CRYSTALLIZATION

ENGLAND:	BUTTERWORTH & CO. (PUBLISHERS) LTD. LONDON: 88 Kingsway, W.C.2
AFRICA:	BUTTERWORTH & CO. (AFRICA) LTD. DURBAN: 33/35 Beach Grove
AUSTRALIA:	BUTTERWORTH & CO. (AUSTRALIA) LTD. SYDNEY: 6–8 O'Connell Street MELBOURNE: 430 Bourke Street BRISBANE: 240 Queen Street
CANADA:	BUTTERWORTH & CO. (CANADA) LTD. TORONTO: 1367 Danforth Avenue, 6
NEW ZEALAND:	BUTTERWORTH & CO. (NEW ZEALAND) LTD' WELLINGTON: 49/51 Ballance Street AUCKLAND: 35 High Street
U.S.A.:	BUTTERWORTH INC. WASHINGTON, D.C.: 7235 Wisconsin Avenue, 14

CRYSTALLIZATION

J. W. MULLIN
B.Sc., Ph.D., F.R.I.C., M.I.Chem.E.
Department of Chemical Engineering,
University College, London

LONDON
BUTTERWORTHS
1961

Printed in Great Britain at the Pitman Press, Bath

PREFACE

CRYSTALLIZATION must surely rank as the oldest unit operation, in the chemical engineering sense. Sodium chloride, for example, has been manufactured by this process since the dawn of civilization. Today there are few sections of the chemical industry that do not, at some stage, utilize crystallization as a method of production, purification or recovery of solid material. Apart from being one of the best and cheapest methods available for the production of pure solids from impure solutions, crystallization has the additional advantage of giving an end product which has many desirable properties. Uniform crystals have good flow, handling and packaging characteristics; they also have an attractive appearance, and this latter property alone can be a very important sales factor.

The industrial applications of crystallization are not necessarily confined to the production of pure solid substances. In recent years large-scale purification techniques have been developed for substances that are normally liquid at room temperature. The petroleum industry, for example, in which distillation has long held pride of place as the major processing operation, is turning its attention most keenly to low-temperature crystallization as a method for the separation of 'difficult' liquid hydrocarbon mixtures.

It is rather surprising that few books, indeed none in the English language, have been devoted to a general treatment of crystallization practice, in view of its importance and extensive industrial application. One reason for this lack of attention could easily be that crystallization is still referred to as more of an art than a science. There is undoubtedly some truth in this old adage, as anyone who has designed and subsequently operated a crystallizer will know, but it cannot be denied that nowadays there is a considerable amount of science associated with the art.

Despite the large number of advances that have been made in recent years in crystallization technology there is still plenty of evidence of the reluctance to talk about crystallization as a process divorced from considerations of the actual substance being crystallized. To some extent this state of affairs is similar to that which existed in the field of distillation some decades ago when little attempt had been made to correlate the highly specialized techniques developed, more or less independently, for the processing of such commodities as coal tar, alcohol and petroleum products. The transformation from an 'art' to a 'science' was eventually made when it came to be recognized that the key factor which unified distillation design methods lay in the equilibrium physical properties of the working systems.

There is a growing trend today towards a unified approach to crystallization problems, but there is still some way to go before crystallization ceases to be the Cinderella of the unit operations. More data, particularly of the applied kind, should be published. In this age of prolific outputs of technical literature such a recommendation is not made lightly, but there is a real deficiency of this type of published information. There is, at the same time, a wealth of knowledge and experience retained in the process industries, much of it empirical but none the less valuable when collected and correlated.

The object of this book is to outline the more important aspects of crystallization theory and practice, together with some closely allied topics. The book is intended to serve process chemists and engineers, and it should prove of interest to students of chemical engineering and chemical technology. Whilst many of the techniques and operations have been described with reference to specific processes or industries, an attempt has been made to treat the subject matter in as general a manner as possible in order to emphasize the unit operational nature of crystallization. Particular attention has been paid to the newer and more recently developed processing methods, even where these have not as yet proved adaptable to the large-scale manufacture of crystals.

Chapter 1 consists of a brief introduction to the fundamental concepts of the crystalline state and crystallography. A short selective bibliography is given at the end of this section for the benefit of those who wish to read further into this specialized field. The choice of subject matter for Chapters 2–8 was largely dictated by the scope of the unit operation of crystallization and its dependence on solution properties and phase equilibria. Chapters 7 and 8 are devoted solely to industrial crystallization and problems associated with crystal production. Chapter 9 deals with the sizing and grading of crystals—operations which are invariably associated, in one way or another, with the manufacture of crystalline materials. Several tables of solubility and heat of solution data have been compiled and included as an Appendix.

My thanks are due to the Editors of *Chemical Engineering Practice* for permission to include some of the material and many of the diagrams previously published by me in Volume 6 of their 12-volume series. I am indebted to Professor M. B. Donald, who first suggested that I should write on this subject, and to many of my colleagues, past and present, for helpful discussions in connection with this work. I would also like to take this opportunity of acknowledging my indebtedness to my wife for the valuable assistance and encouragement she gave me during the preparation of the manuscript.

London,
1960

J. W. MULLIN

CONTENTS

APPENDIX

1

THE CRYSTALLINE STATE

THE three general states of matter—gaseous, liquid and solid—represent very different degrees of atomic or molecular mobility. In the gaseous state, the molecules are in constant, vigorous and random motion; a mass of gas takes the shape of its container, is readily compressed and exhibits a low viscosity. In the liquid state, random molecular motion is much more restricted. The volume occupied by a liquid is limited; a liquid only takes the shape of the occupied part of its container, and its free surface is flat, except in those regions where it comes into contact with the container walls. A liquid exhibits a much higher viscosity than a gas and is less easily compressed. In the solid state, molecular motion is confined to an oscillation about a fixed position, and the rigid structure generally resists compression very strongly; in fact it will often fracture when subjected to a deforming force.

Some substances, such as wax, pitch and glass, which possess the outward appearance of being in the solid state, yield and flow under pressure, and they are sometimes regarded as highly viscous liquids. Solids may be crystalline or amorphous, and the crystalline state differs from the amorphous state in the regular arrangement of the constituent molecules, atoms or ions into some fixed and rigid pattern known as a lattice. Actually, many of the substances which were once considered to be amorphous have now been shown, by X-ray analysis, to exhibit some degree of regular molecular arrangement, but the term 'crystalline' is most frequently used to indicate a high degree of internal regularity, resulting in the development of definite external crystal faces.

As molecular motion in a gas or liquid is free and random, the physical properties of these fluids are the same no matter in what direction they are measured. In other words, they are *isotropic*. True amorphous solids, because of the random arrangement of their constituent molecules, are also isotropic. Most crystals, however, are *anisotropic*; their mechanical, electrical, magnetic and optical properties can vary according to the direction in which they are measured. Crystals belonging to the cubic system are the exception to this rule; their highly symmetrical internal arrangement renders them isotropic. Anisotropy is most readily detected by refractive index measurements, and the striking phenomenon of double refraction exhibited by a clear crystal of Iceland spar (calcite) is probably the best known example.

Liquid Crystals

Before considering the type of crystal with which everyone is familiar, namely the solid crystalline body, it is worth while mentioning a state of matter which possesses the flow properties of a liquid yet exhibits some of the properties of the crystalline state.

Although liquids are generally isotropic, some 200 cases are known of substances which exhibit anisotropy in the liquid state at temperatures just above their melting point. These liquids bear the unfortunate name 'liquid crystals'; the term is inapt because the word 'crystal' implies the existence of a rigid space lattice. Lattice formation is not possible in the liquid state, but some form of molecular orientation can occur with certain types of molecules under certain conditions. Accordingly, the name 'anisotropic liquid' is preferred to 'liquid crystal'. The name 'mesomorphic state' was proposed by Friedel (1922) to indicate that anisotropic liquids are intermediate between the true liquid and crystalline solid states.

Among the better known examples of anisotropic liquids are *p*-azoxyphenetole, *p*-azoxyanisole, cholesteryl benzoate, ammonium oleate and sodium stearate. These substances exhibit a sharp melting point, but they melt to form a turbid liquid. On further heating, the liquid suddenly becomes clear at some fixed temperature. On cooling, the reverse processes occur at the same temperatures as before. It is in the turbid liquid stage that anisotropy is exhibited. The changes in physical state occurring with change in temperature for the case of *p*-azoxyphenetole are:

$$\underset{\text{(anisotropic)}}{\text{Solid}} \overset{137^\circ\text{C}}{\rightleftharpoons} \underset{\substack{\text{(anisotropic,}\\ \text{mesomorphic)}}}{\text{Turbid liquid}} \overset{167^\circ\text{C}}{\rightleftharpoons} \underset{\text{(isotropic)}}{\text{Clear liquid}}$$

The simplest representation of the phenomenon is given by Bose's swarm theory, according to which molecules orientate into a number of groups in parallel formation (*Figure 1.1*). In many respects this is rather similar to the

Figure 1.1. Isotropic and anisotropic liquids: (a) isotropic: molecules in random arrangement; (b) anisotropic: molecules aligned into swarms

behaviour of a large number of logs floating down a river. Substances which can exist in the mesomorphic state are generally organic compounds, often aromatic, with elongated molecules. Properties such as double refraction and the production of interference colours in polarized light are attributed to the scattering of light at the boundaries of these swarms.

The mesomorphic state is conveniently divided into two main classes. The *smectic* (soap-like) state is characterized by an oily nature, and the flow of such liquids occurs by a gliding movement of thin layers over one another. Liquids in the *nematic* (thread-like) state flow like normal viscous liquids, but mobile threads can often be observed within the liquid layer. A third class, in which strong optical activity is exhibited, is known as the *cholesteric* state; some workers regard this state as a special case of the nematic. The name arises from the fact that cholesteryl compounds form the majority of known examples.

For further information on this subject, reference should be made to the original literature (see Bibliography).

Crystalline Solids

The true solid crystal comprises a rigid lattice of molecules, atoms or ions, the locations of which are characteristic of the substance. The regularity of the internal structure of this solid body results in the crystal having a characteristic shape; smooth surfaces or faces develop as a crystal grows, and the

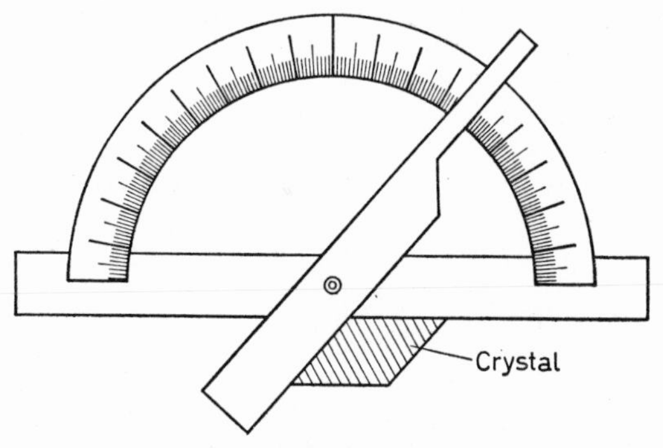

Figure 1.2. Simple contact goniometer

planes of these faces are parallel to atomic planes in the lattice. Very rarely, however, do any two crystals of a given substance look identical; in fact, any two given crystals often look completely different in both size and external shape. In a way this is not very surprising as many crystals, especially the natural minerals, have grown under different conditions. Few natural crystals have grown 'free'; most have grown under some restraint resulting in stunted growth in one direction and exaggerated growth in another.

This state of affairs prevented the general classification of crystals for centuries. The first advance in the science of crystallography came when Steno (1669) observed a unique property of all quartz crystals. He found that the angle between any two given faces on a quartz crystal was constant, irrespective of the relative sizes of these faces. This fact was confirmed later by other workers, and the Law of Constant Interfacial Angles was proposed by Haüy (1784): the angles between corresponding faces of all crystals of a given substance are constant. They may vary in size, and the development of the various faces (the crystal habit) may differ considerably, but the interfacial angles do not vary; they are characteristic of the substance.

Crystal angles are measured with an instrument known as a goniometer. The simple contact goniometer (*Figure 1.2*), consisting of an arm pivoted on a protractor, can only be used on moderately large crystals, and precision greater than $\pm$ 0·5° is rarely possible. This type of instrument is seldom used nowadays. The reflecting goniometer (*Figure 1.3*), which was developed by Wollaston (1809), is a more versatile and accurate apparatus. A crystal is mounted at the centre of a graduated turntable, a beam of light from an illuminated slit being reflected from one face of the crystal. The reflection is

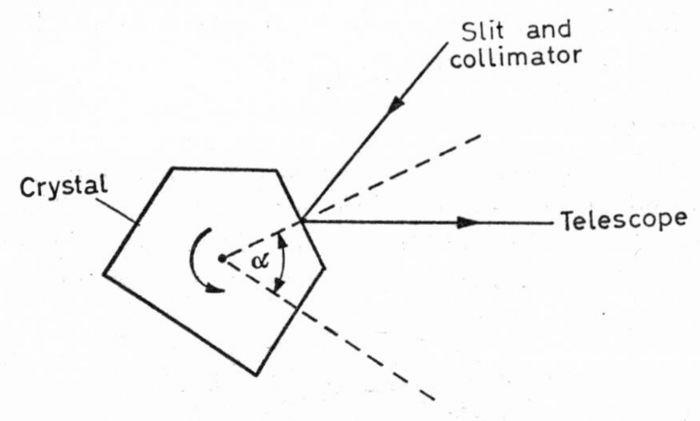

Figure 1.3. Reflecting goniometer

observed in a telescope and read on the graduated scale. The turntable is then rotated until the reflection from the next face of the crystal is observed in the telescope, and a second reading is taken from the scale. The difference α between the two readings is the angle between the normals to the two faces, and the interfacial angle is therefore $180 - \alpha°$.

Crystal Symmetry

Many of the geometric shapes that appear in the crystalline state are readily recognized as being to some degree symmetrical, and this fact can be used as a means of crystal classification. The three simple elements of symmetry that can be considered are:

1. Symmetry about a point (a *centre* of symmetry)
2. Symmetry about a line (an *axis* of symmetry)
3. Symmetry about a plane (a *plane* of symmetry)

It must be remembered, however, that while some crystals may possess a centre and several different axes and planes of symmetry, others may have no element of symmetry at all.

A crystal possesses a centre of symmetry when every point on the surface of the crystal has an identical point on the opposite side of the centre, equidistant from it. A perfect cube is a good example of a body having a centre of symmetry (at its mass centre).

If a crystal is rotated through 360° about any given axis, it obviously returns to its original position. If, however, the crystal *appears* to have reached its original position more than once during its complete rotation, the chosen axis is an axis of symmetry. If the crystal has to be rotated through 180° (360/2) before coming into coincidence with its original position, the axis is one of 2-fold symmetry (called a diad axis). If it has to be rotated through 120° (360/3), 90° (360/4) or 60° (360/6) the axes are of 3-fold symmetry (triad axis), 4-fold symmetry (tetrad axis) and 6-fold symmetry (hexad axis), respectively. These are the only axes of symmetry possible in the crystalline state.

A cube, for instance, has 13 axes of symmetry: 6 diad axes through opposite edges, 4 triad axes through opposite corners and 3 tetrad axes

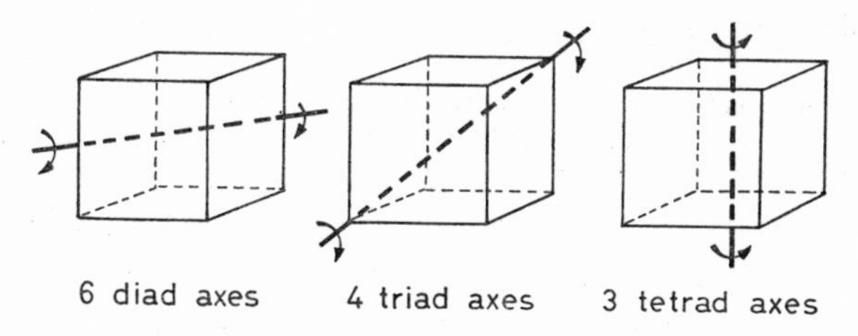

Figure 1.4. The thirteen axes of symmetry in a cube

through opposite faces. One each of these axes of symmetry are shown in *Figure 1.4.*

The third simple type is symmetry about a plane. A plane of symmetry bisects a solid object in such a manner that one half becomes the mirror image of the other half in the given plane. This type of symmetry is quite common and is often the only type exhibited by a crystal. A cube has 9 planes of symmetry: 3 rectangular planes each parallel to two faces, and 6 diagonal planes passing through opposite edges, as shown in *Figure 1.5.*

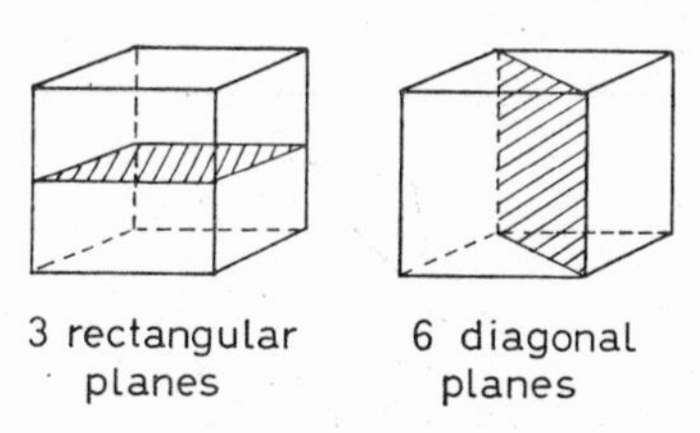

Figure 1.5. The nine planes of symmetry in a cube

It can be seen, therefore, that the cube is a highly symmetrical body, as it possesses 23 elements of symmetry (a centre, 9 planes and 13 axes). An octahedron also has the same 23 elements of symmetry; so despite the difference in outward appearance there is a definite crystallographic relationship between these two forms. *Figure 1.6* indicates the passage from the cubic (hexahedral) to the octahedral form, and *vice versa*, by a progressive and symmetrical removal of the corners. The intermediate solid forms shown (truncated cube, truncated octahedron and cubo-octahedron) are three of the 13 Archimedian semi-regular solids which are called *combination forms*, i.e. combinations of a cube and an octahedron. Crystals exhibiting combination forms are commonly encountered. The tetrahedron is also related to the cube and octahedron; in fact these three forms belong to the five regular solids of geometry. The other two (the regular dodecahedron and icosahedron) do not occur in the crystalline state. The rhombic dodecahedron, however, is frequently found, particularly in crystals of garnet. *Table 1.1* lists the properties of the six regular and semi-regular forms most often encountered in crystals. The Euler relationship is useful for calculating the number of faces, edges and corners of any polyhedron:

$$E = F + C - 2$$

This relation states that the number of edges is two less than the sum of the number of faces and corners.

Table 1.1. Properties of some Regular and Semi-regular Forms found in the Crystalline State

Form	*Faces*	*Edges*	*Corners*	*Edges at a corner*	*Elements of Symmetry*		
					Centre	*Planes*	*Axes*
Regular solids							
Tetrahedron .	4	6	4	3	*No*	6	7
Hexahedron (cube) . .	6	12	8	3	*Yes*	9	13
Octahedron .	8	12	6	4	*Yes*	9	13
Semi-regular solids							
Truncated cube	14	36	24	3	*Yes*	9	13
Truncated octahedron .	14	36	24	3	*Yes*	9	13
Cubo-octahedron	14	24	12	4	*Yes*	9	13

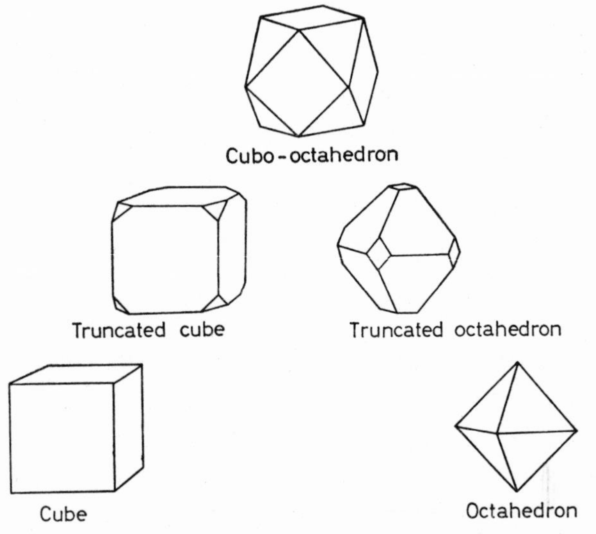

Figure 1.6. Combination forms of cube and octahedron

A fourth element of symmetry which is exhibited by some crystals is known by the names 'compound, or alternating, symmetry', or symmetry about a 'rotation-reflection axis' or 'axis of rotatory inversion'. This type of symmetry obtains when one crystal face can be related to another by performing two operations: (*a*) rotation about an axis, and (*b*) reflection in a plane at right angles to the axis, or inversion about the centre. *Figure 1.7*

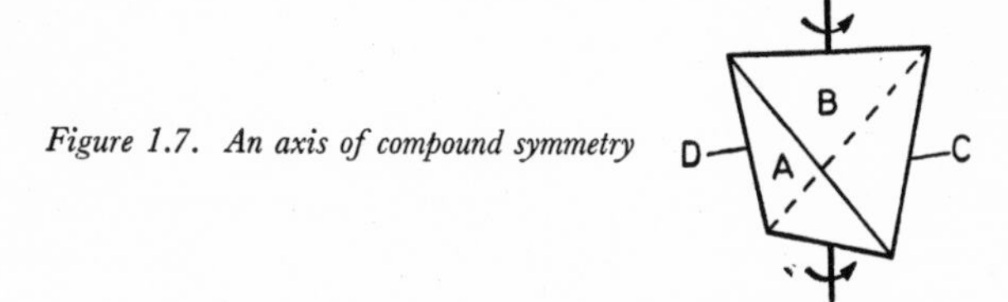

Figure 1.7. An axis of compound symmetry

illustrates the case of a tetrahedron, where the four faces are marked *A*, *B*, *C* and *D*. Face *A* can be transformed into face *B* after rotation through 90°, followed by an inversion. This procedure can be repeated four times, so the chosen axis is a compound axis of 4-fold symmetry.

Crystal Systems

There are only 32 possible combinations of the above-mentioned elements of symmetry, including the asymmetric state (no elements of symmetry), and these are called the 32 *point groups* or *classes*. All but one or two of these classes have been observed in crystalline bodies. For convenience these 32 classes are grouped into seven *systems* which are known by the following names: regular (5 possible classes), tetragonal (7), orthorhombic (3), monoclinic (3), triclinic (2), trigonal (5) and hexagonal (7).

The first six of these systems can be described with reference to three axes, x, y and z. The z axis is vertical, the x axis is directed from front to back and the y axis from right to left, as shown in *Figure 1.8a*. The angle between the

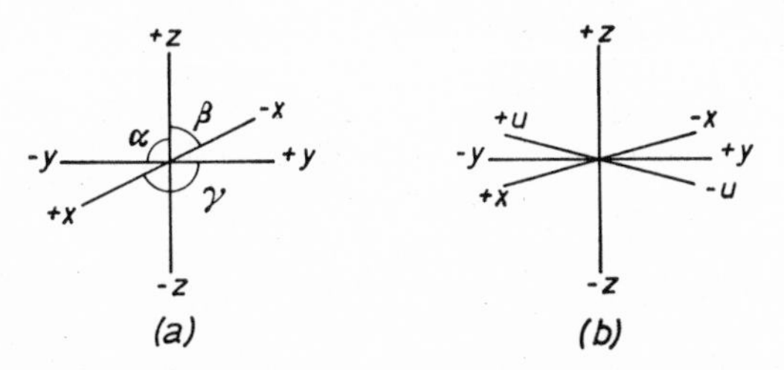

Figure 1.8. Crystallographic axes for describing the seven crystal systems:

(*a*) *three axes* $\widehat{yz} = \alpha$; $\widehat{xz} = \beta$; $\widehat{xy} = \gamma$;

(*b*) *four axes (hexagonal system)* $xy = \widehat{yu} = \widehat{ux} = 60°$ (120°)

axes y and z is denoted by α, that between x and z by β, and that between x and y by γ. Four axes are required to describe the hexagonal system; the z axis is vertical and perpendicular to the other three axes (x, y and u) which are co-planar and inclined at 60° (or 120°) to each other, as shown in *Figure 1.8b*. Some workers prefer to describe the trigonal system with reference to four axes. Descriptions of the seven crystal systems, together with some of the other names occasionally employed, are given in *Table 1.2.*

Table 1.2. The Seven Crystal Systems

System	*Other names*	*Angles between axes*	*Length of axes*	*Examples*
REGULAR	Cubic Octahedral Isometric Tesseral	$\alpha = \beta = \gamma = 90°$	$x = y = z$	Sodium chloride Potassium chloride Alums Diamond
TETRAGONAL	Pyramidal Quadratic	$\alpha = \beta = \gamma = 90°$	$x = y \neq z$	Rutile Zircon Nickel sulphate
ORTHO-RHOMBIC	Rhombic Prismatic Isoclinic Trimetric	$\alpha = \beta = \gamma = 90°$	$x \neq y \neq z$	Potassium permanganate Silver nitrate Iodine α-sulphur
MONO-CLINIC	Mono-symmetric Clinorhombic Oblique	$\alpha = \beta = 90° \neq \gamma$	$x \neq y \neq z$	Potassium chlorate Sucrose Oxalic acid β-sulphur
TRICLINIC	Anorthic Asymmetric	$\alpha \neq \beta \neq \gamma \neq 90°$	$x \neq y \neq z$	Potassium dichromate Copper sulphate
TRIGONAL	Rhombo-hedral	$\alpha = \beta = \gamma \neq 90°$	$x = y = z$	Sodium nitrate Ruby Sapphire
HEXAGONAL	None	z axis is perpendicular to the x, y and u axes which are inclined at 60° to each other	$x = y = u \neq z$	Silver iodide Graphite Water (ice) Potassium nitrate

For the regular, tetragonal and orthorhombic systems, the three axes x, y and z are mutually perpendicular. The systems differ in the relative lengths of these axes: in the regular system they are all equal, in the orthorhombic system they are all unequal while, in the tetragonal system, two are equal and the third different. The three axes are all unequal in the monoclinic and triclinic systems; in the former, two of the angles are 90° and one angle different, and in the latter all three angles are unequal and none equal to 90°. Sometimes the limitation 'not equal to 30°, 60° or 90°' is also applied to the triclinic system. In the trigonal system, three equal axes intersect at equal angles, but the angles are not 90°. The hexagonal system is described with reference to four axes. The axis of 6-fold symmetry (hexad axis) is usually chosen as the z axis, and the other three equal-length axes, located in a plane at 90° to the z axis, intersect each other at 60° (or 120°).

Each crystal system contains several classes which exhibit only a partial symmetry; for instance, only one-half or one-quarter of the maximum number of faces permitted by the symmetry may have been developed. The *holohedral* class is that which has the maximum number of similar faces, i.e. possesses the highest degree of symmetry. In the *hemihedral* class only half this number of faces have been developed, and in the *tetartohedral* class only one-quarter have been developed. For example, the regular tetrahedron (4 faces) is the hemihedral form of the holohedral octahedron (8 faces) and the wedge-shaped sphenoid is the hemihedral form of the tetragonal bipyramid (*Figure 1.9*).

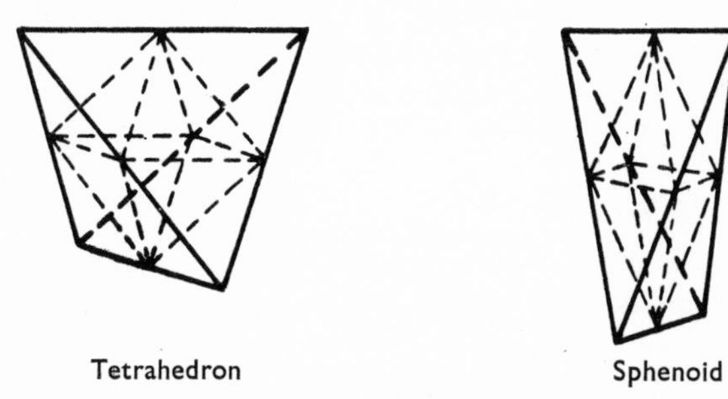

Figure 1.9. Hemihedral forms of the octahedron and tetragonal bipyramid

It has been mentioned above that crystals exhibiting combination forms are often encountered. The simplest forms of any crystal system are the prism and the pyramid. The cube, for instance, is the prism form of the regular system and the octahedron is the pyramidal form, and some combinations of these two forms have been indicated in *Figure 1.6*. Two simple combination forms in the tetragonal system are shown in *Figure 1.10*.

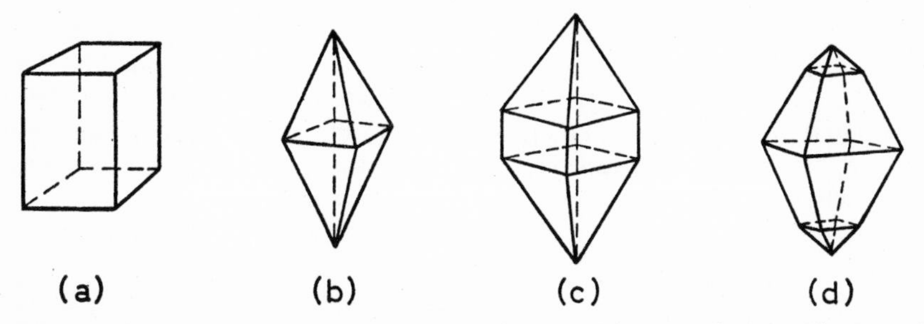

Figure 1.10. Simple combination forms in the tetragonal system: (a) tetragonal prism; (b) tetragonal bipyramid; (c) combination of prism and bipyramid; (d) combination of two bipyramids

Figures 1.10a and *b* are the tetragonal prism and bipyramid, respectively. *Figure 1.10c* shows a tetragonal prism which is terminated by two tetragonal pyramids, and *Figure 1.10d* the combination of two different tetragonal bipyramids. It frequently happens that a crystal develops a group of faces which intersect to form a series of parallel edges; such a set of faces is said to constitute a *zone*. In *Figure 1.10b*, for instance, the four prism faces make a zone.

The crystal system favoured by a substance is to some extent dependent on the atomic or molecular complexity of the substance. More than 80 per cent of the crystalline elements and very simple inorganic compounds belong to the regular and hexagonal systems. As the constituent molecules become more complex the orthorhombic and monoclinic systems are favoured; about 80 per cent of the known crystalline organic substances and 60 per cent of the natural minerals belong to these systems.

Miller Indices

All the faces of a crystal can be described and numbered in terms of their axial intercepts. The axes referred to here are the crystallographic axes (usually 3), which are chosen arbitrarily; one or more of these axes may be axes of symmetry or parallel to them, but three convenient crystal edges can be used if desired. It is best if the three axes are mutually perpendicular, but this cannot always be arranged. On the other hand, some crystals require four axes for indexing purposes.

If, for example, three crystallographic axes have been decided upon, a plane which is inclined to all three axes is chosen as the standard or *parametral plane*. It is sometimes possible to choose one of the crystal faces to act as the parametral plane. The intercepts X, Y and Z of this plane on the axes x, y and z are called parameters a, b and c. The ratios of the parameters $a:b$ and $b:c$ are called the axial ratios, and by convention the values of the parameters are reduced so that the value of b is unity.

W. H. Miller (1839) suggested that each face of a crystal could be represented by the indices h, k and l, defined by

$$h = \frac{a}{X}, \quad k = \frac{b}{Y} \quad \text{and} \quad l = \frac{c}{Z}$$

For the parametral plane, the axial intercepts X, Y and Z are the parameters a, b and c, so the indices h, k and l are a/a, b/b and c/c, i.e. 1, 1 and 1. This is usually written (111). The indices for the other faces of the crystal are calculated from the values of their respective intercepts X, Y and Z, and these intercepts can always be represented by ma, nb and pc, where m, n and p are small whole numbers or infinity (Haüy's Law of Rational Intercepts).

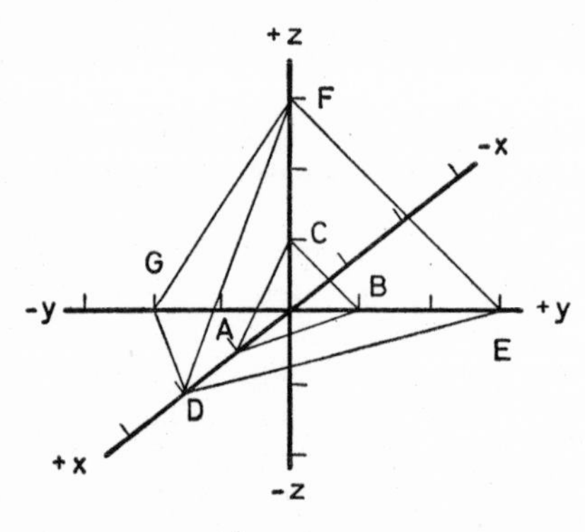

Figure 1.11. Intercepts of planes on the crystallographic axes

The procedure for allotting face indices is indicated in *Figure 1.11* where equal divisions are made on the x, y and z axes. The parametral plane ABC, with axial intercepts of $0A = a$, $0B = b$ and $0C = c$, respectively, is indexed

(111) as described above. Plane DEF has axial intercepts $X = 0D = 2a$, $Y = 0E = 3b$ and $Z = 0F = 3c$; so the indices for this face can be calculated as

$$h = a/X = a/2a = \tfrac{1}{2}$$
$$k = b/Y = b/3b = \tfrac{1}{3}$$
$$l = c/Z = c/3c = \tfrac{1}{3}$$

Hence $h:k:l = \frac{1}{2}:\frac{1}{3}:\frac{1}{3}$, and multiplying through by six, $h:k:l = 3:2:2$. Face DEF, therefore, is indexed (322). Similarly, face DFG which has axial intercepts of $X = 2a$, $Y = -2b$ and $Z = 3c$, gives $h:k:l = \frac{1}{2}:-\frac{1}{2}:\frac{1}{3} = 3:-3:2$ or $(3\bar{3}2)$. Thus the Miller indices of a face are inversely proportional to its axial intercepts.

Figure 1.12 shows two simple crystals belonging to the regular system. As there is no inclined face in the cube, no face can be chosen as the parametral plane (111). The intercepts Y and Z of face A on the axes y and z are at

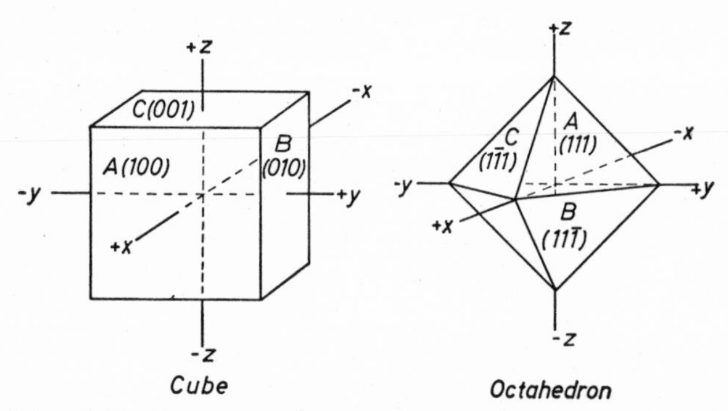

Figure 1.12. Two simple crystals belonging to the regular system, showing the use of Miller indices

infinity, so the indices h, k and l for this face will be a/a, b/∞ and c/∞, or (100). Similarly, faces B and C are designated (010) and (001), respectively. For the octahedron, face A is chosen arbitrarily as the parametral plane, so it is designated (111). As the crystal belongs to the regular system, the axial intercepts made by the other faces are all equal in magnitude, but not in sign, to the parametral intercepts a, b and c. For instance, the intercept of face B on the z axis is negative, so this face is designated $(11\bar{1})$. Similarly, face C is designated $(1\bar{1}1)$, and the unmarked D face is $(1\bar{1}\bar{1})$.

Space Lattices

The external development of smooth faces on a crystal arises from some regularity in the internal arrangement of the constituent ions, atoms or molecules. Any account of the crystalline state, therefore, should include some reference to the internal structure of crystals. It is beyond the scope of this book to deal in any detail with this large topic, but a brief description will be given of the concept of the space lattice. For further information reference should be made to the specialized works listed in the Bibliography.

It is well known that some crystals can be split by cleavage into smaller crystals which bear a distinct resemblance in shape to the parent body. Whilst there is clearly a mechanical limit to the number of times that this process can be repeated, eighteenth century investigators, Hooke and Haüy in particular, were led to the conclusion that all crystals are built up from a large number of minute units, each shaped in a manner similar to the larger crystal. This hypothesis constituted a very important step forward in the science of crystallography because its logical extension led to the modern concept of the space lattice.

A space lattice is a regular arrangement of points in three dimensions, each point representing a structural unit, e.g. an atom or a molecule. The whole structure is homogeneous, i.e. every point in the lattice has an environment identical to every other point. For instance, if a line is drawn between any two points it will, when produced in both directions, pass through other points in the lattice spaced in an identical manner to the chosen pair. Another way in which this homogeneity can be visualized is to imagine an observer located within the structure; he would get the same view of his surroundings from any of the points in the lattice.

By geometrical reasoning, Bravais (1848) came to the conclusion that there were only fourteen possible basic types of lattice that could give the above environmental identity. These fourteen unit cells can be classified into seven groups based on their symmetry, and these seven groups correspond to the seven crystal systems listed in *Table 1.2*. The fourteen Bravais lattices are given in *Table 1.3*. The three cubic lattices are illustrated in *Figure 1.13*;

Table 1.3. The Fourteen Bravais Lattices

Type of Symmetry	*Lattice*	*Corresponding Crystal System*
CUBIC	Cube Body-centred cube Face-centred cube	REGULAR
TETRAGONAL	Square prism Body-centred square prism	TETRAGONAL
ORTHORHOMBIC	Rectangular prism Body-centred rectangular prism Rhombic prism Body-centred rhombic prism	ORTHORHOMBIC
MONOCLINIC	Monoclinic parallelepiped Clinorhombic prism	MONOCLINIC
TRICLINIC	Triclinic parallelepiped	TRICLINIC
RHOMBOIDAL	Rhombohedron	TRIGONAL
HEXAGONAL	Hexagonal prism	HEXAGONAL

the first comprises eight elementary particles arranged at the corners of a cube, the second consists of a cubic structure with a ninth particle located at the centre of the cube, and the third of a cube with six extra particles, each located on a face of the cube.

The points in any lattice can be arranged to lie on a large number of different planes, called lattice planes, some of which will contain more points per unit area than others. The external faces of a crystal are parallel to lattice planes, and the most commonly occurring faces will be those which correspond to planes containing a high density of points, usually referred to as a high reticular density. Cleavage also occurs along lattice planes.

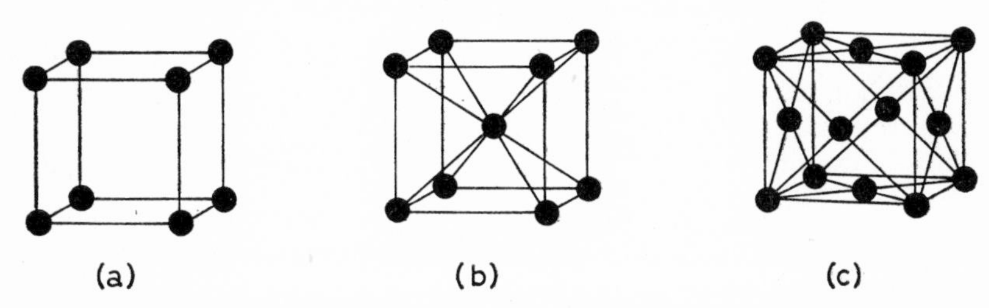

Figure 1.13. The three cubic lattices: (a) cube; (b) body-centred cube; (c) face-centred cube

Although there are only fourteen basic lattices, interpenetration of lattices can occur in actual crystals, and it has been deduced that 230 combinations are possible which still result in the identity of environment of any given point. These combinations are the 230 space groups which are further divided into the 32 point groups, or classes, mentioned above in connection with the seven crystal systems.

Isomorphism and Polymorphism

Two or more substances which crystallize in almost identical forms are said to be *isomorphous* (Greek: 'of equal form'). This is not a contradiction of Haüy's law because these crystals do show small, but quite definite, differences in their respective interfacial angles. Isomorphs are often chemically similar and can then be represented by similar chemical formulae; this statement is one form of Mitscherlich's Law of Isomorphism which is now recognized only as a broad generalization. One group of compounds which obey and illustrate Mitscherlich's law is represented by the formula $M_2'SO_4 \cdot M_2'''(SO_4)_3 \cdot 24H_2O$ (the alums) where M' represents a univalent radical (e.g. K or NH_4) and M''' represents a tervalent radical (e.g. Al, Cr or Fe). Many phosphates and arsenates, sulphates and selenates are also isomorphous.

Sometimes isomorphous substances can crystallize together out of a solution to form 'mixed crystals' or, as they are better termed, solid solutions. In such cases, the composition of the homogeneous solid phase which is deposited follows no fixed pattern; it depends largely on the relative concentrations and solubilities of the substances in the original solvent. For instance, chrome alum, $K_2SO_4 \cdot Cr_2(SO_4)_3 \cdot 24H_2O$ (purple), and potash alum, $K_2SO_4 \cdot Al_2(SO_4)_3 \cdot 24H_2O$ (colourless), crystallize from their respective aqueous solutions as regular octahedra. When an aqueous solution containing both salts is crystallized, regular octahedra are again formed, but the colour of the crystals (which are now homogeneous solid solutions) can vary from almost colourless to deep purple, depending on the proportions of the two alums in the crystallizing solution.

Another phenomenon often shown by isomorphs is the formation of overgrowth crystals. For example, if a crystal of chrome alum (octahedral) is placed in a saturated solution of potash alum, it will grow in a regular manner such that the purple core is covered with a continuous colourless overgrowth. In a similar manner an overgrowth crystal of nickel sulphate, $NiSO_4 \cdot 7H_2O$ (green), and zinc sulphate, $ZnSO_4 \cdot 7H_2O$ (colourless), can be prepared.

There have been many 'rules' and 'tests' proposed for the phenomenon of isomorphism, but in view of the large number of known exceptions to these it is now recognized that the only general property of isomorphism is that crystals of the different substances shall show very close similarity. All the other properties, including those mentioned above, are merely confirmatory and not necessarily shown by all isomorphs.

A substance capable of crystallizing into different, but chemically identical crystalline forms is said to exhibit *polymorphism.* Dimorphous and trimorphous substances are commonly known, e.g.

Carbon: graphite (hexagonal)
diamond (regular)

Silicon dioxide: cristobalite (regular)
tridymite (hexagonal)
quartz (trigonal)

The term *allotropy* instead of polymorphism is often used when the substance is an element.

The different crystalline forms exhibited by one substance may result from a variation in the crystallization temperature or a change of solvent. Sulphur, for instance, crystallizes in the form of orthorhombic crystals (αS) from a carbon disulphide solution, and of monoclinic crystals (βS) from the melt. In this particular case the two crystalline forms are inter-convertible: β-sulphur cooled below 95·5° C changes to the α form. This interconversion between two crystal forms at a definite transition temperature is called *enantiotropy* (Greek: 'change into opposite') and is accompanied by a change in volume. Ammonium nitrate (melting point 169·2° C) exhibits four enantiotropic changes between − 18 and 125° C, as shown below.

$$\text{liquid} \underset{169\cdot6^\circ}{\rightleftharpoons} \overset{\text{(I)}}{\text{cubic}} \underset{125\cdot2^\circ}{\rightleftharpoons} \overset{\text{(II)}}{\text{trigonal}} \underset{84\cdot2^\circ}{\rightleftharpoons} \overset{\text{(III)}}{\text{orthorhombic}} \underset{32\cdot3^\circ}{\rightleftharpoons} \overset{\text{(IV)}}{\text{orthorhombic}} \underset{-18^\circ}{\rightleftharpoons} \overset{\text{(V)}}{\text{tetragonal}}$$

The transitions from forms II to III and IV to V result in volume increases, the changes from I to II and III to IV are accompanied by a decrease in volume. These volume changes frequently cause difficulty in the processing and storage of ammonium nitrate. The salt can readily burst a metal container into which it has been cast when change II to III occurs. The drying of ammonium nitrate crystals must be carried out within fixed temperature limits, e.g. 40 to 80° C, otherwise the crystals can disintegrate when a transition temperature is reached.

When polymorphs are not interconvertible, the crystal forms are said to be *monotropic*; graphite and diamond are monotropic forms of carbon. The term *isopolymorphism* is used when each of the polymorphous forms of one substance are respectively isomorphous with the polymorphous forms of another substance. For instance, the regular and orthorhombic polymorphs of arsenious oxide, As_2O_3, are respectively isomorphous with the regular and orthorhombic polymorphs of antimony trioxide, Sb_2O_3. These two oxides are thus said to be isodimorphous.

Enantiomorphism

Two crystals of the same substance which are the mirror images of each other are said to be enantiomorphous (Greek: 'of opposite form'). These crystals have no planes of symmetry at all. Most enantiomorphous substances exhibit the property of optical activity, i.e. they are capable of rotating the plane of polarized light; one form will rotate it to the left (laevo-rotatory or *l*-form) and the other to the right (dextro-rotatory or *d*-form). Tartaric acid and certain sugars are well known examples of optically active substances.

Enantiomorphous substances are not necessarily optically active, but all known optically active substances are capable of being crystallized into enantiomorphous forms. In some cases the solution of an optically active crystal is also optically active, indicating that the actual molecules of the substance are enantiomorphous. In other cases solution or fusion destroy the optical activity, indicating that enantiomorphism was confined to the crystal structure only.

Optical activity has been associated with compounds which possess one or more atoms around which different elements or groups are arranged asymmetrically, so that the molecule can exist in mirror image forms. In organic compounds the presence of an asymmetric carbon atom often favours optical activity. Tartaric acid offers a good example of this, and three possible arrangements of the tartaric acid molecule are shown in *Figure 1.14*.

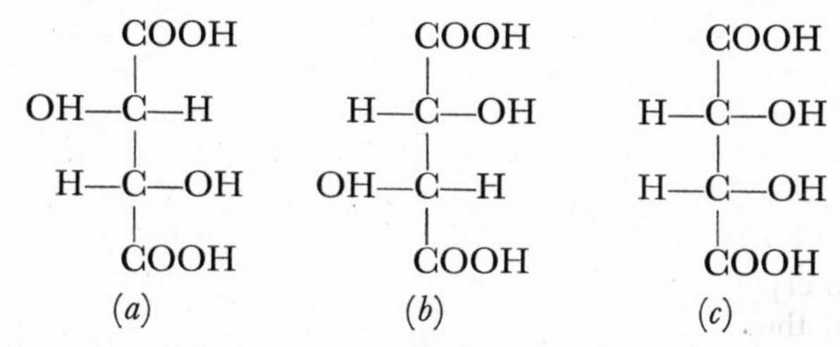

Figure 1.14. The tartaric acid molecule: (a) and (b) optically active forms; (c) meso-tartaric acid, optically inactive

The (*a*) and (*b*) forms are mirror images of each other; both contain two asymmetric carbon atoms and both are optically active; one will be the *d*-form and the other the *l*-form. There are two asymmetric carbon atoms in formula (*c*) but this form (*meso*-tartaric acid) is optically inactive; the potential optical activity of one-half of the molecule is compensated by the opposite potential optical activity of the other.

The case of tartaric acid serves to illustrate another property known as *racemism*. A mixture of crystalline *d*- and *l*-tartaric acids dissolved in water can, if mixed in the right proportions, produce an optically inactive solution. Crystallization of this solution will yield crystals of optically inactive racemic acid which are different in form from the *d*- and *l*-crystals. There is, however, a difference between a racematic and a *meso*-form of a substance; the former can be resolved into *d*- and *l*-forms while the latter cannot.

A racemate can be resolved in a number of ways. Pasteur (1848) found that crystals of the sodium ammonium salt of racemic acid,

$$Na \cdot NH_4 \cdot C_4H_4O_6 \cdot H_2O$$

deposited from aqueous solution, consisted of two clearly different types, one being the mirror image of the other. The *d*- and *l*-forms were easily separated by hand picking. Bacterial attack was also shown by Pasteur to be effective in the resolution of racemic acid. *Penicillium Glaucum* allowed to grow in a dilute solution of sodium ammonium racemate destroys the *d*-form but, apart from being a rather wasteful process, the attack is not always completely selective. A racematic is best resolved by forming a salt or ester with an optically active base (usually an alkaloid) or alcohol. For example, a racemate of an acidic substance A with say the dextro form of an optically active base B will give

$$dl\text{A} + d\text{B} \rightarrow d\text{A} \cdot d\text{B} + l\text{A} \cdot d\text{B}$$

and the two salts $d\text{A} \cdot d\text{B}$ and $l\text{A} \cdot d\text{B}$ can be separated by fractional crystallization.

Crystal Habit

Although crystals can be classified according to the seven general systems (*Table 1.1*), the relative sizes of the faces of a particular crystal can vary considerably. This variation is called a modification of habit. The crystals may grow more rapidly, or be stunted, in one direction; thus an elongated growth of the prismatic habit gives a needle-shaped crystal (acicular habit) while a stunted growth gives a flat plate-like crystal (tabular, platy or flaky habit). Nearly all manufactured and natural crystals are distorted to some degree, and this fact frequently leads to a misunderstanding of the term 'symmetry'. Perfect geometric symmetry is rarely observed in crystals, but crystallographic symmetry is readily detected by means of a goniometer.

Figure 1.15 shows three different habits of a crystal belonging to the hexagonal system. The centre diagram (*b*) shows a crystal with a predominant prismatic habit. This combination-form crystal is terminated by hexagonal pyramids and two flat faces perpendicular to the vertical axis; these flat faces are called *pinacoids*. A stunted growth in the vertical direction (or elongated growth in the directions of the other axes) results in a tabular crystal (*a*); excessively flattened crystals are usually called plates or flakes.

An elongated growth in the vertical direction yields a needle or acicular crystal (*c*); flattened needle crystals are usually called blades.

The relative growths of the faces of a crystal can be altered, and often controlled, by a number of factors. Rapid crystallization, such as that produced by the sudden cooling or seeding of a supersaturated solution, generally results in the formation of needle crystals. Impurities in the crystallizing solution often stunt the growth of a crystal in certain directions,

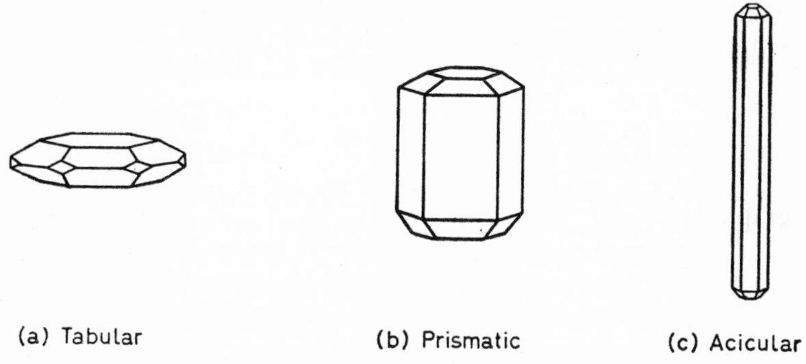

Figure 1.15. Crystal habit illustrated on a hexagonal crystal

while crystallization from solutions of the given substance in different solvents may result in a change of habit. The degree of supersaturation or supercooling of a solution or melt often exerts a considerable influence on the crystal habit, and so can the state of agitation of the system. These and other factors affecting the control of habit are discussed in Chapter 5.

Dendrites

Rapid crystallization from supercooled melts, supersaturated solutions and vapours frequently produces tree-like formations called dendrites, the growth of which is indicated in *Figure 1.16*. The main crystal stem grows quite

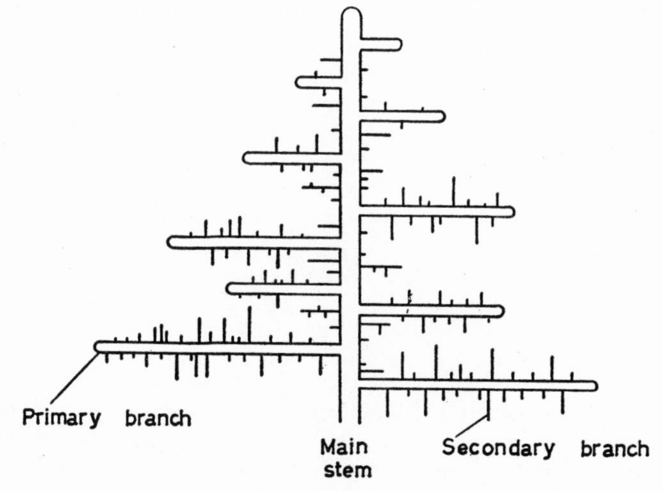

Figure 1.16. Dendritic growth

rapidly in a supercooled system that has been seeded, and at a later stage primary branches grow at a slower rate out of the stem, often at right angles to it. In certain cases, small secondary branches may grow slowly out of the primaries. Eventually branching ceases and the pattern becomes filled in with crystalline material.

Most metals crystallize from the molten state in this manner, but because of the filling-in process the final crystalline mass may show little outward appearance of dendrite formation. The fascinating patterns of snow crystals are good examples of dendritic growth, and the frosting of windows often affords a visual observation of this phenomenon occurring in two dimensions. The growth of a dendrite can be observed quite easily under a microscope by seeding a drop of a saturated solution on the slide.

Dendrites form most commonly during the early stages of crystallization; at later stages a more normal uniform growth takes place and the pattern may be obliterated. Dendritic growth occurs quite readily in thin liquid layers, probably because of the high rate of evaporative cooling, while agitation tends to suppress this type of growth. Dendrite formation is favoured by substances which have high latent heats of crystallization and low heat conductivities. Although many theories have been proposed to account for dendritic growth, none, so far, is completely acceptable. Buckley (see Bibliography) presents a detailed critical account of these theories.

Composite Crystals

Most crystalline natural minerals, and many crystals produced industrially, exhibit some form of aggregation or intergrowth, and prevention of the formation of these composite crystals is one of the problems of large-scale crystallization. The presence of aggregates in a crystalline mass spoils the appearance of the product and interferes with its free-flowing nature. More important, however, aggregation is often indicative of impurity because crystal clusters readily retain impure mother liquor and resist efficient washing.

Composite crystals may occur in simple symmetrical forms or in random clusters. The simplest form of aggregate results from the phenomenon

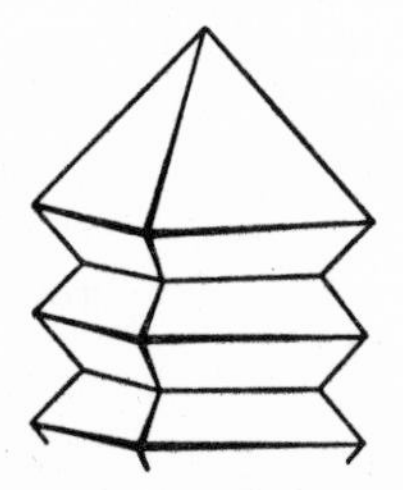

Figure 1.17. Parallel growth on a crystal of potash alum

known as *parallel growth*; individual forms of the same substance grow on the top of one another in such a manner that all corresponding faces and edges of the individuals are parallel. Potash alum, $K_2SO_4 \cdot Al_2(SO_4)_3 \cdot 24H_2O$, exhibits this type of growth; *Figure 1.17* shows a typical structure in which regular octahedra are piled on top of one another in a column symmetrical about the vertical axis. Parallel growth is often associated with isomorphs; for instance, parallel growths of one alum can be formed on the crystals of another, but this property is no longer regarded as an infallible test for isomorphism.

Another composite crystal frequently encountered is known as a *twin* or a *macle*; it appears to be composed of two intergrown individuals, similar in form, joined symmetrically about an axis (a twin axis) or a plane (a twin plane). A twin axis is a possible crystal edge and a twin plane is a possible crystal face. Many types of twins may be formed in simple shapes such as a V, +, L and so forth, or they may show an interpenetration giving the appearance that one individual has passed completely through the other (*Figure 1.18*). Partial interpenetration (*Figure 1.19*) can also occur. In some

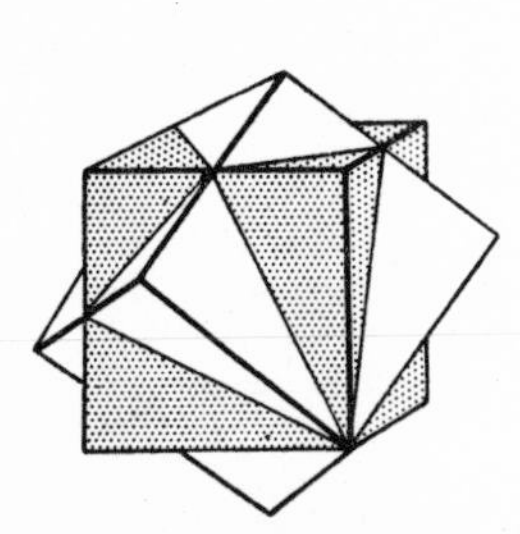

Figure 1.18. Interpenetrant twin of two cubes (e.g. fluorspar)

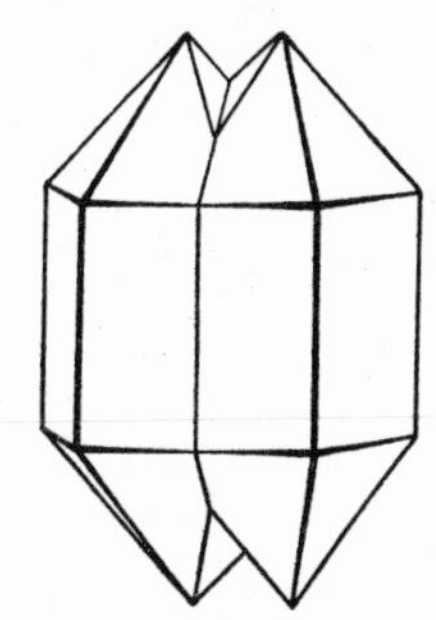

Figure 1.19. Partial interpenetrant twin (e.g. quartz)

cases, a twin crystal may present the outward appearance of a form which possesses a higher degree of symmetry than that of the individuals, and this is known as *mimetic twinning*. A typical example of this behaviour is orthorhombic potassium sulphate which can form a twin looking almost identical with a hexagonal bipyramid.

Parallel growth and twinning (or even triplet formation) are generally encountered when crystallization has been allowed to take place in an undisturbed medium. Although twins of individuals belonging to most of the seven crystal systems are known, twinning occurs most frequently when the crystals belong to the orthorhombic or monoclinic systems. Certain impurities in the crystallizing medium can cause twin formation even though an appreciable agitation is effected; this is one of the problems encountered in the commercial crystallization of sugar.

The formation of crystal clusters, aggregates or conglomerates which possess no symmetrical properties is probably more frequently encountered in large-scale crystallization than the formation of twins. Relatively little is still known about the growth of these irregular crystal masses, but among the factors which often favour their formation are poor agitation, the presence of certain impurities in the crystalling solution, seeding at high degrees of supersaturation and the presence of too many seed crystals, leading to conditions of overcrowding in the crystallizer.

BIBLIOGRAPHY

BRIGGS, D. B., *The Study of Crystals*, 1930. London; Dent
BUCKLEY, H. E., *Crystal Growth*, 1952. London; Chapman & Hall
BUNN, C. W., *Chemical Crystallography*, 1945. Oxford; Clarendon Press
HARTSHORNE, N. H. and STUART, A., *Crystals and the Polarising Microscope*, 3rd. Ed., 1960. London; Arnold
PARTINGTON, J. R., *An Advanced Treatise on Physical Chemistry*, Vol. III, *The Properties of Solids*, 1952. London; Longmans Green
PHILLIPS, F. C., *Introduction to Crystallography*, 2nd Ed., 1956. London; Longmans Green
SMITH, G. F. H., *Gemstones*, revised by F. C. Phillips, 1958. London; Methuen
TUTTON, A. E. H., *Crystallography and Practical Crystal Measurement*, Vols. I and II, 2nd Ed., 1922. London; Macmillan
WELLS, A. F., *The Third Dimension in Chemistry*, 1956. Oxford; Clarendon Press
'Symposium on Liquid Crystals,' *Trans. Faraday Soc.* 29 (1933) 881, 1060

2

SOLUTIONS AND SOLUBILITY

A SOLUTION is a homogeneous mixture of two or more substances; it may be gaseous, liquid or solid, and its constituents are usually called solvents and solutes. There is no particular reason why any one component of a solution should be termed the solvent, but it is conventional to give this name to the component present in excess. Many cases exist, however, where considerable confusion can arise. For example, a salt such as potassium nitrate fuses in the presence of small amounts of water at a much lower temperature than the pure salt does. The term 'solvent' for water can hardly be justified in such cases. It may seem strange to refer to 'a solution of water in potassium nitrate', yet this would be the correct description. It has been suggested[1] that fusion is nothing more than an extreme case of liquefaction by solution, so it may be said that when a salt dissolves in water, the salt does, in fact, melt.

Due to the widespread and often indiscriminate use of the word 'melt', it is difficult to give a precise definition of the term. Strictly speaking, a melt is the liquid phase of a pure substance which is solid at normal temperatures. In its general application the term also includes homogeneous liquid mixtures of two or more substances which solidify on cooling. Thus α-naphthol (m.p. 96° C) in the liquid state is a melt. Homogeneous liquid mixtures consisting of say α-naphthol and β-naphthol (m.p. 122° C), or α-naphthol, β-naphthol and naphthalene (m.p. 80° C), would also be considered to be melts, while liquid mixtures containing say α-naphthol and benzene, or α-naphthol, β-naphthol and ethyl alcohol could be classified as solutions. It must be pointed out, however, that no rigid definition is possible; the KNO_3—H_2O system quoted above, and the many well known cases of hydrated salts dissolving in their own water of crystallization at elevated temperatures would in all probability be considered as melts.

Solubility Diagrams

The composition of a solution can be expressed in a number of ways. One is to state the weight of solute present in a given volume of solution; for example, the concentration of a certain salt solution may be quoted as 20 g/l. Whilst this sort of expression may be convenient for the analytical laboratory it can be most inconvenient in industrial practice, since it is necessary to know the density of the solution before the relative weights of solute and solvent can be determined. Volume measures can be rather misleading as they are dependent upon temperature. Another method is to state the weight of solute present in a given weight of solution, but here again the relative weights of solute and solvent are not expressed precisely.

The solubility of a solute is most conveniently stated as the parts by weight per part (or 100 parts) by weight of solvent. To avoid confusion in the case of hydrated salts dissolved in water the solute concentration should always refer to the *anhydrous* salt. No difficulty will then arise in cases where several hydrated forms can exist over the temperature range considered.

All the above methods of solubility expression can lead to the use of the term 'percentage concentration', and unless precisely defined this term can be very misleading. For instance, a 10 per cent aqueous solution of sodium sulphate could, without further definition, be taken to mean any one of the following:

10 lb. of Na_2SO_4 (anhyd.) in 100 lb. of water
10 lb. of Na_2SO_4 (anhyd.) in 100 lb. of solution
10 lb. of $Na_2SO_4 \cdot 10H_2O$ in 100 lb. of water
10 lb. of $Na_2SO_4 \cdot 10H_2O$ in 100 lb. of solution

To show how misleading this loose form of expression can be, let 10 lb. of Na_2SO_4 (anhyd.) in 100 lb. of water be the correct description of the solution concentration. This would then be equivalent to

9·1 lb. of Na_2SO_4 (anhyd.) in 100 lb. of solution
20·6 lb. of $Na_2SO_4 \cdot 10H_2O$ in 100 lb. of solution
26·0 lb. of $Na_2SO_4 \cdot 10H_2O$ in 100 lb. of solution

Solubility data may also be recorded in terms of equivalent, molar or molal quantities; a normal solution (N) contains one g-equivalent of the solute per litre of solution, while a molar solution (M) contains one mole per litre. These expressions are particularly useful in laboratory practice, but both are temperature-dependent; the normality and molarity of a given solution decreases with an increase in temperature. A molal solution (m) contains one mole (or lb.-mol.) of solute per 1,000 g (or 1,000 lb.) of solvent, and concentrations are often expressed in terms of molality when phase changes occur in the solute-solvent system over a given temperature range.

Concentrations expressed as moles (or lb.-mol.) of solute per mole (or lb.-mol.) of mixture are frequently used in industrial practice, especially for multicomponent liquid mixtures. The mole fraction x of a particular component in a mixture of several substances is given by

$$x_1 = \frac{m_1/M_1}{m_1/M_1 + m_2/M_2 + m_3/M_3 + \dots} \quad (1)$$

and similarly

$$x_2 = \frac{m_2/M_2}{m_1/M_1 + m_2/M_2 + m_3/M_3 + \dots} \quad (2)$$

where m is the mass of a particular component, and M its molecular weight. For any mixture the sum of all the mole fractions is unity. The term 'mole percentage' ($100\,x$) is also used.

As a simple illustration of the use of these weight compositions, take the case of an aqueous solution of ammonium chloride (mol. wt. = 53·5) containing 31 lb. of NH_4Cl per 100 lb. of solution; this can be expressed in the following ways:

(*a*) *weight composition*
$= 31 \times 100/(100 - 31) = 45$ lb. NH_4Cl/100 lb. water

(*b*) *molal composition:*

$$NH_4Cl = 31/53{\cdot}5 \qquad = 0{\cdot}58 \text{ lb.-mole}$$
$$H_2O = (100 - 31)/18 \qquad = 3{\cdot}83 \text{ lb.-mole}$$
$$4{\cdot}41 \text{ lb.-mole}$$

molality $= 0{\cdot}58 \times 100/(100 - 31) = 8{\cdot}4$

(*c*) *mole fractions:*

$$NH_4Cl = 0{\cdot}58/4{\cdot}41 \qquad = 0{\cdot}13$$
$$H_2O = 3{\cdot}83/4{\cdot}41 \qquad = 0{\cdot}87$$
$$1{\cdot}00$$

In the majority of cases the solubility of a solute in a solvent increases with an increase in temperature, but there are a few well known exceptions to this rule. Some typical solubility curves for various salts in water are shown in

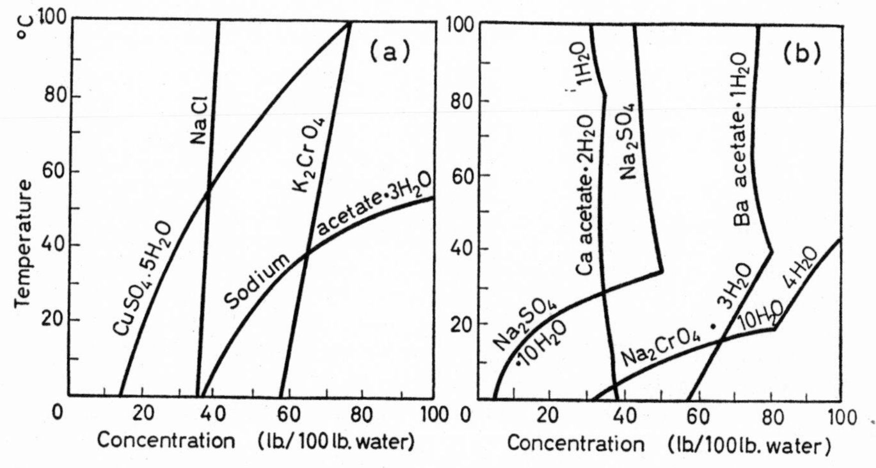

Figure 2.1. Solubility curves for some salts in water: (a) smooth curves; (b) indicating occurrence of phase changes

Figure 2.1, where all concentrations are expressed as lb. of anhydrous substance per 100 lb. of water. In *Figure 2.1a*, sodium chloride is a good example of a salt whose solubility increases only slightly with an increase in temperature, while sodium acetate shows a fairly rapid increase.

The solubility characteristics of a solute–solvent system have a considerable influence on the choice of a method of crystallization. It would be useless, for instance, to cool a hot saturated solution of sodium chloride in the hope of depositing crystals in any quantity; cooling from say 90 to 20° C would only produce about 7 lb. of NaCl for every 100 lb. of water present. The yield could be increased, however, by removing some of the water by evaporation, and this is what is done in practice. On the other hand, a direct cooling–crystallization operation would be adequate for a salt such as copper sulphate; cooling from 90 to 20° C would produce about 44 lb. of

$CuSO_4$ for every 100 lb. of water present in the original solution. As the stable phase of copper sulphate at 20° C is the pentahydrate, the actual crystal yield would be about 69 lb. of $CuSO_4 \cdot 5H_2O$ for every 100 lb. of water present initially.

Not all solubility curves are smooth, as can be seen in *Figure 2.1b*. A discontinuity in the solubility curve denotes a phase change. For example, the solid phase deposited from an aqueous solution of sodium sulphate below 32·4° C will consist of the decahydrate, whereas the solid deposited above this temperature will consist of the anhydrous salt. The solubility curves for two different phases meet at the transition point, and a system may show a number of these points. For instance, three forms of ferrous sulphate may be deposited from aqueous solution, depending upon the temperature: $FeSO_4 \cdot 7H_2O$ up to 56° C, $FeSO_4 \cdot 4H_2O$ from 56° to 64° C and $FeSO_4 \cdot H_2O$ above 64° C. In the case of sodium hydroxide no less than six hydrates and the anhydrous substance can be deposited from aqueous solution between − 24° and 62° C.

Referring back to the case of sodium sulphate (*Figure 2.1b*) it can be seen that above 32·4° C, when the anhydrous salt is the stable form, the solubility decreases with an increase in temperature. This negative solubility effect, or inverted solubility as it is sometimes called, is also exhibited by substances such as anhydrous sodium sulphite, calcium sulphate (gypsum), calcium, barium and strontium acetates, calcium hydroxide, etc. These substances can cause trouble in certain types of crystallizer by causing a deposition of scale on the heat transfer surfaces.

The general trend of a solubility curve can be predicted from Le Chatelier's Principle which, for the present purpose, can be stated as: when a system in equilibrium is subjected to a change in temperature or pressure, the system will adjust itself to a new equilibrium state in order to relieve the effect of the change. Most solutes dissolve in their near-saturated solutions with an absorption of heat (endothermic heat of solution); so an increase in the temperature of these solutions results in an increase in the solubility. An inverted solubility effect occurs when the solute dissolves in its near-saturated solution with an evolution of heat (exothermic heat of solution). Strictly speaking, solubility is also a function of pressure, but the effect is quite negligible in the systems encountered in crystallization from the liquid phase.

Many equations have been proposed for the correlation and prediction of solubility data, but none has been found to be of general applicability. In any case, an experimentally determined solubility is undoubtedly preferred to an estimated value. Nevertheless, there are two equations which are commonly used to express the influence of temperature on solubility, viz.

$$c = A + Bt + Ct^2 + \dots \tag{3}$$

and

$$\log x = \frac{a}{T} + b \tag{4}$$

In equation (3), c = mass of solute per given mass of solvent, t = temperature in °C or °F. Values of the constants A, B and C, etc. for a number of common solute–solvent systems may be found in the literature. The solubility in equation 4 is expressed as the mole fraction x of non-solvated solute in the solution, $T = °K$, and a and b are constants. This latter relationship can be extremely useful. The graphical interpolation and extrapolation of solubility values or the estimation of transition points from the conventional solubility plots, such as those shown in *Figure 2.1*, can prove difficult and unreliable when only a few experimental values are available, especially when the points lie on one or more different curved lines. However, if equation 4 can be applied—and it does apply reasonably well to a large number of systems—these difficulties can be minimized.

Solubility data can be plotted, in accordance with equation 4, in the manner shown in *Figure 2.2*. Mole fractions x are recorded on the logarithmic

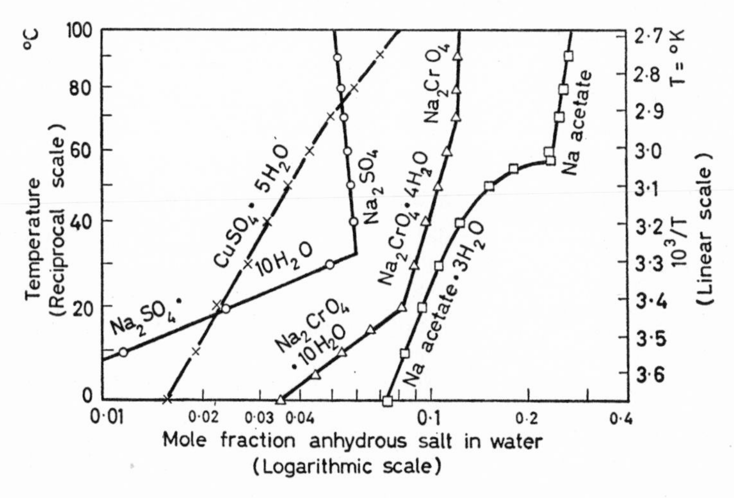

Figure 2.2. Alternative method for the graphical representation of solubility data

abscissa and the values of $10^3/T$ (because $1/T$ in the range 273–373° K is a rather small quantity) are recorded on the right-hand linear ordinate scale. Alternatively, special log-reciprocal graph paper can be used and the temperature in °C can be plotted direct. The left-hand ordinate in *Figure 2.2* is marked off on the reciprocal scale.

In *Figure 2.1* the solubility of $CuSO_4$ over the temperature range 0–100° C is represented by a smooth curve, while the solubilities of Na_2SO_4 and Na_2CrO_4 are represented by smooth curves which intersect at transition points. Several advantages of the log x versus $1/T$ plot shown in *Figure 2.2* immediately become apparent. First, the data for the above three salts lie on a series of straight lines. Secondly, and probably not so important, the data for the highly soluble salts sodium chromate and acetate can be recorded for the complete range 0–100° C on the same graph as that used for the data of the less soluble salts. Thirdly, the existence of transition points can be more clearly detected.

It is easier, for example, to produce the two straight lines for sodium sulphate in *Figure 2.2* to meet at 32·4° C than it is to extend the two corresponding curves in *Figure 2.1*. The two straight lines for $CuSO_4$ intersect at about 67° C indicating a phase transition at this temperature; this transition between two different crystalline forms of the pentahydrate was not detected in *Figure 2.1*. Incidentally, the transition $CuSO_4 \cdot 5H_2O \rightleftharpoons CuSO_4 \cdot 3H_2O$ occurs at 95·9° C. Only two of the transitions for the sodium chromate system are indicated in *Figure 2.2*. There are actually three transition points in this system: $10H_2O \rightleftharpoons 6H_2O$ (19·6° C), $6H_2O \rightleftharpoons 4H_2O$ (26·0° C) and $4H_2O \rightleftharpoons$ anhydrous (64·8° C).

The solubility data for sodium acetate are included in *Figure 2.2* to illustrate the fact that straight lines do not always result from this method of plotting. Curved lines are often obtained for highly soluble substances, or in regions where the temperature coefficient of solubility is high, or in cases where several hydrates can exist over a narrow range of temperature. It is possible, of course, that the curved portion of the sodium acetate line in the region of about 40–58° C is really a series of straight lines representing hydrates other than the trihydrate, but no evidence to support this view is available.

Saturation and Supersaturation

A solution which is in equilibrium with the solid phase is said to be saturated with respect to that solid. However, it is relatively easy to prepare a solution containing more dissolved solid than that represented by saturation condition, and such a solution is said to be supersaturated. Uncontaminated solutions in clean containers, cooled slowly without disturbance in a dust-free atmosphere, can readily be made to show appreciable degrees

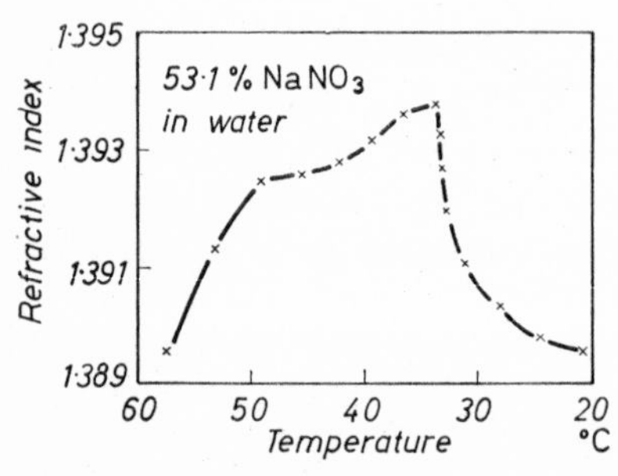

Figure 2.3. Showing the change in refractive index of an aqueous salt solution on cooling (*After* H. A. MIERS[3])

of supersaturation. The state of supersaturation is an essential feature of all crystallization operations. Wilhelm Ostwald (1897) first introduced the terms 'labile' (unstable) and 'metastable' supersaturation; they refer to supersaturated solutions in which spontaneous deposition of the solid phase, in the absence of solid nuclei, will and will not occur, respectively.

MIERS[2–5] carried out extensive researches into the relationship between supersaturation and spontaneous crystallization. He measured the refractive indices of concentrated aqueous salt solutions during a cooling process[3], and *Figure 2.3* shows one typical result: the change in refractive index of a 53·1 per cent solution of sodium nitrate in water with decrease in temperature.

A glass prism of known refractive index was immersed in the solution and the refractive index was measured by the method of total reflection within the prism using a goniometer. As the initially unsaturated solution was cooled, with stirring, from about 60° C, the refractive index increased considerably until a few small crystals appeared in the solution. The refractive index continued to increase at a slightly slower rate and reached a maximum value at about 36° C. At this point a copious separation of fine crystals occurred, accompanied by a sudden drop in the refractive index with no change in temperature. Further cooling reduced the refractive index to an approximately constant value at about 20° C. A 53 per cent solution of sodium nitrate is saturated at about 50° C, but the spontaneous deposition of crystals did not occur until the temperature of the solution had fallen to 36° C. Similar results were obtained with other concentrations of sodium nitrate and with aqueous solutions of other salts, with and without stirring. The degree of supercooling necessary to bring about a sudden reduction of the refractive index of the solution, and simultaneously cause a copious separation of crystals, was found to be lower for an agitated solution than for a quiescent one.

Miers's results can be represented in another manner, as shown diagrammatically in *Figure 2.4.* The lower continuous line is the normal solubility

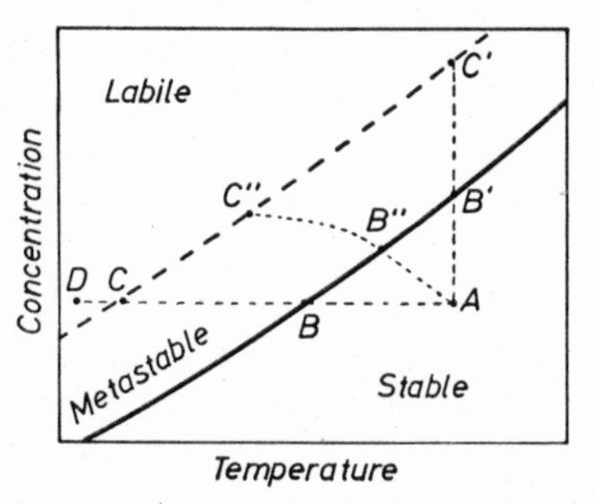

Figure 2.4. The solubility–supersolubility diagram

curve for the salt concerned. Temperatures and concentrations at which spontaneous crystallization occurs are represented by the upper broken curve, generally referred to as the supersolubility curve. This curve is not as well defined as the solubility curve and its position in the diagram depends, among other things, on the degree of agitation of the solution. Some workers prefer to picture the supersolubility 'curve' as a region or narrow band located in the supersaturated zone.

However, in spite of the fact that the supersolubility curve is ill-defined, there is no doubt that a region of metastability exists in the supersaturated region above the solubility curve. The diagram is therefore divided into three zones, one well defined and the other two variable to some degree:

(1) The stable (unsaturated) zone where crystallization is impossible.
(2) The metastable (supersaturated) zone, between the solubility and supersolubility curves, where spontaneous crystallization is improbable. However, if a crystal seed were placed in such a metastable solution, growth would occur on it.

(3) The unstable or labile (supersaturated) zone, where spontaneous crystallization is probable, but not inevitable.

If a solution represented by point A in *Figure 2.4* is cooled without loss of solvent (line ABC), spontaneous crystallization cannot occur until conditions represented by point C are reached. At this point, crystallization may be spontaneous or it may be induced by seeding, agitation or by mechanical shock. Further cooling to some point D may be necessary before crystallization can be induced, especially with very soluble substances such as sodium

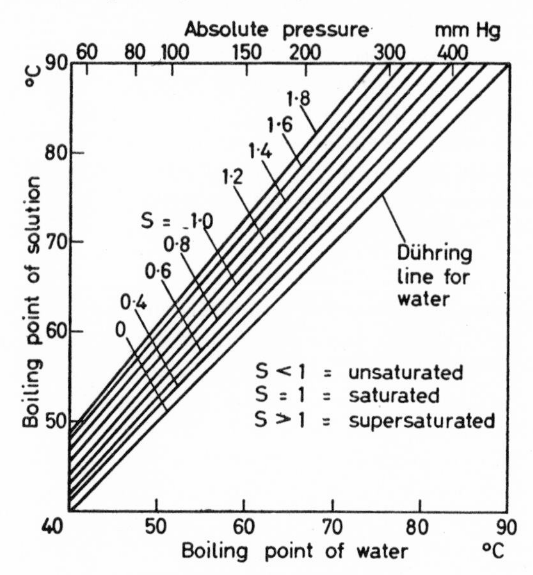

Figure 2.5. Dühring-type plot showing constant supersaturation lines (range $S = 0$ to 1·8) for aqueous solutions of sucrose. (*After* A. L. HOLVEN[6])

thiosulphate. Although the tendency to crystallize increases once the labile zone is penetrated, the solution may have become so highly viscous as to prevent crystallization and would set to a glass.

Supersaturation can also be achieved by removing some of the solvent from the solution by evaporation. Line $AB'C'$ represents such an operation carried out at constant temperature. Penetration beyond the supersolubility curve into the labile zone rarely happens, as the surface from which evaporation takes place is usually supersaturated to a greater degree than the bulk of the solution. Crystals which appear on this surface eventually fall into the solution and seed it, often before conditions represented by point C' are reached in the bulk of the solution. In practice, a combination of cooling and evaporation is employed, and such an operation is represented by the line $AB''C''$ in *Figure 2.4.*

A coefficient or degree of supersaturation S can be defined by

$$S = \frac{c}{c^*} \tag{5}$$

where c is the actual concentration of the substance in the solution (e.g. parts/100 parts of solvent) at some given temperature, and c^* the normal equilibrium saturation concentration in the pure solvent at the same temperature. The term 'percentage supersaturation' ($100\,S$) may also be employed.

HOLVEN[6] has devised a method for the determination of the degree of supersaturation of a solution from a knowledge of its boiling point elevation. He suggested that it should be possible to apply the principles of Dühring's rule—The boiling point of a solution is a linear function of the boiling point of the pure solvent at the same pressure—to the boiling points of aqueous sucrose solutions at various degrees of supersaturation. It was found that over the range of pressures normally encountered in sugar boiling practice a plot of the boiling point of the sugar solution, for a given degree of supersaturation, against the boiling point of water at the same absolute pressure yielded a straight line. Holven used these findings to develop an automatic method for recording and controlling the degree of supersaturation in sugar boilers. *Figure 2.5* shows a Dühring-type plot of the boiling points, over a range of reduced pressures, of water and pure sucrose solutions at various degrees of supersaturation.

Theoretical Crystal Yield

If the solubility data for a substance in a particular solvent are known, it is a simple matter to calculate the maximum yield of pure crystals that could be obtained by cooling or evaporating a given solution. The calculated yield will be a maximum because the assumption has to be made that the final mother liquor in contact with the deposited crystals will be just saturated. Generally, some degree of supersaturation may be expected, but this cannot be estimated. The yield will refer only to the quantity of pure crystals deposited from the solution, but the actual yield of solid material may be slightly higher than that calculated because crystal masses invariably retain some mother liquor even after filtration. When the crystals are dried they become coated with a layer of material which is frequently of a lower grade than that in the bulk of the crystals. Impure dry crystal masses produced commercially are very often the result of inadequate mother liquor removal.

Washing on a filter helps to reduce the amount of mother liquor retained by a mass of crystals, but there is always the danger of reducing the final yield by dissolution during the washing operation. If the crystals are readily soluble in the working solvent, another liquid in which the substance is relatively insoluble may be used. Alternatively, a wash consisting of a cold, near-saturated solution of the pure substance in the working solvent may be employed. The efficiency of washing depends largely on the shape and size of the crystals (see Chapter 7).

The calculation of the yield for the case of crystallization by cooling is quite straightforward if the initial concentration and the solubility of the substance at the lower temperature are known. The calculation can be complicated slightly if some of the solvent is lost, deliberately or accidentally, during the cooling process, or if the substance itself removes some of the solvent, e.g. by taking up water of crystallization. All these possibilities are taken into account in the following equations which may be used to calculate the maximum yields of pure crystals under a variety of conditions.

Let C_1 = initial solution concentration (lb. anhydrous salt/lb. solvent)
C_2 = final solution concentration (lb. anhydrous salt/lb. solvent)
W = initial weight of solvent (lb.)
V = solvent lost by evaporation (lb. per lb. of original solvent)
R = ratio of molecular weights of hydrate and anhydrous salt
Y = crystal yield (lb.)

Substance crystallizes unchanged (e.g. anhydrous salt)

Total loss of solvent: $$Y = WC_1 \quad (6)$$

No loss of solvent: $$Y = W(C_1 - C_2) \quad (7)$$

Partial loss of solvent: $$Y = W[C_1 - C_2(1 - V)] \quad (8)$$

Substance crystallizes as a solvate

Total loss of free solvent: $$Y = WRC_1 \quad (9)$$

No loss of solvent: $$Y = \frac{WR(C_1 - C_2)}{1 - C_2(R - 1)} \quad (10)$$

Partial loss of solvent: $$Y = \frac{WR[C_1 - C_2(1 - V)]}{1 - C_2(R - 1)} \quad (11)$$

Equation 11 can, of course, be used as the general equation for all cases.

Example

Calculate the theoretical yield of pure crystals that could be obtained from a solution containing 1,000 lb. of sodium sulphate (mol. wt. = 142) in 5,000 lb. of water by cooling to 50° F. The solubility of sodium sulphate at 50° F is 8·9 parts of anhydrous salt per 100 parts of water, and the deposited crystals will consist of the decahydrate (mol. wt. = 322). Assume that 2 per cent of the water will be lost by evaporation during the cooling process.

Solution

R = 322/142 = 2·27
C_1 = 0·2 lb. Na_2SO_4 per lb. water
C_2 = 0·089 lb. Na_2SO_4 per lb. water
W = 5,000 lb. water
V = 0·02 lb. per lb. of water present initially.

Substituting these values in equation 11:

$$Y = \frac{2{\cdot}27 \times 5{,}000[0{\cdot}2 - 0{\cdot}089(1 - 0{\cdot}02)]}{1 - 0{\cdot}089(2{\cdot}27 - 1)}$$

Yield = 1,445 lb. $Na_2SO_4 \cdot 10H_2O$

Determination of Solubility

The technique of solubility determination is basically simple, but it demands a high degree of practical skill and accuracy. A solution saturated with a given solid solute, as defined above, is one which is in equilibrium with the solid phase, and it is the achievement of true equilibrium that presents one of the biggest experimental difficulties. Prolonged and intimate contact is required between excess solid and solution at a constant temperature, generally for several hours. In certain cases contact for many days may be necessary; viscous solutions and systems at relatively low temperatures require long contact times. Both the solute and solvent should be of the highest purity possible, unless for some reason solubility data for commercial substances are required, and the solute particles should be small enough to facilitate fairly rapid dissolution. Very small particles, however, cannot settle readily in viscous solutions, and it is generally found that a close-sieved sample around 20-mesh size is suitable for most purposes.

Once equilibrium has been attained, the mixture is allowed to stand for an hour or more, still at constant temperature, to enable any finely dispersed solids to settle. A sample of the clear supernatant liquid is carefully withdrawn by means of a warmed pipette, and a weighed quantity (*not* a measured volume) of the sample is analysed. One point on the solubility curve can then be plotted. The analysis of solutions can be effected by conventional volumetric or gravimetric techniques, but in certain cases colorimetric methods or measurements of refractive index, density, conductivity, etc. may yield a more rapid, and sometimes a more reliable, result.

There is, however, one further step to be taken in order to complete the information, and that is to determine the composition of the solid phase that was in equilibrium with the solution at the given temperature. The stable phase can change appreciably over quite short ranges of temperature, especially in hydrated systems. For example, in the determination of the solubility of sodium carbonate in water over the temperature range 0 to 100° C it would be found that the stable solid phase is $Na_2CO_3 \cdot 10H_2O$ up to 32·0° C, $Na_2CO_3 \cdot 7H_2O$ between 32·0° and 35·4° C, and $Na_2CO_3 \cdot H_2O$ above 35·4° C. In order to determine the composition of the solid phase, a sample is taken from the container in which the equilibrium state was achieved and dried carefully at the temperature of the experiment, before commencing the analysis. The 'wet residue' and 'synthetic complex' methods described in Chapter 4 provide alternative procedures for the determination of solid phase composition.

A check can be made on a solubility determination at a given temperature by approaching equilibrium in two different ways: (*a*) from the unsaturated state and (*b*) from the supersaturated state. In the first method a quantity of solid, in excess of the amount required to saturate the solvent at the given temperature, is added to the solvent and the two are agitated until equilibrium is reached. In the second method the same quantities of solute and solvent are mixed, but the system is then heated above the required temperature, if solubility increases with temperature, so that most of the solid is dissolved. The solution is agitated for a long period at the required temperature and the excess solid is deposited. If the two solubility determinations agree it can be reasonably assumed that the result represents the true equilibrium saturation concentration at the given temperature.

Constant temperature control during the experimental procedure is essential though its limits vary according to the system under investigation and the required precision of the solubility. Far greater care has to be taken when the solubility changes appreciably with a change in temperature. In the determination of the solubility of, say, sodium chloride in water at 30° C (*Figure 2.1a*), a variation of ± 0·5° in the experimental temperature would allow for a potential precision of about 0·1 per cent in the solubility, but the same temperature variation would only allow for a potential precision of about 5 per cent in the case of sodium sulphate at 30° C (*Figure 2.1b*). It should be quite obvious that the thermometer used in the thermostat must be accurately calibrated with reference to a standard thermometer.

A simple apparatus for the determination of solubility consists of a thermostatically controlled water bath (or oil bath for temperatures above about 80° C), a stoppered flask or test tube of about 50 cm^3 capacity and a device for shaking the flask while it is immersed in the bath. Alternatively, the flask can be fitted with a stirrer; a very compact apparatus operated on this principle has been described in detail by PURDON and SLATER[7].

The sampling tube, or pipette, can be left standing in a stoppered tube immersed in the bath so that it attains the same temperature as the solution under investigation. A filter-tip consisting of a sintered glass end-piece, or more simply a piece of glass wool, can be provided on the sampling tube to prevent finely dispersed solids being drawn up with the sample of solution. Some sampling pipettes, e.g. the Landolt type, can be used as the weighing vessel. ZIMMERMAN[8] has made an extensive review of the literature up to 1950 on the subject of experimental solubility determination.

An apparatus for the direct and rapid measurement of the saturation temperature of solutions has been devised by DAUNCEY and STILL[9]. This new technique is based on an optical effect caused by the slight change in

Figure 2.6. Appearance of the illuminated slit during saturation temperature measurements: (a) unsaturated; (b) saturated; (c) supersaturated solutions in contact with the flat crystal face. (*After* L. A. DAUNCEY *and* J. E. STILL[9])

concentration, and therefore in refractive index, occurring in a layer of solution immediately in contact with a crystal that is either growing or dissolving. The saturation cell is a small Perspex container fitted with a stirrer (or the solution may be passed continuously through it), a calibrated thermometer and a holder for a medium-sized crystal. The cell is placed in a thermostatically controlled water bath also made of Perspex. A beam of light from an optical slit is directed on to an edge of the crystal, and the appearance of the slit when viewed from behind the crystal will take the form of one of the three sketches shown in *Figure 2.6.* The light is bent into an obtuse angle when the solution is unsaturated (*a*), and an acute angle when it is supersaturated (*c*). As soon as it is determined that the solution near the crystal face is unsaturated or supersaturated, the temperature controller is adjusted until view (*b*) is obtained.

Dauncy and Still worked with ethylenediamine hydrogen tartrate and ammonium dihydrogen phosphate and reported that points on the solubility curves could be plotted at the rate of 8–10 per hour. Wise and Nicholson[10] have adapted this method and applied it with considerable success in the determination of sugar solubilities. Kelly[11] has reviewed these and other techniques in some detail.

Effect of Particle Size on Solubility

If the particles of a solute suspended in a solvent are small enough, concentrations greater than that represented by the normal solubility of the substance can temporarily be obtained. Numerous attempts have been made to correlate the relationship between particle size and solubility, the first being due to Ostwald (1900) who derived thermodynamically an equation which was later corrected by Freundlich (1909) in the form

$$\ln \frac{c_1}{c_2} = \frac{2M\sigma}{RT\rho}\left(\frac{1}{r_1} - \frac{1}{r_2}\right) \qquad (12)$$

where

c_1 and c_2 = solubilities of spherical particles of radius r_1 and r_2, respectively
R = gas constant
T = absolute temperature
ρ = density of the solid
M = molecular weight of the solid in solution
σ = surface energy of the solid particle in contact with the solution

If the normal equilibrium solubility of a substance, i.e. the solubility of large particles with flat surfaces ($r \rightarrow \infty$), is denoted by c^*, the solubility c_r of a particle of radius r can be expressed as

$$\ln \frac{c_r}{c^*} = \frac{2M\sigma}{RT\rho r} \qquad (13)$$

and this relationship has been found[12] to hold fairly well for dilute aqueous solutions of gypsum ($CaSO_4 \cdot 2H_2O$) for particle sizes 0·5 to 50μ.

The Ostwald–Freundlich equation involves a number of assumptions which are not strictly valid. For instance, both the density of the solid and the solid–liquid surface energy are assumed to be independent of particle size. In addition, the particles are considered to be spherical, and no account is taken of any dissociation of the solid in solution. Jones[13, 14] modified the equation to include cases of particles with various geometric shapes and to allow for degrees of ionization or dissociation of the dissolved solid. Later[15] this modified equation was used to provide a general theory of supersaturation in which a critical size of the crystal nucleus necessary for growth to commence was specified. The Jones equation, which allowed for ionization effects, was rather complex and solubilities could not be calculated directly from it. DUNDON and MACK[16] presented a simpler relationship which can be written in a form similar to equation 12:

$$\ln \frac{c_1}{c_2} = \frac{2M\sigma}{\boldsymbol{R}T\rho}\left(\frac{1}{r_1} - \frac{1}{r_2}\right)(1 - \alpha + n\alpha) \tag{14}$$

where α is the degree of dissociation, n the number of ions formed from the dissociation of one molecule. If large particles are considered ($r_1 \rightarrow \infty$), $c_1 \rightarrow c^*$ and equation 14 becomes

$$\ln \frac{c_r}{c^*} = \frac{2M\sigma}{\boldsymbol{R}T\rho r}(1 - \alpha + n\alpha) \tag{15}$$

HULETT[17, 18] investigated the effect of fine grinding on the solubilities in water of various substances and reported a 19 per cent increase in the solubility of gypsum for $0{\cdot}4\mu$ particles and 81 per cent increase for $0{\cdot}1\mu$ particles of $BaSO_4$. DUNDON[19] measured the increases in the solubilities of many other salts and found an apparent relationship between the surface energy σ, molecular volume (mol. wt./density) and hardness of the crystalline substances (see *Table 2.1*).

Table 2.1. Particle Size and Solubility Increase. (After DUNDON[19])

Substance	*Particle size* μ	*Solubility increase* per cent	*Molecular volume*	*Surface energy* erg/cm²	*Hardness (Mohs)*
PbI_2	0·4	2	74·8	130	v. soft
$CaSO_4{\cdot}2H_2O$	0·2–0·5	4–12·5	74·2	370	1·6–2
Hg_2CrO_4	0·3	10	60·1	575	1·6–2
PbF_2	0·3	9	29·7	900	1·6–2
$SrSO_4$	0·25	26	46·4	1,400	3–3·5
$BaSO_4$	0·1	80	52·0	1,250	2·5–3·5
CaF_2	0·3	18	24·6	2,500	2·5–4

Equations 12–15 suffer from the serious defect that they postulate a continual increase of solubility with reduction in particle size. To overcome this anomaly KNAPP[20] considered that the particles carried a small surface electric charge, so that the total surface energy would be the sum of the normal surface energy and the electric charge. From these considerations he derived, for the case of isolated charged spheres, an equation written in the form

$$c_r = c^* \exp\left(\frac{a}{r} - \frac{b}{r^4}\right) \tag{16}$$

where $a = 2\sigma M/\boldsymbol{R}T\rho$ and $b = Q^2M/8\pi K\boldsymbol{R}T\rho$, Q being the electric charge on the particle and K the dielectric constant of the solid. Equation 14 gives a curve of the type shown in *Figure 2.7*. As the particle size is reduced, the solubility increases above the normal solubility until some maximum value is reached. On account of the repulsion between the charged particles, any further reduction of particle size would result in a reduction of solubility. The work of DUNDON and MACK[16, 19] appears to afford some confirmation of Knapp's postulation. Solubility determination with $CaSO_4 \cdot 2H_2O$ in the size range 0·2–0·5μ, for instance, showed a maximum at about 0·3μ[19].

HARBURY[21] proposed a modified form of equation 15 to allow for the fact that the surface energy σ is not independent of particle radius at the very low values of r encountered in crystal nuclei (0·01–0·001μ). He also suggested that σ should be replaced by a quantity σ' which, though being of the same dimensions as σ, no longer represents the specific free surface energy of a macro crystal lattice. In point of fact σ' was really a 'catch-all' for several correction factors, and at extremely low values of r, σ' is only a small fraction

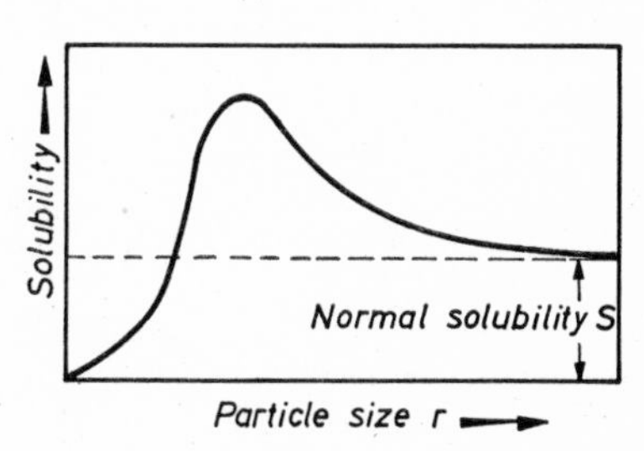

Figure 2.7. The effect of particle size on solubility. (After L. F. KNAPP[20])

of σ. Harbury measured the maximum increases in the solubilities of a number of fairly soluble inorganic salts, such as KNO_3, $KClO_3$, $K_2Cr_2O_7$ and $Na_2SO_4 \cdot 10H_2O$ in water. Values of the ratio c_{max}: c^* were found to be 2·75, 3·0, 6·0 and 13·0, respectively. He suggested that this ratio was a function of a quantity which he called the 'molar surface' of the substance, defined as $(M/\rho)^{\frac{2}{3}}$. For the above-mentioned inorganic salts it was found that

$$\ln \frac{c_{max}}{c^*} \simeq 3{\cdot}3(M/\rho)^{\frac{2}{3}} \qquad (17)$$

Effect of Impurities on Solubility

Pure solutions are rarely encountered outside the analytical laboratory. Industrial solutions are almost invariably impure, and the presence of an impurity can have a considerable effect on the solubility characteristics of a system. If to a saturated binary solution of A (a solid solute) and B (a liquid solvent) a third component C (also soluble in B) is added, one of four conditions can result. First—this is comparatively rare—nothing may happen: the system remains in its original saturated state. Secondly, component C may combine or react chemically with A by forming a complex or compound, thus altering the whole nature of the system. These two cases will not be considered here.

The third and fourth possibilities, however, are extremely important. The presence of component C may make the solution undersaturated or supersaturated with respect to solute A. In the former case, A would be precipitated or 'salted-out'. In the latter, the solution would be capable of dissolving more A or, in other words, A would be 'salted-in'.

Take, for example, the case of a solution of salol (phenyl salicylate) in ethyl alcohol at 25° C. At this temperature a saturated solution contains 54 parts of salol per 100 parts of ethyl alcohol. If, however, the aqueous ethanol azeotrope had been used as the solvent (95·5 per cent ethyl alcohol by weight), a saturated solution at 25° C would contain only 31 parts of salol per 100 parts of solvent (i.e. 32·5 parts/100 parts of ethyl alcohol present). Here the presence of a small quantity of water, which may be considered as an impurity, causes a reduction in the solubility of salol.

On the other hand, a saturated solution of salicylic acid in benzene at 25° C contains 1 part of salicylic acid per 100 parts of benzene, while a saturated solution at the same temperature in a solvent containing 95 parts of benzene per 5 parts of acetone contains 4 parts per 100 parts of solvent (i.e. 4·25 parts/100 parts of benzene present). In this case, the presence of the 'impurity' (acetone) causes an increase in the solubility of the salicylic acid.

The addition of a soluble salt to a saturated aqueous solution of another salt will generally result in the precipitation of some of the latter if the two salts have an ion in common. Salting-out due to the common ion effect can be explained by the *solubility product* principle first proposed by Nernst (1889). In a saturated solution of a sparingly soluble salt, MR, dissociation may be assumed to be complete because of the very low electrolyte concentration. If one molecule of MR dissociates into n positive ions and n' negative ions according to the equation

$$M_nR_{n'} \rightleftharpoons nM^+ + n'R^-$$

then

$$(c_{M^+})^n(c_{R^-})^{n'} = k_s \tag{18}$$

where c_{M^+} and c_{R^-} are the ionic concentrations (e.g. g-ions/litre) and k_s is the solubility product. If the solubility of a uni-univalent salt ($n = n' = 1$) at a given temperature is denoted by s^* (e.g. mole/litre) then

$$c_{M^+} = c_{R^-} = s^*$$

and

$$(c_{M^+})(c_{R^-}) = k_s \tag{19}$$

$$= s^{*2} \tag{20}$$

The addition of a soluble salt N^+R^- to a saturated solution of M^+R^- would result in a reduction in the solubility of salt M^+R^-. The lower solubility, s, produced by the addition of e g-ions/litre of R^- ions is given by

$$s = \tfrac{1}{2}(e^2 - 4s^{*2})^{\frac{1}{2}} - \tfrac{1}{2}e \tag{21}$$

$$= \tfrac{1}{2}(e^2 - 4k_s)^{\frac{1}{2}} - \tfrac{1}{2}e \tag{22}$$

The common ion effect is of great importance in gravimetric analysis when an almost complete precipitation of one substance is essential.

Strictly speaking, the simple solubility product principle can only be applied to solutions of sparingly soluble salts (about 0·01 mole/litre or less), and a more fundamental approach involves the use of the chemical potential and activity concepts.

(For a complete account of these concepts and their application to aqueous solutions, reference should be made to specialized works, e.g. J. H. Hildebrand and R. L. Scott, *Solubility of Non-electrolytes*, 3rd edition, 1950 (New York; Reinhold) or S. Glasstone, *Textbook of Physical Chemistry*, 2nd edition, 1953 (London, Macmillan), 954–974.)

If one molecule of a solute dissociates in solution according to the equation

$$M_nR_{n'} \rightleftharpoons nM^+ + n'R^-$$

then the chemical potential μ of the undissociated solid in equilibrium with the saturated solution must be equal to the sum of the chemical potentials of the ions in solution, thus

$$\mu_{M_nR_{n'}} = n\mu_{M^+} + n'\mu_{R^-} \quad (23)$$
$$= n\boldsymbol{R}T \ln a_{M^+} + n\boldsymbol{R}T \ln a_{R^-} \quad (24)$$

where a_{M^+} and a_{R^-} are the activities of the M^+ and R^- ions and $\boldsymbol{R}$ is the gas constant. As long as the solution is saturated with respect to MR, the chemical potential of the solid $\mu_{M_nR_{n'}}$ will remain constant, therefore

$$n\boldsymbol{R}T \ln a_{M^+} + n\boldsymbol{R}T \ln a_{R^-} = \text{constant} \quad (25)$$

and

$$(a_{M^+})^n(a_{R^-})^{n'} = \text{constant} = K_s \quad (26)$$

where K_s is the activity product or the activity solubility product. For a uni-univalent salt

$$(a_{M^+})(a_{R^-}) = K_s \quad (27)$$

The activity, a, of an ion may be expressed in terms of ionic concentration, c, and the activity coefficient, f, by the relationship

$$a = cf \quad (28)$$

Therefore,

$$(c_{M^+}f_{M^+})(c_{R^-}f_{R^-}) = K_s \quad (29)$$

For a sparingly soluble salt the activity coefficients f_{M^+} and f_{R^-} are almost equal to unity; so equation 29 reduces to equation 19.

A number of cases which appear anomalous when the simple solubility product is used can be explained when activity coefficients are taken into account. For instance, the addition of a common ion generally decreases the solubility of a salt, but cases are known where large additions of a salt with a common ion result in increased solubility. The reason for this is that a large increase in ionic concentration can bring about a reduction in the activity coefficients. Thus from equation 29 an increase in c_{R^-} will result in a decrease in c_{M^+}, i.e. precipitation of MR, if f_{M^+} and f_{R^-} remain fairly constant, but an increase in c_{R^-} to a value which reduces both f_{M^+} and f_{R^-} must result in an increase in c_{M^+} if K_s is to remain constant. The addition of a salt without a common ion usually increases the solubility; this again is the result of the increased ionic concentration reducing the activity coefficients, but the salting-in effect is much less pronounced than the salting-out effect considered above.

The salting-out effect of an electrolyte produced when it is added to an aqueous solution of a non-electrolyte, at one given temperature, can often be represented by the equation

$$\log \frac{s}{s^*} = kC \quad (30)$$

where s^* and s are the equilibrium saturation concentrations (mole/litre) of the non-electrolyte in pure water and a salt solution of concentration C (mole/litre), respectively. The constant k is called the salting parameter, and it refers to one particular electrolyte and its effect on one particular non-electrolyte. This type of relationship will often apply with a reasonable accuracy for low non-electrolyte concentrations and electrolyte concentrations up to 4 or 5 mole/litre.

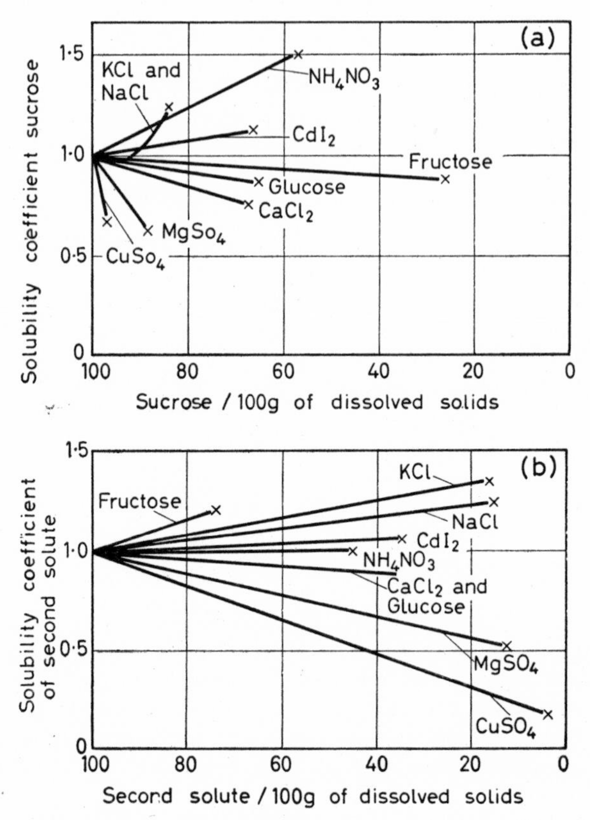

Figure 2.8. Influence of (a) second solute on the solubility coefficient of sucrose, (b) sucrose on the solubility coefficient of the second solute. (*After* F. H. C. KELLY[23])

Occasionally the presence of an electrolyte increases the solubility of a non-electrolyte in water (negative value of k), and this salting-in effect is exhibited by several salts with large anions or cations, which themselves are very soluble in water. Sodium benzoate and sodium p-toluenesulphonate are good examples of these 'hydrotropic' salts, and the phenomenon of salting-in is sometimes referred to as 'hydrotropism'. Values of the salting parameter k for three salts applied to benzoic acid in aqueous solution are: NaCl (0·17), KCl (0·14) and sodium benzoate (– 0·22). LONG and MCDEVIT[22] have made a comprehensive review of salting-in and salting-out phenomena.

The salting effect of an impurity can also be considered in terms of a simple solubility ratio c/c^*, where c and c^* are the equilibrium saturation concentrations (e.g. parts/100 parts of solvent) of the given solute in the impure solution and pure solvent, respectively, at one given temperature. KELLY[23, 24] used this approach and called the solubility ratio the 'solubility coefficient'. *Figure 2.8a* shows the influence of certain substances on the solubility of sucrose in water; the solubility coefficient of sucrose is plotted against the 'purity' of the sucrose solution. *Figure 2.8b* shows the effect of sucrose on the solubilities of several other substances. Kelly concluded that solutes which form a hydrate have a salting-out effect on sucrose within the temperature ranges over which the hydrates are stable, while solutes which do not form a hydrate have a salting-in effect. Solutes which increased the solubility of sucrose also had their own solubilities increased, and *vice versa*, although fructose and ammonium nitrate proved exceptions to this generalization.

When considerable quantities of a soluble impurity are present in, or deliberately added to, a binary solution, the system is best analysed with reference to a ternary equilibrium diagram (see Chapter 4).

Choice of a Solvent

For the industrial crystallization of inorganic substances from solution the use of water as the solvent is almost exclusive. This fact is quite understandable because, apart from the relative ease with which a very large number of chemical compounds dissolve in it, water is readily available, cheap and quite innocuous. For these reasons water is used whenever possible in the industrial crystallization of organic compounds. Yet there are many cases in this particular field where for a variety of reasons some other solvent must be employed.

The selection of the best solvent for a given crystallization operation is not always an easy matter. Many factors must be considered and some compromise must inevitably be made; several undesirable characteristics may have to be accepted to secure the aid of one important solvent property. There are several hundred organic liquids that are potentially capable of acting as crystallization solvents, but outside the laboratory the list can be shortened to a few dozen selected from the following groups: acetic acid and its esters; lower alcohols and ketones; ethers; chlorinated hydrocarbons, e.g. chloroform and carbon tetrachloride; benzene, toluene, xylene and light petroleum fractions.

Occasionally, a mixture of two or more solvents will be found to possess the best properties for crystallization purposes. Common binary solvent mixtures that have proved useful include alcohol–water, alcohol–ketone, alcohol–ether, alcohol–carbon tetrachloride, alcohol–chloroform, alcohol–benzene, alcohol–toluene, alcohol–xylene, benzene–hexane and hexane–trichlorethylene. Sulphuric acid, of various strength, can often be used to advantage as a solvent for certain organic compounds.

Broadly speaking, the factors that have to be considered when choosing a solvent can be grouped under the headings: solvent properties, economic aspects and industrial hazards. Several well-known publications[25–29] deal comprehensively with these aspects. The following brief notes indicate the main points in the selection process.

Solvent Power

The solute to be crystallized should be capable of being dissolved fairly easily in the solvent; it should also be capable of being deposited easily from the solution in the desired crystalline form after cooling, evaporation or dilution. It is important to remember that the crystal habit can often be changed by changing the solvent; for example, one solvent may favour the precipitation of needle crystals while another may induce prisms.

There are many exceptions to the frequently quoted rule that 'like dissolves like', but this rough empiricism can serve as a useful guide. Solvents may be classified as being polar or non-polar; the former description is given to liquids which have high dielectric constants, e.g. water, acids, alcohols, while the latter refers to liquids of low dielectric constant, e.g. aromatic hydrocarbons. A non-polar solute (e.g. anthracene) is generally more soluble in a non-polar solvent (e.g. benzene) than in a polar solvent (e.g. water). However, close chemical similarity between solute and solvent should be avoided because their mutual solubility will in all probability be high, and crystallization may prove difficult or uneconomical.

The 'power' of a solvent is usually expressed as the mass of solute that can be dissolved in a given mass of pure solvent at one specified temperature. Water, for example, is a more powerful solvent at 20° C for calcium chloride than say *n*-propanol (75 and 16 g/100 g of solvent, respectively). At the same temperature *n*-propanol is a more powerful solvent than water for say benzoic acid (42·5 and 0·29 g/100 g of solvent, respectively).

In cooling–crystallization the temperature coefficient of solubility is another important factor to be considered. For example, at 20° C water is a more powerful solvent for potassium sulphate (11 g/100 g of water) than for potassium chlorate (7 g/100 g), but the converse is true at 80° C (K_2SO_4 = 21 g, $KClO_3$ = 39 g/100 g). Thus, on cooling the saturated solutions from 80 to 20° C, about 82 per cent of the dissolved $KClO_3$ is deposited compared with about 47 per cent of the K_2SO_4.

Both the solvent power and the temperature coefficient of solubility must be considered when choosing a solvent for a cooling–crystallization process; the former quantity decides the volume of the crystallizer, while the latter determines the solute yield. It frequently happens, especially in aqueous organic systems, that a low solubility is combined with a high temperature coefficient of solubility. For example, the solubilities of salicylic acid in water at 20 and 80° C are 0·20 and 2·26 g/100 g, respectively. Therefore, on cooling from 80 to 20° C, most of the dissolved solute (91 per cent) is deposited, and consequently the solute yield is high. However, on account of the low solubility even at the higher temperature, such a large crystallizer would be required to give a reasonable through-put that water could not be considered as a solvent for salicylic acid.

Potassium chromate is an example of a solute with a reasonably high solubility in water but a low temperature coefficient of solubility (61·7 and

72·1 g/100 g at 20 and 80° C, respectively). The low yield on cooling (about 14 per cent) makes it necessary to effect crystallization in some other manner, such as a combination of cooling and evaporation, thus increasing the cost of the operation.

Purity

No deleterious impurity, dissolved or suspended, should be introduced into the crystallizing system. The solvent, therefore, should be as clean and as pure as possible. No colouring matter should be permitted to affect the appearance of the final crystals, and no residual odours should remain in the product after drying. This latter problem is often encountered after crystallization from petroleum and coal tar solvents. If no previous experience has been obtained with a solvent, simple laboratory trials should be made, but due caution should be exercised as laboratory filtration, washing and drying techniques generally prove to be far more efficient than the corresponding large-scale operations.

Chemical Reactivity

The solvent should be stable under all foreseeable operating conditions; it should neither decompose nor oxidize, and it must not attack any of the materials of construction of the plant. When organic solvents are being used, care must be taken in choosing the correct gasket materials; most common types of rubber, for example, swell and disintegrate after prolonged contact with hydrocarbons such as benzene and toluene.

The solute and solvent should not be capable of reacting together chemically; glacial acetic acid, for instance, would hardly be chosen if the solute were liable to be acetylated. Solvate formation, however, may be permitted under certain circumstances. Hydrated crystals are frequently desired as end products, but should the anhydrous substance be required the drying process may prove difficult and expensive. Methanol, ethanol, benzene and acetic acid are also known to form solvates with certain substances, and the loss of solvent on drying imposes an additional cost on the process.

The removal of a solvent from a solvate by drying at elevated temperatures generally results in the destruction of the crystalline nature of the product. Freeze drying, i.e. sublimation of the solvent, may prevent this undesirable feature but the costs of the operation are invariably prohibitive. Occasionally, the difficulties may be overcome by washing the crystalline mass with a liquid which acts as a solvent for the mother solvent and a non-solvent for the solute.

Handling and Processing

Highly viscous solvents are not generally conducive to efficient crystallization, filtration and washing operations. In general, therefore, solvents of low viscosity are preferred. If the solvent recovery process involves distillation, a reasonably volatile solvent is desirable, the latent heat of vaporization should be low and possible azeotrope formation should be avoided. On the other hand, the loss of a solvent with a high vapour pressure from filters and other processing equipment can be considerable and may prove both costly and hazardous. Solvents with freezing points above about $-5°$ C present wintertime storage and transportation difficulties. Benzene (f.p. 5° C) and glacial acetic acid (f.p. 16° C) are good examples of commonly used solvents which suffer from this defect.

Inflammability and Toxicity

A large proportion of the organic solvents employed in crystallization processes are inflammable, and their use necessitates stringent operating conditions. Two of the most important properties of an inflammable solvent are the flash point and explosive limits; the former is defined as the temperature at which the mixture of air and vapour above the liquid can be ignited by means of a spark, and the latter generally refers to the percentages by volume in air between which the air-vapour mixture will explode in a confined space. Diethyl ether is an example of a solvent with a very low flash point ($-30°$ C) and wide range of explosive limits (1–50 per cent).

Rigidly enforced safety precautions should be taken on the plant, and operating personnel must be made fully aware of the potential dangers. Costly flame-proof lights and electrical equipment have to be installed, and all vessels and pipelines should be earthed as a precaution against the build-up of static electricity. Maintenance tools should be of the non-ferrous type, and operators should wear rubber-soled footwear to avoid the accidental production of a spark. Adequate fire-fighting equipment is necessary and safe solvent storage areas must be provided.

All organic solvents are toxic to a greater or lesser degree; the prolonged inhalation of almost any vapour will produce some harmful effect on a human being. Some solvents are acute poisons; some have a cumulative poisoning effect while others produce narcosis or intoxication on inhalation, or dermatitis when contacted with the skin. Details on these aspects and on the maximum vapour concentrations permitted in working areas can be obtained from the specialized reference books.

Adequate ventilation should be provided in the plant, and all vessels and containers should be leak-free. Many of the precautions discussed above with regard to inflammable solvents should be enforced, particularly those concerning the operating personnel.

REFERENCES

[1] GUTHRIE, F., Salt solutions and attached water, *Phil. Mag.* 18 (1884) 22, 105
[2] MIERS, H. A., Variation in angles of alum crystals, *Phil. Trans.* A202 (1904) 459
[3] MIERS, H. A. and ISAAC, F., Refractive indices of crystallizing solutions, *J. chem. Soc.* 89 (1906) 413
[4] MIERS, H. A. and ISAAC, F., The spontaneous crystallization of binary mixtures, *Proc. roy. Soc.* A79 (1907) 322
[5] MIERS, H. A., The growth of crystals in supersaturated liquids, *J. Inst. Metals* 37 (1927) 331
[6] HOLVEN, A. L., Supersaturation in sugar boiling operations, *Industr. Engng Chem.* 34 (1942) 1234
[7] PURDON, F. F. and SLATER, V. W., *Aqueous Solution and the Phase Diagram*, 1946. London; Arnold

[8] ZIMMERMAN, H. K., The experimental determination of solubilities, *Chem. Rev.* 51 (1952) 25
[9] DAUNCEY, L. A. and STILL, J. E., Apparatus for the direct measurement of the saturation temperatures of solutions, *J. appl. Chem.* 2 (1952) 399
[10] WISE, W. S. and NICHOLSON, E. B., Solubilities and heats of crystallization of sucrose and methyl α-*d*-glucoside in aqueous solution, *J. chem. Soc.* (1955) 2714
[11] KELLY, F. H. C., The solubility of sucrose in impure solutions, in *Principles of Sugar Technology* Vol. 2, ed. P. Honig, 1959. Amsterdam; Elsevier
[12] JONES, W. J. and PARTINGTON, J. R., Experiments on supersaturated solutions, *J. chem. Soc.* 107 (1915) 1019
[13] JONES, W. J., Über die Grösse der Oberflächenenergie fester Stoffe, *Z. phys. Chem.* 82 (1913) 448
[14] JONES, W. J.,Über die Beziehung zwischen geometrischer Form und Dampfdruck, Löslichkeit und Formenstabilität, *Ann. Phys. Lpz.* 41 (1913) 441
[15] JONES, W. J. and PARTINGTON, J. R., A theory of supersaturation, *Phil. Mag.* 29 (1915) 35
[16] DUNDON, M. L. and MACK, E., The solubility and surface energy of calcium sulphate, *J. Amer. chem. Soc.* 45 (1923) 2479
[17] HULETT, G. A., Beziehungen zwischen Oberflächenspannung und Löslichkeit, *Z. phys. Chem.* 37 (1901) 385
[18] HULETT, G. A., The solubility of gypsum as affected by particle size and crystallographic surface, *J. Amer. chem. Soc.* 27 (1905) 49
[19] DUNDON, M. L., The surface energy of several salts, *J. Amer. chem. Soc.* 45 (1923) 2658
[20] KNAPP, L. F., The solubility of small particles and the stability of colloids, *Trans. Faraday Soc.* 17 (1922) 457
[21] HARBURY, L., Solubility and melting point as functions of particle size, *J. phys. Chem.* 50 (1946) 190
[22] LONG, F. A. and MCDEVIT, W. F., Activity coefficients of non-electrolyte solutes in aqueous salt solutions, *Chem. Rev.* 51 (1952) 119
[23] KELLY, F. H. C., Phase equilibria in sugar solutions (Parts I–V, Ternary Systems), *J. appl. Chem.* 4 (1954) 401
[24] KELLY, F. H. C., Phase equilibria in sugar solutions (Parts VI–X, Quaternary and Quinary System), *J. appl. Chem.* 5 (1955) 66, 120, 170
[25] DURRANS, T. H., *Solvents*, 1930. London; Chapman & Hall
[26] JORDAN, O., *The Technology of Solvents* (Transl. A. D. Whitehead), 1940. London; Leonard Hill
[27] MELLAN, I., *Industrial Solvents*, 2nd Edition, 1950. New York; Reinhold
[28] SAX, N. I., *Handbook of Dangerous Materials*, 1951. New York; Reinhold
[29] *Toxicity of Industrial Organic Solvents*, 1937. London; H.M. Stationery Office

3

PHYSICAL AND THERMAL PROPERTIES

PHYSICAL PROPERTIES

Density

THE concentration of commercial solutions is often expressed in terms of a density measurement. Density is defined as mass per unit volume, and the most common expression of this quantity encountered in the scientific literature has the units g/cm^3. In industrial and engineering practice the units lb./ft.^{3}, kg/m^3 and lb./gal. (U.S. and Imperial) are frequently employed. It must be remembered, however, that the density of a solution is affected by temperature as well as by concentration.

The ratio of the density of a liquid at one temperature to the density of water at another is known as the *specific gravity* of the liquid; thus for complete definition all expressions of specific gravity must be accompanied by a temperature designation. For instance, the specific gravity of a solution quoted as 1·23 (20°/20° C) refers to the ratio of the densities of the solution and water both at 20° C, while 1·23 (20°/4° C) refers to the solution at 20° C and water at 4° C. At 4° C (39·2° F) water exhibits its maximum density of 1·000 g/cm^3; thus a specific gravity expressed with reference to water at 4° C is numerically equal to the density of the solution in g/cm^3 at the given temperature.

Liquid densities are most conveniently measured by the use of a hydrometer which consists of a hollow cylinder or float, loaded with lead shot or mercury, on which is mounted a graduated stem. The depth to which the stem sinks in the liquid gives a measure of the density. As density varies with temperature, hydrometers are marked with the temperature at which they were calibrated. Liquid density can also be determined by the specific gravity bottle, pyknometer or Westphal balance methods, details of which are given in most textbooks of practical physics.

For reasons of convenience, several density scales have been developed for industrial use, and hydrometers graduated in these scales are readily available. In Great Britain, for instance, the *Twaddell* scale is widely employed, especially in the acid and alkali industries. It is used for liquids heavier than water (pure water at 4° C = 0° Tw), and the relationship between specific gravity and degrees Twaddell is given by

$$\text{S.G.} = 1 + (^\circ\text{Tw}/200) \qquad (1)$$

or

$$^\circ\text{Tw} = 200\,(\text{S.G.} - 1) \qquad (2)$$

On account of its simplicity and easy convertibility into and from values of specific gravity, the Twaddell scale has also gained popularity in many parts of the world.

The *Beaumé* (or Baumé) scale was developed in the salt industry; pure water was taken as 0° Bé, each degree representing 1 per cent of dissolved salt. Unfortunately, no reference temperature was originally specified, and as percentage composition can be interpreted in many ways, as many as twenty different Beaumé scales have been in use at one time or another. Several of these are still in use, but the two most widely used are the *Rational* and the *U.S.* scales. The relationship between these and specific gravity are

(*a*) *liquids lighter than water:*

Rational °Bé = (144·3/S.G.) − 144·3 [Water = 0° at 15° C]
U.S. °Bé = (140/S.G.) − 130 [Water = 10° at 60° F]

(*b*) *liquids heavier than water:*

Rational °Bé = 144·3 − (144·3/S.G.) [Water = 0° at 15° C]
U.S. °Bé = 145 − (145/S.G.) [Water = 0° at 60° F]

Thus, there is little difference between the two 'heavy' scales, but an appreciable difference between the two 'light' scales.

A Beaumé-type density scale was proposed by the American Petroleum Institute for liquids lighter than water. This is now known as the *A.P.I.* scale and is used in the petroleum industries of most countries. Water at 60° F is 10° A.P.I. The conversion between degrees A.P.I. and specific gravity is given by

$$°\text{A.P.I.} = (141{\cdot}5/\text{S.G.}) - 131{\cdot}5 \qquad (3)$$

Several other density scales are still in fairly common use in certain industries. These are best described in tabular form, as shown below; the terms 'light' and 'heavy' refer to the scale conversions for liquids lighter or heavier than water, respectively. The number of degrees on a particular scale is denoted by N.

Table 3.1. Industrial Density Scales

Scale	*Reference temperature*	*Specific Gravity*	
		Light	*Heavy*
Balling	17·5° C	200/(200 + N)	200/(200 − N)
Beck	12·5° C	170/(170 + N)	170/(170 − N)
Brix	60° F	400/(400 + N)	400/(400 − N)
Cartier	12·5° C	136·8/(126·1 + N)	136·8/(126·1 − N)
Fisher	60° F	*as Brix scale*	

The heavy Brix or Fisher scale quoted above is not widely used, but another Brix scale is used throughout the sugar industry. On this scale a solution containing N per cent sugar by weight is said to have a density of N degrees Brix. A density scale known by the names Salinometer, Salometer or Salimeter is used in the salt industry. Water is 0° Sal and a saturated aqueous solution of sodium chloride at 20° C (26·5 parts salt/100 parts of solution by weight) is 100° Sal.

Table 3.2 compares some of the more common density scales used in industrial practice.

Table 3.2. Comparison between various Density Scales used in Industrial Practice
(*Reference temperature* = 60° F)

Specific gravity	°Tw	*Degrees Beaumé*		*Degrees A.P.I.*	lb./gal. (*Imp.*)	lb./gal. (*U.S.*)	lb./ft.^{3}
		Rational	*U.S.*				
0·50	—	144·3	150·0	151·5	5·0	4·17	31·2
0·55	—	118·0	124·5	125·8	5·5	4·58	34·3
0·60	—	96·2	103·3	104·3	6·0	4·99	37·4
0·65	—	77·7	85·4	86·2	6·5	5·41	40·5
0·70	—	61·9	70·0	70·6	7·0	5·83	43·6
0·75	—	48·1	56·7	57·2	7·5	6·24	46·7
0·80	—	36·1	45·0	45·4	8·0	6·66	49·8
0·85	—	25·4	34·7	35·0	8·5	7·08	52·9
0·90	—	16·0	25·5	25·7	9·0	7·49	56·1
0·95	—	7·6	17·4	17·5	9·5	7·91	59·2
1·00	—	0·0	10·0	10·0	10·0	8·33	62·3
1·00	0	0·0	0·0	10·0	10·0	8·33	62·3
1·05	10	6·7	6·9	3·3	10·5	8·75	65·4
1·10	20	13·1	13·2	—	11·0	9·16	68·5
1·15	30	18·8	18·9	—	11·5	9·58	71·7
1·20	40	24·1	24·2	—	12·0	10·00	74·8
1·25	50	28·8	29·0	—	12·5	10·41	77·9
1·30	60	33·2	33·5	—	13·0	10·83	81·0
1·35	70	37·4	37·6	—	13·5	11·25	84·1
1·40	80	41·2	41·4	—	14·0	11·66	87·3
1·45	90	44·7	45·0	—	14·5	12·08	90·4
1·5	100	48·1	48·3	—	15·0	12·50	93·5

Viscosity

The performance of many items of equipment encountered in crystallization practice is often profoundly affected by the flow properties of the liquid media. Heat transfer, for example, may be severely impeded in 'thick' sluggish liquors or magmas; crystallization may occur only with difficulty, and filtration and washing of the crystalline product may be impaired.

A measure of the resistance to flow exhibited by a fluid moving over itself is given by the property known as viscosity, which is defined as a shearing stress per unit area per unit velocity gradient within the fluid. The dimensions of viscosity, therefore, are: mass/length . time, and the common c.g.s. unit is the poise (1 g/cm · sec). As this unit is rather large, the centipoise is more often used; 1 centipoise = 0·01 poise = 2·42 lb./ft. · h. Viscosity, η, so defined, is sometimes referred to as the 'dynamic' or 'absolute' viscosity or 'coefficient of viscosity'. The kinematic viscosity, ν, of a fluid is equal to η/ρ where ρ is its density. The dimensions of ν are length^2/time and the common units are stokes, centistokes and ft.^{2}/h: 1 centistokes = 0·01 stokes ≡ 0·0388 ft.^{2}/h.

The viscosity of a liquid decreases with increasing temperature, and for many liquids the relationship

$$\eta = A \cdot \exp(-B/T) \quad (4)$$

holds reasonably well. A and B are constants, and the temperature T is in degrees absolute. Plots of log η against $1/T$, or log η against log T, usually yield fairly straight lines, and this fact may be utilized in estimating the viscosity of a liquid at some temperature if values of η are known at two other temperatures. *Table 3.3* gives some values of the viscosity of several common solvents over a range of temperature.

Table 3.3. Absolute Viscosities (centipoise) of some Common Solvents

Solvent	*Temperature* °C					
	0	20	40	60	80	100
Water	1·79	1·00	0·655	0·468	0·355	0·281
Acetone	0·389	0·322	0·261	—	—	—
Benzene	—	0·654	0·492	0·396	0·318	—
Toluene	0·76	0·587	0·471	0·380	0·310	0·250
o-Xylene	1·11	0·825	0·625	0·502	0·405	0·345
Carbon tetrachloride	1·37	0·975	0·746	0·595	—	—
Methanol	0·810	0·592	0·456	0·350	—	—
Ethanol	1·77	1·19	0·826	0·605	—	—
n-Propanol	4·20	2·56	1·40	0·925	0·645	0·443
n-Butanol	5·14	2·95	1·77	1·15	0·762	0·540

In general, solid electrolytes and non-electrolytes increase the viscosity of water, although a few exceptions to this rule are known. Occasionally the increase in viscosity is considerable, as shown by the system sucrose-water[1] (*Table 3.4*). At low temperatures the viscosity of sucrose solutions increases

Table 3.4. Absolute Viscosities (centipoise) of Aqueous Solutions of Sucrose (After E. HATSCHEK[1], *from data of Bingham and Jackson)*

Temperature °C	g *Sucrose per* 100 g *of Solution*		
	20	40	60
0	3·804	14·77	238
10	2·652	9·794	109·8
20	1·960	6·200	56·5
30	1·504	4·382	33·78
40	1·193	3·249	21·28
50	0·970	2·497	14·01
60	0·808	1·982	9·83
70	0·685	1·608	7·15
80	0·590	1·334	5·40
90	—	1·123	4·15
100	—	0·960	3·34

greatly with an increase in concentration, while at high concentration the decrease in viscosity with increasing temperature is extremely rapid. *Figure 3.1a* shows an example of a solute which decreases the viscosity of the solvent; in this system (KI–water) a minimum viscosity is obtained. Several other potassium and ammonium salts also exhibit a similar behaviour. *Figure 3.1b* shows the effect of concentration and temperature on the system ethanol–water; this system exhibits a maximum viscosity. Unfortunately, no completely reliable method is available for the prediction of the viscosities of solutions or liquid mixtures.

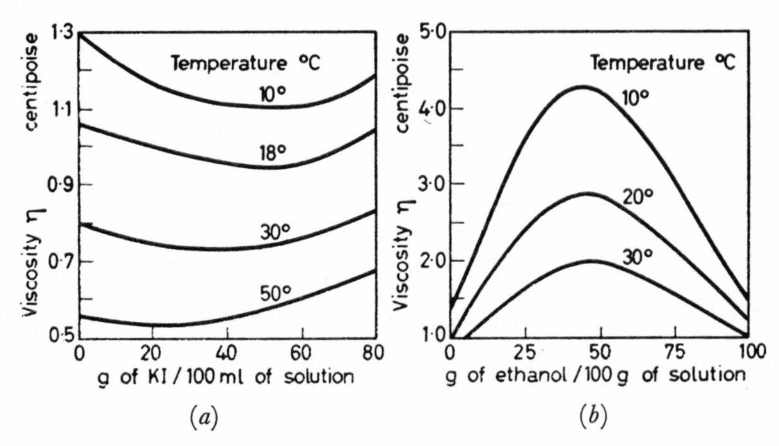

Figure 3.1. Aqueous solutions exhibiting (a) minimum, (b) maximum viscosities (After E. HATSCHEK[1])

The viscosity characteristics of liquids can be altered considerably by the presence of finely dispersed solid particles, especially of colloidal size. Einstein (1911) suggested that the viscosity of a suspension of rigid spherical particles in a liquid, when the distance between the spheres is much greater than their diameter, would be governed by the equation

$$\eta_s = \eta_0(1 + 2{\cdot}5\phi) \tag{5}$$

where η_s = viscosity of the disperse system, η_0 = viscosity of the pure dispersion medium, and ϕ = ratio of the volume of the dispersed particles to the total volume of the disperse system. The Einstein equation applies reasonably well to lyophobic sols and very dilute suspensions. For moderately concentrated lyophilic sols the following modifications of the Einstein equation have been proposed by Hatschek (1916), Kunitz (1926) and Guth and Simha (1936):

$$\eta_s = \eta_0(1 + 2{\cdot}5\phi + 14{\cdot}1\phi^2) \quad \text{(Guth and Simha)} \tag{6}$$

$$\eta_s = \eta_0\left(\frac{1 + 0{\cdot}5\phi}{1 - \phi^4}\right) \quad \text{(Kunitz)} \tag{7}$$

$$\eta_s = \frac{\eta_0}{1 - \phi^{\frac{1}{3}}} \quad \text{(Hatschek)} \tag{8}$$

Numerous instruments have been devised for the measurement of viscosity. Many of these are based on the flow of fluid through a capillary tube, and Poiseuille's law can be applied in the form

$$\eta = \frac{\pi r^4 \Delta P}{8lV} \tag{9}$$

where ΔP = the pressure drop across the capillary of length l and radius r, V = volume of fluid flowing in unit time. One simple type of *tube viscometer*[2] is shown in *Figure 3.2.* The liquid under test is sucked into leg B until the

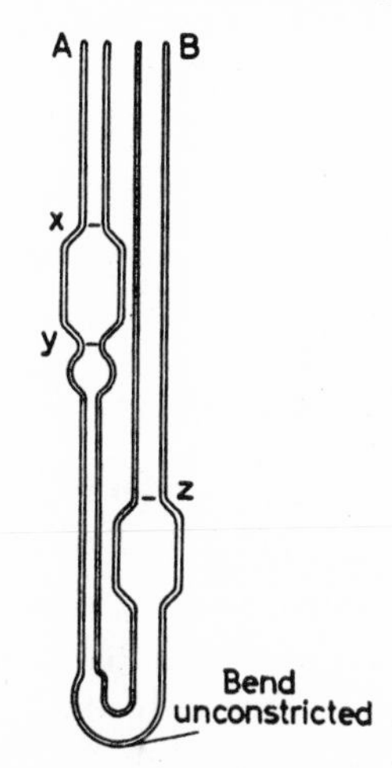

Figure 3.2. Simple tube viscometer
(*After* British Pharmacopœia[2])

level in this leg reaches mark z. The tube is arranged truly vertical, the temperature of the liquid is measured and kept constant. The liquid is then sucked up into leg A to a point 1 cm above mark x, and the time t (sec) for the meniscus to fall from x to y recorded. The mean of several measurements is taken. The kinematic viscosity of the liquid ν (centistokes), can be calculated from the equation

$$\nu = kt \tag{10}$$

where k is a constant for the apparatus, determined by measurements on a liquid of known viscosity, e.g. water.

The falling-sphere method of viscosity determination also has many applications, and Stokes' law may be applied in the form

$$\eta = \frac{(\rho_s - \rho_l)d^2 g}{18u} \tag{11}$$

where d, ρ_s and u are the diameter, density and terminal velocity, respectively, of a solid sphere falling in the liquid of density ρ_l. A simple *falling-sphere viscometer*[2] is shown in *Figure 3.3.* The liquid under test is contained in the inner tube of 3·2 mm diam. and about 450 mm length. The central portion of the tube contains two reference marks, 150 mm apart. The tube is held truly vertical, the temperature of the liquid is measured and kept constant. A $\frac{1}{16}$ in. diam. steel ball, free from rust and previously warmed to the test temperature, is allowed to fall down the tube, the time of its passage between the two reference marks being measured. A mean of several measurements should be taken. The viscosity of the liquid can then be calculated from equation 11.

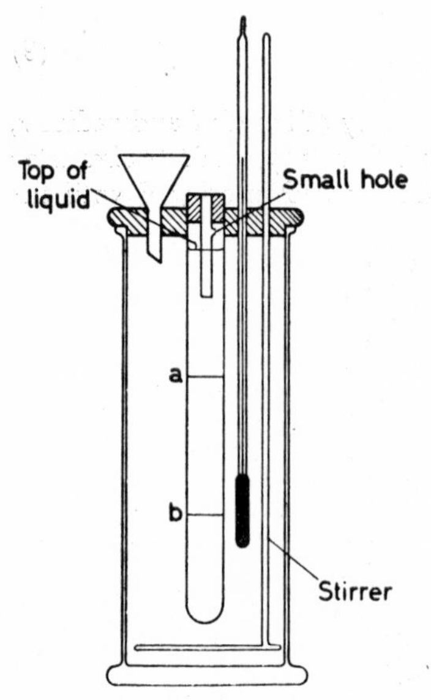

Figure 3.3. Falling-sphere viscometer. (*From* British Pharmacopœia[2], *by courtesy of* General Medical Council, London)

Several viscometers are based on the concentric-cylinder method. The liquid under test is contained in the annulus between two vertical co-axial cylinders; one cylinder can be made to rotate at a constant speed, and the couple required to prevent the other cylinder rotating can be measured. For detailed information on practical viscometry reference should be made to the publications of HATSCHEK[1] and BARR[3].

THERMAL PROPERTIES

Before the heating or cooling requirements for a crystallization operation can be determined, a knowledge of many thermal properties of solid, liquid and gaseous systems may be required. A solution can be crystallized by cooling or evaporation, or by a combination of both processes. In the case of cooling, the heat which has to be removed from the system is the sum of the sensible heat and the heat of crystallization (exothermic in most cases). Before the heating requirements for concentration processes can be calculated, the latent heat of vaporization of the solvent and the heat of crystallization of the solute must be known. In the case of crystallization from the melt, a value of the latent heat of fusion, amongst other heat quantities, is needed.

The supersaturation of a solution, and hence its crystallization, can also be achieved by the addition of a 'diluent' in the form of a liquid which is miscible in all proportions with the original solvent. The solute should be sparingly soluble in the diluent. This method of crystallization is referred to as 'salting-out' or 'dilution crystallization'. Water is the most widely used diluent in the crystallization of organic compounds from alcoholic solution, but alcohols can be used as diluents in the crystallization of inorganic salts from aqueous solution. When two solvents are mixed, heat is generally liberated; hence a knowledge of the heat of mixing is required for the heat balance calculation of a salting-out process.

Units of Heat

The four heat energy units in common use are the gramme-calorie (cal), the kilogram-calorie (kcal), the British Thermal Unit (B.t.u.) and centigrade heat unit (c.h.u.). The precise definitions of these units are made, by international agreement, with reference to the units of mechanical energy. Thus, 1 cal = 4·18605 international joules = 4·18674 absolute joules. However, for use in industrial chemical practice, the older but less precise definitions are more convenient. These definitions are based on the heat energy required to raise the temperature of a unit mass of water by one degree:

1 cal = 1 g water raised through 1° C (or 1° K)
1 kcal = 1 kg water raised through 1° C (or 1° K)
1 c.h.u. = 1 lb. water raised through 1° C (or 1° K)
1 B.t.u. = 1 lb. water raised through 1° F (or 1° R)

The centigrade heat unit is sometimes called the pound-calorie unit (p.c.u.). *Table 3.5* indicates the equivalent values of these heat units.

Table 3.5. Equivalent Values of the Four Common Heat Energy Units

cal	kcal	B.t.u.	c.h.u.
1	0·001	0·00397	0·00221
1,000	1	3·97	2·21
252	0·252	1	0·556
453·6	0·4536	1·8	1

Heat Capacity

The amount of heat energy associated with a given temperature change in a given system is a function of the chemical and physical states of the system. A measure of this heat energy can be expressed in terms of the quantity known as the heat capacity. The term 'specific heat' is often used synonymously with heat capacity, but strictly speaking the specific heat of a substance is the ratio of its heat capacity to the heat capacity of an equal mass of water, usually at the reference temperature of 15° C. Heat capacities may be expressed on a mass or molal basis. As a one-degree temperature range is considered, the Celsius (centigrade), Kelvin, Fahrenheit or Rankine scales can be employed. The equivalent heat capacity units are

mass heat capacity, c

1 cal/g °C (or °K) = 1 B.t.u./lb. °F (or °R)
= 1 c.h.u./lb. °C (or °K)

molal heat capacity, C

1 cal/mole °C (or °K) = 1 B.t.u./lb.-mol. °F (or °R)
= 1 c.h.u./lb.-mol. °C (or °K)

For gases, two heat capacities have to be considered, at constant pressure, C_p, and at constant volume, C_v. The value of the ratio of these two quantities, $C_p/C_v = \gamma$, varies from about 1·67 for mono-atomic gases (e.g. He) to about 1·3 for tri-atomic gases (e.g. CO_2). For liquids and solids there is little difference between C_p and C_v, i.e. $\gamma \sim 1$, and it is usual to find C_p values only quoted in the literature.

Solids

Values of the heat capacities of solid substances near normal atmospheric temperature can be estimated with a reasonable degree of accuracy by combining two empirical rules. The first of these, due to Dulong and Petit, applies to solid elemental substances and may be written

$$\text{mass heat capacity} \times \text{atomic weight} = \text{atomic heat} \simeq 6{\cdot}2$$

The second rule, due to Kopp, applies to solid compounds and may be expressed by

$$\text{molal heat capacity} = \text{sum of the atomic heats of the constituent atoms.}$$

In applying these rules, the following exceptions to the approximation 'atomic heat $\simeq$ 6·2' must be noted

$$\begin{array}{llll} C = 1{\cdot}8 & H = 2{\cdot}3 & B = 2{\cdot}7 & Si = 3{\cdot}8 \\ O = 4{\cdot}0 & F = 5{\cdot}0 & S = 5{\cdot}4 & [H_2O] = 9{\cdot}8 \end{array}$$

The last substance, $[H_2O]$, refers to water as ice or as water of crystallization in solid substances. Obviously a reliable measured value of a heat capacity is preferable to an estimated value, but in the absence of the former Kopp's rule can prove extremely useful. A few calculated and observed values of C_p are compared in *Table 3.6*.

Table 3.6. Estimated (Kopp's Rule) and Observed Values of C_p for Several Solid Substances at Room Temperature

Solid	*Formula*	*Calculation*	C_p (cal/mole °C)	
			Calc.	*Obs.*
Sodium chloride	$NaCl$	6·2 + 6·2	12·4	12·4
Magnesium sulphate	$MgSO_4 \cdot 7H_2O$	6·2 + 5·4 + 4(4·0) + 7(9·8)	96·2	89·5
Iodobenzene	C_6H_5I	6(1·8) + 5(2·3) + 6·2	28·5	24·6
Naphthalene	$C_{10}H_8$	10(1·8) + 8(2·3)	36·4	37·6
Potassium sulphate	K_2SO_4	2(6·2) + 5·4 + 4(4·0)	33·8	30·6
Oxalic acid	$C_2H_2O_4 \cdot 2H_2O$	2(1·8) + 6(2·3) + 4(4·0) + 2(9·8)	43·8	43·5

Many inorganic solids have values of mass heat capacity, c_p, in the range 0·1–0·3 cal/g °C while many organic solids have values in the range 0·2–0·5. In general, the heat capacity increases slightly with an increase in temperature. For example, the values of c_p at 0° and 100° C for sodium chloride are 0·21 and 0·22 cal/g °C, respectively, and the corresponding values for anthracene are 0·30 and 0·35.

Liquids

Although several methods have been proposed[4] for the estimation of the heat capacity of a liquid, none is completely reliable. However, one recent method[5] worthy of mention is based on the additivity of the heat capacity contributions $[C_p]$ of the various atomic groupings in the molecules of organic liquids. *Table 3.7* gives some values of $[C_p]$, and the following examples illustrate the use of the method; the heat capacity values (cal/mole °C) in parentheses denote values obtained experimentally at 20° C.

Methyl alcohol ($CH_3 \cdot OH$) $9{\cdot}9 + 11{\cdot}0 = 20{\cdot}9$ (19·5)

Toluene ($C_6H_5 \cdot CH_3$) $30{\cdot}5 + 9{\cdot}9 = 40{\cdot}4$ (36·8)

*iso*Butyl acetate $CH_3 \cdot COO \cdot CH \langle {CH_3 \atop CH_2 \cdot CH_3}$

$$= 3(9{\cdot}9) + 14{\cdot}5 + 5{\cdot}4 + 6{\cdot}3 = 55{\cdot}9 \ (53{\cdot}3)$$

The heat capacity of a substance in the liquid state is generally higher than that in the solid state. A large number of organic liquids have mass heat capacity values in the range 0·4–0·6 cal/g °C at about room temperature. The heat capacity of a liquid usually increases with increasing temperature; for example, the values of c_p for ethyl alcohol at 0°, 20° and 60° C are 0·54, 0·56 and 0·68 cal/g °C, those for benzene at 20°, 40° and 60° C are 0·40, 0·42 and 0·45. Water is an exceptional case; it has a very high heat capacity and exhibits a minimum value at 30° C. The values of C_p for water at 0°, 15°, 30° and 100° C are 1·008, 1·000, 0·9987 and 1·007 cal/g °C, respectively.

Table 3.7. Contributions of Various Atomic Groups to the Heat Capacity of Organic Liquids at 20° C

(*After* A. I. JOHNSON and C. J. HUANG[5])

Group	$[C_p]$	*Group*	$[C_p]$
C_6H_5—	30·5	—OH	11·0
CH_3—	9·9	—NO_2	15·3
—CH_2—	6·3	—NH_2	15·2
—CH	5·4	—CN	13·9
—COOH	19·1	—Cl	8·6
—COO— (esters)	14·5	—Br	3·7
C=O (ketones)	14·7	—S—	10·6
—H (formates)	3·6	—O— (ethers)	8·4

Liquid Mixtures

There is no reliable method for predicting the heat capacity of liquid mixtures, but in the absence of experimental data the following equation may be used for a mixture of two or more liquids

$$C_{p_{\text{mixt}}} = x_A C_{pA} + x_B C_{pB} + \ldots \quad (12)$$

where x denotes the mole fraction of the given component of the mixture.

For example, the value of c_p for methanol (mol. wt. = 32·0) at 20° C is 0·58 cal/g °C. From equation (12) it can be calculated that an aqueous solution containing 75 mole per cent of methanol (mean mol. wt. of mixture = 28·5) has a mass heat capacity of 0·64 cal/g °C. The experimental value at 20° C is 0·69.

For dilute aqueous solutions of inorganic salts a rough estimate of the mass heat capacity can be made by ignoring the heat capacity contribution of the dissolved substance, i.e.

$$c_p = 1 - X \tag{13}$$

where X = g solute/g water, and c_p = cal/g °C. Thus, solutions containing 5 g NaCl, 10 g KCl and 15 g $CuSO_4$ per 100 g of solution would, by this method, be estimated to have mass heat capacities of 0·95, 0·90 and 0·85 cal/g °C, respectively. Experimental values (25° C) for these solutions are 0·94, 0·91 and 0·83. This estimation method cannot be applied to aqueous solutions of non-electrolytes or acids.

Another rough estimation method for the heat capacity of aqueous solutions is based on the empirical relationship

$$c_p = \frac{1}{\rho} \tag{14}$$

where ρ = density of the solution in g/cm³. For example, at 30° C a 2 per cent aqueous solution of sodium carbonate by weight has a density of 1·016 g/cm³ and a heat capacity of 0·98 cal/g °C ($1/\rho$ = 0·985), while a 20 per cent solution has a density of 1·21 and a heat capacity of 0·86 ($1/\rho$ = 0·83).

Thermal Conductivity

The thermal conductivity, κ, of a substance is defined as the rate of heat transfer by conduction across a unit area, through a layer of unit thickness, under the influence of a unit temperature difference, the direction of heat transmission being normal to the reference area. Fourier's equation for steady conduction may be written as

$$\frac{dq}{dt} = -\kappa A \frac{d\theta}{dx} \tag{15}$$

where q, t, A, θ and x are units of heat, time, area, temperature and length (thickness), respectively. The units of κ, therefore, may be expressed as cal/sec cm² (°C/cm) or B.t.u./h · ft.² (°F/ft.).

An increase in the temperature of a liquid usually results in a slight decrease in thermal conductivity, but water is a notable exception to this generalization. *Table 3.8* gives values of κ for a few pure liquids. In the reference literature, thermal conductivity data, particularly on solutions, are sparse and often conflicting, and no reliable method of estimation has yet been devised. For aqueous methanol and ethanol solutions, however, the

Table 3.8. Thermal Conductivities of Some Pure Liquids

Temperature °F	Thermal Conductivity, κ [B.t.u./h · ft.^{2}(°F/ft.)]					
	Water	*Acetone*	*Benzene*	*Methanol*	*Ethanol*	CCl_4
50	0·34	0·095	0·083	0·12	0·105	0·065
100	0·36	0·088	0·090	0·11	0·098	0·057
200	0·39	—	—	—	—	—

following relationship due to BATES *et al.*[6] may be used with a reasonable degree of accuracy up to about 80° C

$$\kappa_{\text{mixt}} = \frac{\kappa_1 \sinh m_1\delta + \kappa_2 \sinh m_2\delta}{\sinh (100\delta)} \tag{16}$$

where κ_1, κ_2 = thermal conductivities of water and alcohol

m_1, m_2 = mass fractions of water and alcohol

δ = a constant (0·90 for methanol, 0·94 for ethanol)

Freezing and Boiling Points

When a non-volatile solute is dissolved in a solvent, the vapour pressure of the solvent is lowered. Consequently, at any given pressure, the boiling point of a solution is higher and the freezing point lower than that of the pure solvent. For dilute ideal solutions, i.e. such as obey Raoult's law, the boiling point elevation and freezing point depression can be calculated by an equation of the form

$$\Delta T = \frac{mK}{M} \tag{17}$$

where m = mass of solute dissolved in a given mass of pure solvent (usually 1,000 g), M = molecular weight of the solute. When ΔT refers to the freezing point depression, $K = K_f$, the cryoscopic constant, when ΔT refers to the boiling point elevation, $K = K_b$, the ebullioscopic constant. Values of K_f and K_b for several common solvents are given in *Table 3.9*; these, in effect, give the depression in freezing point, or elevation in boiling point, in °C when 1 mole of solute is dissolved, without dissociation or association, in 1,000 g of solvent.

The cryoscopic and ebullioscopic constants can be calculated from values of the latent heats of fusion and vaporization, respectively, by the equation

$$K = \frac{RT^2}{1{,}000l} \tag{18}$$

When $K = K_f$, T refers to the freezing point T_f (°K) and l to the latent heat of fusion, l_f (cal/g). When $K = K_b$, T refers to the boiling point, T_b, and $l = l_{vb}$, the latent heat of vaporization at the boiling point. The gas constant $\boldsymbol{R}$ = 1·987 cal/mole °K.

Table 3.9. Cryoscopic and Ebullioscopic Constants for Some Common Solvents

Solvent	Freezing point °C	Boiling point °C	K_f	K_b
Acetic acid	16·7	118·1	3·9	3·1
Acetone	− 95·5	56·3	2·7	1·7
Aniline	− 6·2	184·5	5·9	3·2
Benzene	5·5	80·1	5·1	2·7
Carbon disulphide	− 108·5	46·3	3·8	2·4
Carbon tetrachloride	− 22·6	76·8	32·0	4·9
Chloroform	− 63·5	61·2	4·8	3·8
Cyclohexane	6·2	80	20·0	2·8
Nitrobenzene	5·7	211	7·0	5·3
Methyl alcohol	− 97·8	64·7	2·6	0·8
Phenol	42·0	181	7·3	3·0
Water	0·0	100·0	1·86	0·52

Equation 18 cannot be applied to concentrated solutions or to aqueous solutions of electrolytes. In these cases, the freezing point depression cannot readily be estimated. The boiling point elevation, however, can be predicted with a reasonable degree of accuracy by means of an empirical rule. Dühring (1878) observed that the boiling point of a solution is a linear

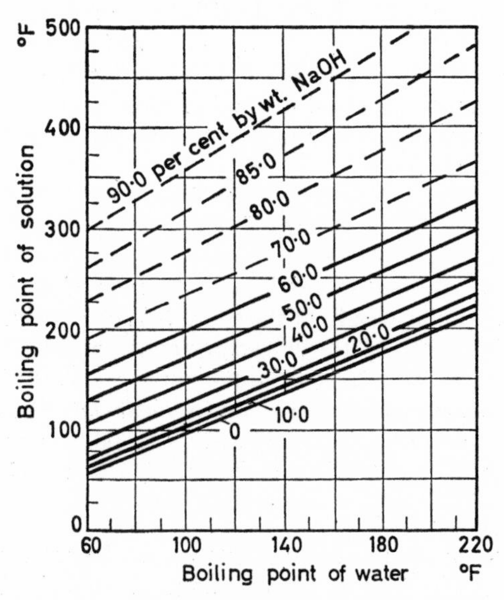

Figure 3.4. Dühring plot for aqueous solutions of sodium hydroxide

function of the boiling point of the pure solvent. Therefore, if the boiling points of solutions of different concentration are plotted against those of the solvent at different pressures, a family of straight lines (not necessarily parallel) will be obtained. A typical Dühring plot for aqueous solutions of sodium hydroxide is given in *Figure 3.4*, from which it can be seen, for example, that at a pressure at which water boils at 180° F, a solution containing 50 per cent by weight of NaOH would boil at about 260° F, i.e. a boiling point elevation of about 80° F.

Melting Point

The melting point of a solid organic substance is frequently adopted as a criterion of purity, but before any reliance can be placed on the test it is absolutely necessary for the experimental procedure to be standardized. Several types of melting point apparatus are available commercially, such as heating blocks on which samples of the material are placed and observed while the temperature of the surface of the block is recorded, but the most widely used method, and the one most frequently recommended, consists of heating a powdered sample of the material in a glass capillary tube located close to the bulb of a thermometer in an agitated bath of liquid.

The best type of glass tube is 0·9–1·2 mm internal diam., about 100 mm long, with walls about 0·1 mm thick. The tube is sealed at one end, and the powdered sample is scraped into the tube and knocked or vibrated down to the closed end to give a compacted layer about 2–4 mm deep. It is then inserted into the liquid bath at about 10° C below the expected melting point and attached close to the middle portion of the thermometer bulb. The bath is continuously agitated and the temperature allowed to rise steadily at about 3° C/min. The melting point of the substance is taken as the temperature at which a definite meniscus is formed in the tube. For pure substances the melting point can be readily and accurately reproduced; for impure substances it is better to record a melting range of temperature.

There are, however, a number of experimental errors that can be encountered if the precise details of the test are not specified. For instance, soft soda glass capillary tubes are most widely used, but several organic substances are extremely sensitive to the presence of traces of alkali. In these cases, much higher melting points can be recorded if Pyrex glass tubes are used. If a substance is known to be alkali-sensitive, e.g. acetylsalicylic acid, the type of glass should be specified before using the melting point as a test for purity. Traces of moisture in the tube will also affect the melting point; thus the capillaries should always be stored in a desiccator.

Some organic substances, especially acetylated compounds, begin to decompose near their melting points, so that it is important not to keep the sample at an elevated temperature for prolonged periods. The insertion of the capillary into the bath at 10° C below the melting point, allowing the temperature to rise at 3°/min, usually eliminates any difficulties; with some unstable compounds an insertion at 5° below the melting point may, however, be necessary. For reproducible results to be obtained the sample should be in a finely divided state, at least finer than 150-mesh. The thermometer, preferably divided in $\frac{1}{10}$° C, should be accurately calibrated over its whole range and, if in constant use, should be checked regularly. The well-known standardization temperatures are the freezing and boiling points of water, but other standards that can be used are the melting points of pure organic substances, such as those indicated in *Table 3.10.*

Table 3.10. Melting Points of Pure Organic Compounds Used for Calibrating Thermometers

Substance	*Melting point* °C
Phenyl salicylate (salol)	42
p-Dichlorbenzene	53
Naphthalene	80
m-Dinitrobenzene	90
Acetamide	114
Benzoic acid	122
Urea	133
Salicylic acid	159
Succinic acid	183
Anthracene	217
p-Nitrobenzoic acid	242
Anthraquinone	285

The liquid used in the heating bath depends on the working temperature; water is quite suitable for melting points from about room temperature to about 70° C; liquid paraffin is widely used but readily darkens at temperatures in excess of about 200° C. Phosphoric acid and glycerol, also used, likewise tend to darken. Sulphuric acid or solutions of potassium sulphate in H_2SO_4 are frequently recommended for heating-bath media, but the potential hazards hardly need any emphasis. In recent years silicone fluids have become available which, though expensive, do not darken and have a prolonged working life.

Latent Heat

When a substance undergoes a phase change, a quantity of heat is transferred between the substance and its surrounding medium. This enthalpy change is called the latent heat, and the following types may be distinguished:

Solid I ⇌ Solid II (latent heat of transition)
Solid ⇌ Liquid (latent heat of fusion)
Solid ⇌ Gas (latent heat of sublimation)
Liquid ⇌ Gas (latent heat of vaporization)

Only the last three of these represent significant quantities of heat energy; heats of transition can for most industrial purposes be ignored. For example, the transformation of monoclinic to orthorhombic sulphur is accompanied by an enthalpy change of about 0·4 B.t.u./lb., whereas the fusion of orthorhombic sulphur is accompanied by an enthalpy change of about 17 B.t.u./lb.

Latent heats, like the other thermal properties, can be expressed on a mass or molar basis, e.g. cal/g or cal/mole. To avoid confusion, all latent heats in this section will be expressed on a molar basis. The relationship between any latent heat, l, and the pressure–volume–temperature conditions of a system is given by the Clapeyron equation

$$\frac{dp}{dT} = \frac{l}{T\Delta v} \tag{19}$$

where dp/dT = rate of change of vapour pressure with absolute temperature, Δv = volume change accompanying the phase change.

A typical temperature-pressure phase diagram for a one-component system is shown in *Figure 3.5.* The sublimation curve AX indicates the increase of the vapour pressure of the solid with an increase in temperature. This is expressed quantitatively by the Clapeyron equation written as

$$\frac{dp}{dT} = \frac{l_s}{T(v_g - v_s)} \tag{20}$$

where v_g and v_s = the molar specific volumes of the vapour and solid, respectively, l_s = latent heat of sublimation. The vaporization curve XB indicates the increase of the vapour pressure of the liquid with an increase in temperature

$$\frac{dp}{dT} = \frac{l_v}{T(v_g - v_l)} \tag{21}$$

where v_l = the specific volume of the liquid, l_v = latent heat vaporization. Curve XB is not a continuation of curve AX; this fact can be confirmed by calculating their slopes dp/dT from equations 20 and 21 at point X.

The fusion line XC indicates the effect of pressure on the melting point of the solid; it can either increase or decrease with an increase in pressure, but the effect is so small that line XC deviates only slightly from the vertical. When the fusion line deviates to the right, as in *Figure 3.5,* the melting point

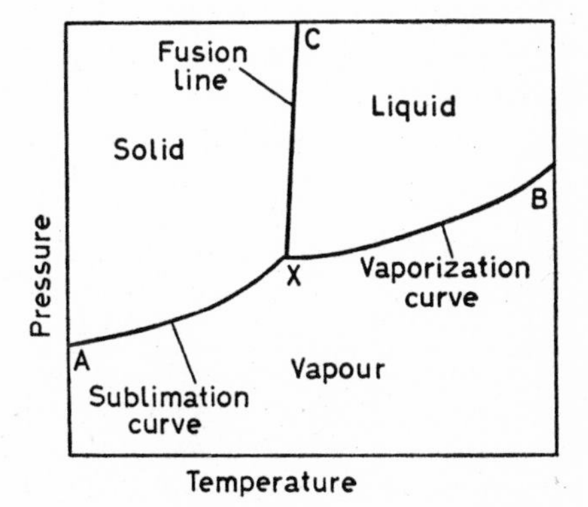

Figure 3.5. Temperature–pressure diagram for a single-component system

increases with an increase in pressure, and the substance contracts on freezing. Most substances behave in this manner. When the line deviates to the left, the melting point decreases with increasing pressure, and the substance expands on freezing. Water and type metals are among the few examples of this behaviour that can be quoted. The equation for the fusion line is

$$\frac{dp}{dT} = \frac{l_f}{T(v_l - v_s)} \tag{22}$$

where l_f is the latent heat of fusion.

The latent heats of sublimation, vaporization and fusion are related by

$$l_s = l_f + l_v \tag{23}$$

but this additivity equation is applicable only at one specific temperature. The variation of a latent heat with temperature can be calculated from the Clausius equation

$$\frac{dl}{dT} - \frac{l}{T} = c_2 - c_1 \tag{24}$$

When $l = l_v$, c_1 and c_2 are the heat capacities of the liquid, just on the point of vaporization, and of the saturated vapour, respectively. Equation 24 can also be used for calculating l_f and l_s, inserting the appropriate values of c.

The specific volumes v_l and v_s are much smaller than v_g; equations 20 and 21 can, therefore, be simplified to

$$\frac{dp}{dT} = \frac{l}{Tv_g} \tag{25}$$

where $l = l_v$ or l_s. From the ideal gas laws, $v_g = \boldsymbol{R}T/p$, so that equation 25 may be written

$$\frac{dp}{p} = \frac{l}{\boldsymbol{R}T^2}\,dT \tag{26}$$

This is the Clausius–Clapeyron equation. If the latent heat is considered to be constant over a small temperature range, T_1 to T_2, equation 26 may be integrated to give

$$\ln\frac{p_2}{p_1} = l\,\frac{(T_2 - T_1)}{\boldsymbol{R}T_1T_2} \tag{27}$$

Equation 27 can be used to estimate latent heats of vaporization and sublimation if vapour pressure data are available, or to estimate vapour pressures from a value of the latent heat. Analysis of sublimation problems, for instance, is frequently difficult owing to the scarcity of published vapour pressure and latent heat data. If two values of vapour pressure are available, however, a considerable amount of information can be derived from equation 27 as illustrated by the following example[7].

Example—Suppose that the only data available on solid anthracene are that its vapour pressures at 210° C and 145° C are 30 and 1 mm Hg, respectively; the vapour pressure at 100° C is required.

Solution—Equation 27 can be used twice, first to calculate a value of l_s then that of the required vapour pressure:

$$\ln\left(\frac{30}{1}\right) = \frac{l_s(483 - 418)}{1{\cdot}99 \times 483 \times 418}$$

$$l_s = 21{,}000 \text{ cal/mole}$$

Substituting this value of l_s in equation 27 again

$$\ln\left(\frac{30}{p_{100°\,C}}\right) = \frac{21{,}000(483 - 373)}{1{\cdot}99 \times 483 \times 373}$$

therefore,

$$p_{100°\,C} = 0{\cdot}048 \text{ mm Hg}$$

Heat of Vaporization

There are several methods available for the estimation of latent heats of vaporization at the atmospheric boiling point of the liquid, l_{vb}. The well known rules due to Trouton (1884) and Kistiakowsky (1923) are only suitable for non-polar liquids, but a more recent method due to GIACALONE[8] is fairly reliable for both polar and non-polar liquids. These three empirical rules may be summarized

$$l_{vb} = 21T_b \quad \text{(Trouton)} \quad (28)$$

$$= (8{\cdot}75 + 4{\cdot}571 \log_{10} T_b)T_b \quad \text{(Kistiakowsky)} \quad (29)$$

$$= \left(\frac{RT_cT_b}{T_c - T_b}\right) \ln P_c \quad \text{(Giacalone)} \quad (30)$$

where T_b = boiling point (°K) of the liquid at 760 mm Hg
T_c = critical temperature (°K) of the liquid
P_c = critical pressure (atm.) of the liquid

Table 3.11 illustrates the use of these three equations and compares calculated and experimental values of l_{vb}. It can be seen that equations (28) and (29) are completely unreliable for polar liquids. Despite its limitations, Trouton's

Table 3.11. Comparison between Estimated and Experimental Values of the Latent Heat of Vaporization of a Liquid at its Boiling Point

Liquid	*Boiling point* 760 mm Hg, °K	*Critical temperature* °K	*Critical pressure* (atm.)	*Latent heat of vaporization at boiling point* cal/mole			
				Equation 28	*Equation* 29	*Equation* 30	*Experimental*
Benzene .	353	563	48·6	7,413	7,200	7,300	7,350
Naphthalene .	491	750	39·2	10,310	10,340	10,320	10,250
Water . .	273	647	218·0	5,733	5,460	9,440	9,708
Ethanol .	351	516	62·7	7,371	7,130	9,050	9,380

rule has the merit of simplicity and when coupled with equation 27 provides a rapid estimation method for the boiling point of a non-polar liquid at pressures both above and below atmospheric; the latent heat of vaporization is assumed to be a constant. A combination of equations 27 and 28 gives

$$\ln \frac{p_2}{p_1} = \frac{21 T_2(T_2 - T_1)}{R T_2 T_1} \tag{31}$$

Example—Benzene boils at 80° C at 760 mm Hg. Estimate its boiling point at a pressure of 200 mm Hg.

Solution—From equation 31

$$\ln \left(\frac{760}{200}\right) = \frac{21 \times 353 \times (353 - T_1)}{1{\cdot}99 \times 353 \times T_1}$$

$$T_1 = 314° \text{ K } (41° \text{ C})$$

The observed boiling point of benzene at 200 mm Hg is 42·0° C.

The latent heat of vaporization of a liquid at some temperature T_1 can be calculated from the latent heat at another temperature T_2 by means of the equation, due to WATSON[9]

$$l_{v_1} = l_{v_2} \left(\frac{T_c - T_1}{T_c - T_2}\right)^{0{\cdot}38} \tag{32}$$

For example, the latent heat of vaporization of benzene at its boiling point (353° K) is 7,350 cal/mole, its critical temperature 563° K. From equation 32 the corresponding value at 25° C (298° K) can be calculated as 8,030 cal/mole, which compares with an experimental result of 8,060 cal/mole.

Heats of Solution and Crystallization

Heat is generally absorbed from the surrounding medium when a solute dissolves in a solvent without reaction (heat of solution), or the solution temperature falls if the dissolution occurs adiabatically. When a solute crystallizes out of its solution, heat is generally liberated (heat of crystallization) and the solution temperature is increased. The reverse cases, viz. heat evolution on dissolution and heat absorption on crystallization, are also encountered, especially with solutes which exhibit an inverted solubility characteristic. The dissolution of an anhydrous salt in water at a temperature at which the hydrated salt is the stable crystalline form frequently leads to the formation of heat energy, owing to the exothermic nature of the hydration process:

$$\underset{\text{(anhydrous)}}{AB} + x\text{H}_2\text{O} \rightarrow \underset{\text{(hydrate)}}{AB\,.\,x\text{H}_2\text{O}}$$

Table 3.12 lists the heats of solution of anhydrous and hydrated magnesium sulphate and sodium carbonate in water to illustrate the effect of water of crystallization.

The enthalpy changes associated with dissolution (ΔH_{soln}) and crystallization (ΔH_{cryst}) are generally recorded as the number of heat units liberated by the system when the process takes place isothermally. According to this system of nomenclature, if an adiabatic operation is considered, the expression $\Delta H_{\text{soln}} = + q$ (heat units per unit mass of solute) means that the solution temperature will increase; $\Delta H_{\text{soln}} = - q$ means that it will fall.

Table 3.12. Heats of Solution of Anhydrous and Hydrated Salts in Water at 18° C and Infinite Dilution

Salt	*Formula*	*Heat of solution* kcal/mole
Magnesium sulphate .	$MgSO_4$	+ 21·1
	$MgSO_4 \cdot H_2O$	+ 14·0
	$MgSO_4 \cdot 2H_2O$	+ 11·7
	$MgSO_4 \cdot 4H_2O$	+ 4·9
	$MgSO_4 \cdot 6H_2O$	+ 0·55
	$MgSO_4 \cdot 7H_2O$	− 3·18
Sodium carbonate .	Na_2CO_3	+ 5·57
	$Na_2CO_3 \cdot H_2O$	+ 2·19
	$Na_2CO_3 \cdot 7H_2O$	− 10·18
	$Na_2CO_3 \cdot 10H_2O$	− 16·22

The magnitude of the heat effect accompanying the dissolution of a solute in a given solute in a solvent or undersaturated solution depends on the quantities of solute and solvent involved, the initial and final concentrations and the temperature at which the dissolution occurs. The standard reference temperature is now generally taken as 25° C, but the older reference temperature of 18° C is still encountered.

In crystallization practice, it is usual to take the heat of crystallization as being equal in magnitude, but opposite in sign, to the heat of solution, i.e.

$$\Delta H_{\text{cryst}} = -\Delta H_{\text{soln}} \tag{33}$$

This assumption, of course, is not strictly correct but the error involved is small. Heats of solution are generally recorded as the enthalpy change associated with the dissolution of a unit quantity of solute in a large excess of pure solvent, i.e. the heat of solution at infinite dilution, $\Delta H^{\infty}_{\text{soln}}$. For most practical purposes the term 'infinite dilution' is taken to mean $< 0{\cdot}01$ mole fraction of solute in the solution.

The heat of crystallization is numerically equal to the heat of solution only when the latter refers to the dissolution of the solute in an almost saturated solution at the specific temperature. The temperature correction may be neglected, but the heat of dilution, ΔH_{dil}, should be taken into account if an accurate value of the heat of crystallization is required, i.e.

$$\Delta H_{\text{cryst}} = -\Delta H^{\infty}_{\text{soln}} + \Delta H_{\text{dil}} \tag{34}$$

Few values of heats of dilution are available in the literature, especially for the concentration ranges generally required, but this quantity is usually only a small fraction of the heat of solution. Furthermore, as the dilution of most aqueous salt solutions is exothermic, i.e. the concentration is endothermic, the true value of the heat of crystallization will be slightly less than that obtained by taking the negative value of the heat of solution alone. Therefore, the calculated quantity of heat to be removed from a crystallizing solution will be slightly greater than the true value; this small error acts as a factor of safety in the design of the heat transfer equipment.

Example—A solution containing 1,000 lb. of Na_2SO_4 in 5,000 lb. of water is cooled from 140 to 50° F in an agitated vessel with an estimated effective weight of mild steel of 1,500 lb. At 50° F the stable crystalline phase is the decahydrate. The heat of solution of $Na_2SO_4 \cdot 10H_2O$ at 18° C and infinite dilution is − 18·75 kcal/mole (− 105 B.t.u./lb.) and the heat capacities of the solution and mild steel are 0·85 and 0·12 B.t.u./lb. °F, respectively. During the cooling process 2 per cent of the water is lost due to evaporation. Calculate the heat to be removed.

Solution—Assume the heat of crystallization of $Na_2SO_4 \cdot 10H_2O$ to be equal and opposite to its heat of solution, $\Delta H_{cryst} = + 105$ B.t.u./lb. The value for the latent heat of evaporation of water will be taken as 1,030 B.t.u./lb. The theoretical crystal yield for this operation has already been calculated on p. 30.

		B.t.u.
Heat of crystallization	= 1,445 × 105	= 151,725
Heat removed from solution	= 6,000 × 90 × 0·85	= 459,000
Heat removed from vessel	= 1,500 × 90 × 0·12	= 16,200
	Total	626,925
Heat lost by evaporation	= 100 × 1,030	= 103,000
	Heat to be removed	523,925 B.t.u.

In the above calculation the heat from the vessel constitutes a very small proportion of the total amount of heat to be removed from the system, but it is included here for the sake of completeness. However, this quantity will not always be negligible, and it is one which is frequently overlooked. In certain cases, radiation and convection losses may also have to be considered, especially when refrigerated systems are involved.

Enthalpy–Concentration Diagrams

The heat effects accompanying a crystallization operation are most frequently determined by making a heat balance over the system, and for a reasonable degree of accuracy many calculations may be necessary, involving a knowledge of heat capacities, heats of crystallization, dilution, vaporization and so on. Much of the burden of calculation, however, can be eased by the use of a graphical technique. Merkel (1929) and Bošnjaković (1932) demonstrated a convenient method for representing enthalpy data for solutions on an enthalpy-concentration (H–x) diagram. McCabe[10] drew attention to the use of the H–x chart for the analysis of several chemical engineering operations, and this approach is now widely used for distillation, evaporation and refrigeration processes, to name but a few.

With regard to applying it to crystallization there are two difficulties. First, enthalpy-concentration diagrams are available, in the literature at least, for only a very few aqueous–inorganic systems. Secondly, the construction of an H–x chart is laborious and would normally be undertaken only if many calculations were to be performed, e.g. on a system of commercial importance. Nevertheless, once an H–x chart is available its use is simple, and a great deal of information can be obtained rapidly. If the concentration x of one component of a binary mixture is expressed as a mass fraction, the enthalpy is expressed as a number of heat units per unit mass of mixture, e.g. B.t.u./lb. Molar units are less frequently used.

The basic rule governing the use of an *H–x* chart is that an adiabatic mixing, or separation, process is represented by a straight line. In *Figure 3.6*,

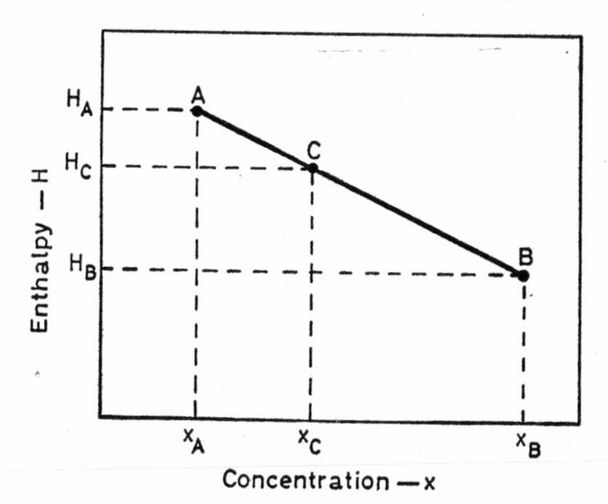

Figure 3.6. An adiabatic mixing process represented on an H–x diagram

points A and B represent the concentrations and enthalpies x_A, H_A and x_B, H_B of two mixtures of the same system. If A is mixed adiabatically with B, the enthalpy and concentration of the resulting mixture is given by point C on the straight line AB. The exact location of point C, which depends on the masses m_A and m_B of the two initial mixtures, can be determined by the mixture rule or lever-arm principle

$$m_A(x_C - x_A) = m_B(x_B - x_C) \tag{35}$$

or

$$x_C = \frac{m_B x_B + m_A x_A}{m_A + m_B} \tag{36}$$

Similarly, if mixture A were to be removed adiabatically from mixture C, the enthalpy and concentration of residue B can be located on the straight line through points A and C by means of the equation

$$x_B = \frac{m_{C'} x_{C'} - m_A x_A}{m_{C'} - m_A} \tag{37}$$

An *H–x* diagram for the system NaOH–water at atmospheric pressure is shown in *Figure 3.7*; this chart[10, 11] is constructed on the basis that the enthalpy of pure water at 32° F is zero. The curved isotherms refer to homogeneous solutions only. The lower right-hand region of the diagram below the saturation curve represents saturated solutions in equilibrium with the various hydrates of NaOH. A simple example will demonstrate the use of this diagram.

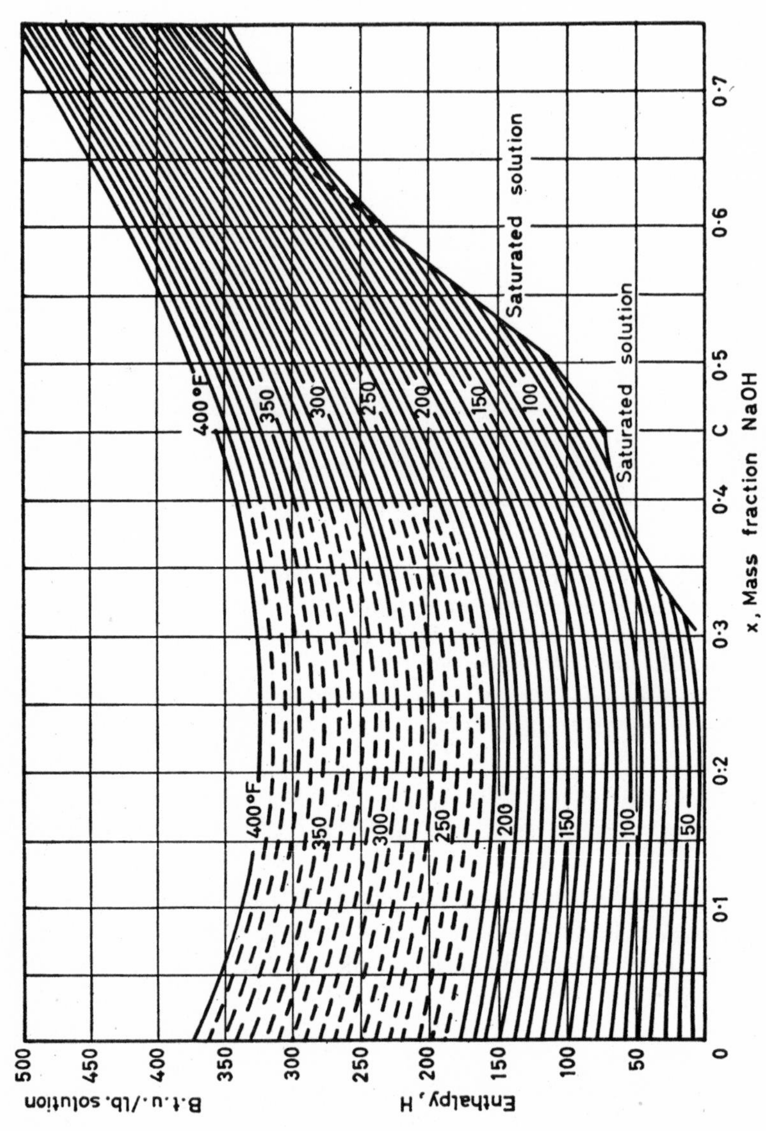

Figure 3.7. Enthalpy-concentration diagram for the system sodium hydroxide–water. (*From* W. L. McCabe *and* J. C. Smith[11], *by courtesy of* McGraw-Hill)

Example—150 lb. of water at 50° F is added to 100 lb. of a 45 per cent solution of NaOH. If the mixing is carried out adiabatically, estimate the temperature of the resulting mixture.

Solution—Locate on *Figure 3.7* point A at $x_A = 0$ on the 50° F isotherm and point B at $x_B = 0{\cdot}45$ on the 60° F isotherm. The required point C representing the final mixture is located, by the mixture rule, at $x_c = 0{\cdot}18$ on the straight line AB. The final temperature is about 110° F.

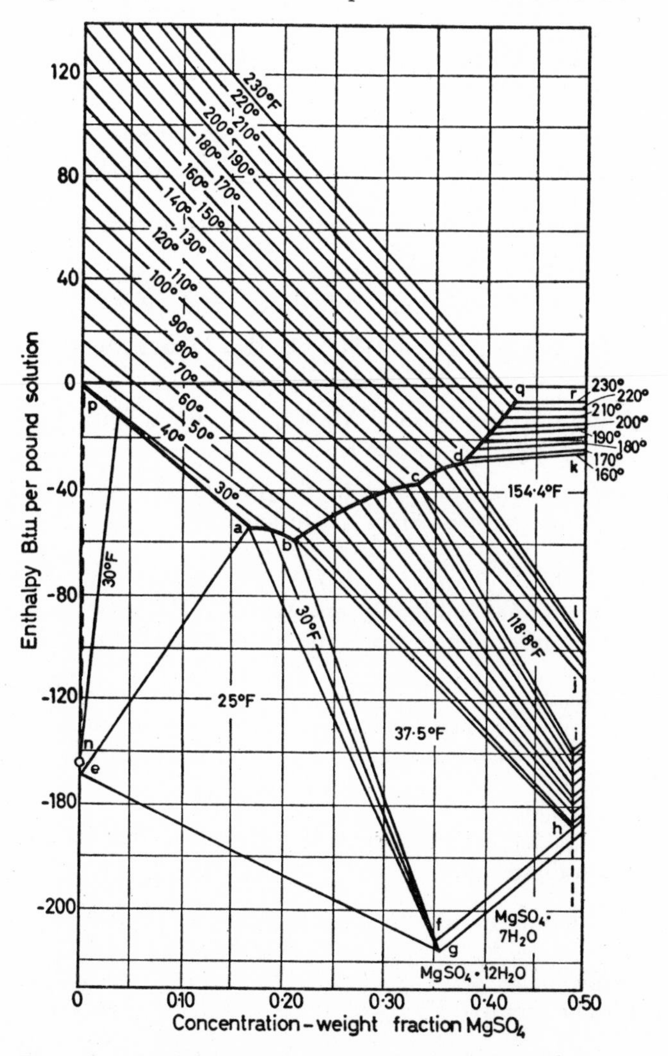

Figure 3.8. Enthalpy-concentration diagram for the system $MgSO_4$–H_2O. (*From* W. L. McCabe[12], *by courtesy of* McGraw-Hill)

The enthalpy-concentration diagram for the system $MgSO_4$–water[12] shown in *Figure 3.8* is more complex than the one described above because enthalpies of liquid and solid phases are recorded on it. The isotherms in the region above curve *pabcdq* represent enthalpies and concentrations of unsaturated aqueous solutions of $MgSO_4$, and the very slight curvature of these isotherms, compared with those in *Figure 3.7*, indicates that the heat of dilution of $MgSO_4$ solutions is very small. Point *p* (zero enthalpy) represents pure water at 32° F, point *n* the enthalpy of pure ice at the same temperature. The portion of the diagram below curve *pabcdq*, which represents liquid–solid systems, can be divided into five polythermal regions

pae	solutions of $MgSO_4$ in equilibrium with pure ice
abfg	equilibrium mixtures of $MgSO_4 \cdot 12H_2O$ and saturated solution
bcih	equilibrium mixtures of $MgSO_4 \cdot 7H_2O$ and saturated solution
cdlj	equilibrium mixtures of $MgSO_4 \cdot 6H_2O$ and saturated solution
dqrk	equilibrium mixtures of $MgSO_4 \cdot H_2O$ and saturated solution

In between these five regions lie four isothermal triangular areas which represent the following conditions.

aef (25° F)	mixtures of ice, cryohydrate *a* and $MgSO_4 \cdot 12H_2O$
bfh (37·5° F)	mixtures of solid $MgSO_4 \cdot 12H_2O$ and $MgSO_4 \cdot 7H_2O$ in a 21 per cent $MgSO_4$ solution
cji (118·8° F)	mixtures of solid $MgSO_4 \cdot 7H_2O$ and $MgSO_4 \cdot 6H_2O$ in a 33 per cent $MgSO_4$ solution
dkl (154·4° F)	mixtures of solid $MgSO_4 \cdot 6H_2O$ and $MgSO_4 \cdot H_2O$ in a 37 per cent $MgSO_4$ solution

The short vertical lines *fg* and *ih* represent the compositions of solid $MgSO_4 \cdot 12H_2O$ (0·359 mass fraction $MgSO_4$) and $MgSO_4 \cdot 7H_2O$ (0·49 mass fraction). The following example demonstrates the use of *Figure 3.8*.

Example—Calculate (*a*) the quantity of heat to be removed and (*b*) the theoretical crystal yield when 5,000 lb. of a 30 per cent solution of $MgSO_4$ by weight at 110° F is cooled to 70° F. Evaporation and radiation losses may be neglected.

Solution—*Figure 3.9* indicates the relevant section—not to scale—of the *H–x* diagram in *Figure 3.8*.

(*a*) Initial solution, *A*	$x_A = 0{\cdot}30,\ H_A = -\ 31$ B.t.u./lb.
Cooled system, *B*	$x_B = 0{\cdot}30,\ H_B = -\ 75$ B.t.u./lb.
Enthalpy change	$\Delta H = -\ 44$ B.t.u./lb.
Heat to be removed	$44 \times 5{,}000 = 220{,}000$ B.t.u.

(*b*) The cooled system *B*, located in the region *bcih* in *Figure 3.8*, comprises $MgSO_4 \cdot 7H_2O$ crystals in equilibrium with solution *S* on curve *bc*. The actual proportions of solid and solution can be calculated by the mixture rule.

Solution composition	$x_S = 0{\cdot}26$
Crystalline phase composition	$x_C = 0{\cdot}49$

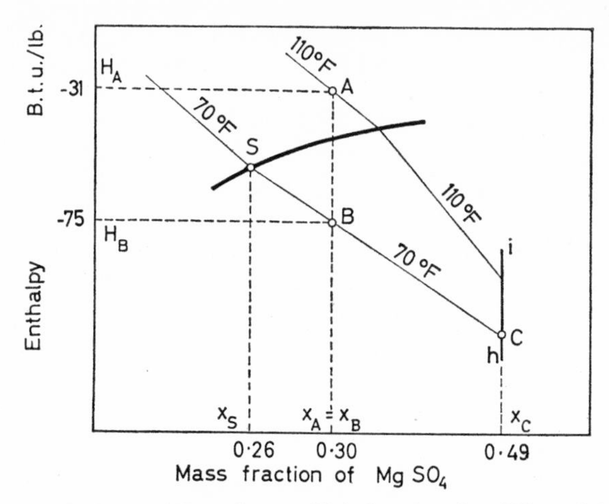

Figure 3.9. Graphical solution of Example on p. 68 (*enlarged portion of Figure 3.8—not to scale*)

Thus,

$$\frac{m_S}{m_C} = \frac{x_C - x_B}{x_B - x_S}$$

$$= \frac{0{\cdot}49 - 0{\cdot}30}{0{\cdot}30 - 0{\cdot}26} = 4{\cdot}75$$

and

$$m_B = m_S + m_C = 5{,}000 \text{ lb.}$$

Therefore, yield

$$m_C = \frac{5{,}000}{5{\cdot}75}$$

$$= 870 \text{ lb. of } MgSO_4 \cdot 7H_2O$$

REFERENCES

1 HATSCHEK, E., *The Viscosity of Liquids*, 1928. London; Bell

2 *British Pharmacopoeia*, General Medical Council, 1958. London; Pharmaceutical Press

3 BARR, G., *Monograph of Viscometry*, 1931. Oxford University Press

4 REID, R. C. and SHERWOOD, T. K., *The Properties of Gases and Liquids*, 1958. New York and London; McGraw-Hill

5 JOHNSON, A. I. and HUANG, C. J., Estimation of heat capacities of organic liquids, *Can. J. Technol.* 33 (1955) 421

6 BATES, O. K., HAZZARD, G. and PALMER, G., Thermal conductivity of liquids, *Industr. Engng Chem.* (*Anal. Ed.*) 10 (1938) 314

7 MULLIN, J. W., Sublimation in theory and practice, *Industr. chem. Mfr* 31 (1955) 540

8 GIACALONE, A., *Gazz. chim. Ital.* 81 (1951) 180 (reported in reference 4)

9 WATSON, K. M., Thermodynamics of the liquid state, *Industr. Engng Chem.*35 (1943) 398

10 McCABE, W. L., The enthalpy-concentration chart—a useful device for chemical engineering calculations, *Trans. Amer. Inst. chem. Engrs* 31 (1935) 129

11 McCABE, W. L. and SMITH, J. C., *Unit Operations of Chemical Engineering*, 1956. New York; McGraw-Hill

12 McCABE, W. L., Crystallization, in *Chemical Engineer's Handbook*, J. H. Perry (Ed.), 1950. New York; McGraw-Hill

4

PHASE EQUILIBRIA

THE amount of information that the simple solubility diagram can yield is strictly limited. For a more complete picture of the behaviour of a given system over a wide range of temperature, pressure and concentration, a phase diagram must be employed. This type of diagram represents graphically, in two or three dimensions, the equilibria between the various phases of a system. The Phase Rule developed by J. Willard Gibbs (1876) relates the number of components, C, phases, P, and degrees of freedom, F, of a system by means of the equation

$$P + F = C + 2$$

These three terms are defined as follows.

The number of *components* of a system is the minimum number of chemical compounds required to express the composition of any phase. In the system water–copper sulphate, for instance, five different chemical compounds can exist, viz. $CuSO_4 \cdot 5H_2O$, $CuSO_4 \cdot 3H_2O$, $CuSO_4 \cdot H_2O$, $CuSO_4$ and H_2O, but for the purpose of applying the Phase Rule there are considered to be only two components, $CuSO_4$ and H_2O, because the composition of each phase can be expressed by the equation

$$CuSO_4 + x\,H_2O \rightleftharpoons CuSO_4 \cdot x\,H_2O$$

Again, in the system represented by the equation

$$CaCO_3 \rightleftharpoons CaO + CO_2$$

three different chemical compounds can exist, but there are only two components because the composition of any phase can be expressed in terms of the compounds CaO and CO_2.

A *phase* is a homogeneous part of a system. Thus any heterogeneous system comprises two or more phases. Any mixture of gases or vapours is a one-phase system. Mixtures of two or more completely miscible liquids or solids are also one-phase systems, but mixtures of two partially miscible liquids or a heterogeneous mixture of two solids are two-phase systems, and so on.

The three variables that can be considered in a system are temperature, pressure and concentration. The number of these variables that may be changed in magnitude without changing the number of phases present is called the number of *degrees of freedom*. In the equilibrium system water–ice–water vapour, $C = 1$, $P = 3$, and from the Phase Rule, $F = 0$. Therefore, in this system there are no degrees of freedom: no alteration may be made in either temperature or pressure (concentration is obviously not a variable in a one-component system) without a change in the number of phases. Such a system is called 'invariant'.

For the system water–water vapour, $C = 1$, $P = 2$ and $F = 1$; so only one variable, pressure or temperature, may be altered independently without changing the number of phases. Such a system is called 'univariant'. The one-phase water vapour system has two degrees of freedom; thus both temperature and pressure may be altered independently without changing the number of phases. Such a system is called 'bivariant'.

Summarizing, it may be said that the physical nature of a system can be expressed in terms of phases, and that the number of phases can be changed by altering one or more of three variables: temperature, pressure or concentration. The chemical nature of a system can be expresses in terms of components, and the number of components is fixed for any given system.

ONE-COMPONENT SYSTEMS

The two variables that can affect the phase equilibria in a one-component, or unary, system are temperature and pressure. The phase diagram for such a system is therefore a temperature–pressure equilibrium diagram. *Figure 4.1* illustrates it on the case of sulphur. This system is chosen because

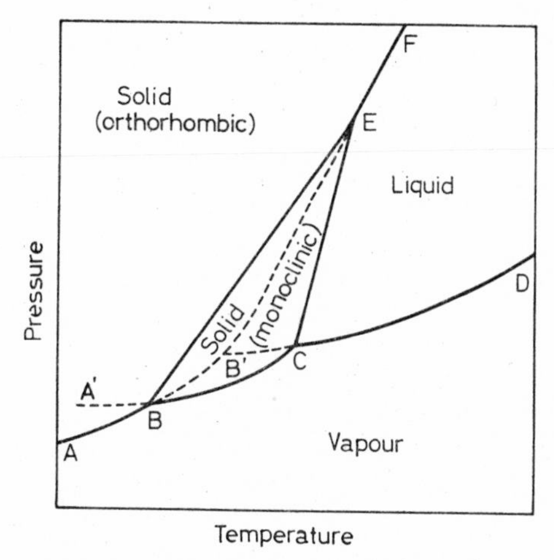

Figure 4.1. Phase diagram for sulphur (not to scale)

it brings out several important points. The diagram (not drawn to scale) indicates the equilibrium relationships between vapour, liquid and two solid forms of sulphur. The area enclosed by the curve *ABEF* is the region in which orthorhombic sulphur is the stable solid form. The areas enclosed by curves *ABCD* and *FECD* indicate the existence of vapour and liquid sulphur, respectively. The 'triangular' area *BEC* represents the region in which monoclinic sulphur is the stable solid form. Curves *AB* and *BC* are the vapour pressure curves for orthorhombic and monoclinic sulphur, respectively, and these curves intersect at the transition point *B*.

Curve *BE* indicates the effect of pressure on the transition temperature for the change orthorhombic S $\rightleftharpoons$ monoclinic S. Point *B*, therefore, is a triple point representing the temperature and pressure (95·5° C and 0·0038 mm Hg) at which orthorhombic sulphur and sulphur vapour can co-exist in stable equilibrium. Curve *EF* indicates the effect of pressure on the melting point of orthorhombic sulphur; point E is a triple point representing the temperature and pressure (151° C and 1,290 atm.) at which orthorhombic and monoclinic sulphur and liquid sulphur are in stable equilibrium. Curve CD is the vapour pressure curve for liquid sulphur, and curve CE indicates the effect of pressure on the melting point of monoclinic sulphur. Point C, therefore, is another triple point (115° C and 0·018 mm Hg) representing the equilibrium between monoclinic and liquid sulphur and sulphur vapour.

The broken lines in *Figure 4.1* represent metastable conditions. If orthorhombic sulphur is heated rapidly beyond 95·5° C the change to the monoclinic form does not occur until a certain time has elapsed; curve BB', a continuation of curve AB, is the vapour pressure curve for metastable orthorhombic sulphur above the transition point. Similarly, if monoclinic sulphur is cooled rapidly below 95·5° C the change to the orthorhombic form does not take place immediately, and curve BA' is the vapour pressure curve for metastable monoclinic sulphur below the transition point. Likewise, curve CB' is the vapour pressure curve for metastable liquid sulphur below the 115° C transition point, and curve $B'E$ the melting point curve for metastable orthorhombic sulphur. Point B', therefore, is a fourth triple point (110° C and 0·013 mm Hg) of the system.

Only three of the four possible phases orthorhombic (solid), monoclinic (solid), liquid and vapour can co-exist in stable equilibrium at any one time, and then only at one of the three 'stable' triple points. This in fact can be deduced from the Phase Rule:

$$3 + \mathrm{F} = 1 + 2$$
$$\mathrm{F} = 0$$

Enantiotropy and Monotropy

A pure substance capable of existing in two different crystalline forms is called dimorphous. The transformation from one form to the other can be reversible or irreversible; in the former case the two crystalline forms are said to be enantiotropic, in the latter, monotropic. These phenomena, already described in Chapter 1, can be demonstrated with reference to the pressure–temperature phase diagram.

Figure 4.2a shows the phase reactions exhibited by two enantiotropic solids, α and β. AB is the vapour pressure curve for the α form, BC for the β form, CD for the liquid. Point B, where the vapour pressure curves of the two solids intersect, is the transition point; the two forms can co-exist in equilibrium under these conditions of temperature and pressure. Point C is a triple point at which vapour, liquid and β solid can co-exist. This point can be considered to be the melting point of the β form.

If the α solid is heated slowly it changes into the β solid and finally melts. The vapour pressure curve ABC is followed. Conversely, if the liquid is cooled slowly, the β form crystallizes out first and then changes into the α form. Rapid heating or cooling, however, can result in a different behaviour. The vapour pressure of the α form can increase along curve BB', a continuation of AB, the α form now being metastable. Similarly, the liquid vapour pressure can fall along curve CB', a continuation of DC, the liquid being metastable. Point B', therefore, is a metastable triple point at which the liquid, vapour and α solid can co-exist in metastable equilibrium.

The type of behaviour described above is well illustrated by the case of sulphur (*Figure 4.1*) where the orthorhombic and monoclinic forms are enantiotropic; the transition point occurs at a lower temperature than the triple point.

Figure 4.2b shows the pressure–temperature curves for a monotropic substance. AB and BC are the vapour pressure curves for the α solid and

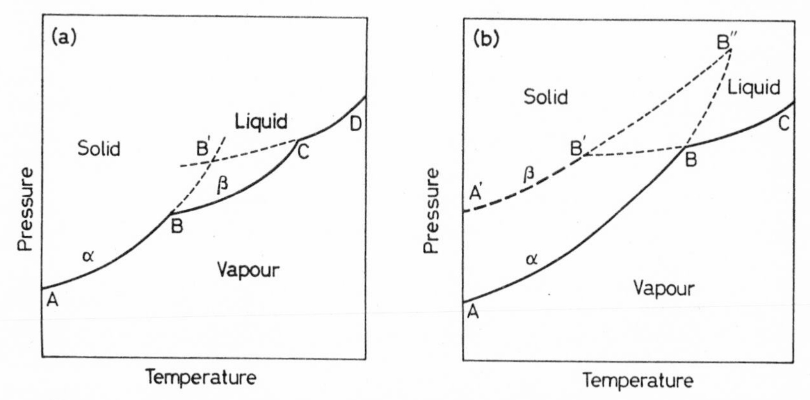

Figure 4.2. Pressure–temperature diagrams for dimorphous substances: (a) enantiotropy; (b) monotropy

liquid, respectively, $A'B'$ is that for the β solid. In this case the vapour pressure curves of the α and β forms do not intersect, so there is no transition point. The solid form with the higher vapour pressure at any given temperature (β in this case) is the metastable form. Curves BB' and BB'' are the vapour pressure curves for the liquid and metastable α solid, so B' is a metastable triple point. If this system did exhibit a true transition point it would lie at point B'', but as this represents a temperature higher than the melting point of the solid it cannot exist.

A typical case of monotropy is the change from white to red phosphorus. Benzophenone is another example of a monotropic substance: the stable melting point is 49° C, whereas the metastable form melts at 29° C.

TWO-COMPONENT SYSTEMS

The three variables that can affect the phase equilibria of a two-component, or binary, system are temperature, pressure and concentration. The behaviour of such a system should, therefore, be represented by a space model with three mutually perpendicular axes of pressure, temperature and concentration. Alternatively, three diagrams with pressure–temperature, pressure–concentration and temperature–concentration axes, respectively, can be employed. However, in most crystallization processes the main interest lies in the liquid and solid phases of a system; a knowledge of the behaviour of the vapour phase is only required when considering sublimation processes. Because pressure has little effect on the equilibria between liquids and solids, the phase changes can be represented on a temperature–concentration diagram; the pressure, usually atmospheric, is ignored. Such a system is said to be 'condensed', and a 'reduced' phase rule can be formulated excluding the pressure variable

$$P + F' = C + 1$$

where F' is the number of degrees of freedom, not including pressure.

Four different types of two-component system will now be considered. Detailed attention is paid to the first type solely to illustrate the information that can be deduced from a phase diagram. It will be noted that the concentration of a solution on a phase diagram is normally given as a mass fraction or mass percentage and not as 'mass of solute per unit mass of solvent', as recommended for the solubility diagram (Chapter 2). Mole fractions and mole percentages are also suitable concentration units for use in phase diagrams.

Simple Eutectic

A typical example of a system in which the components do not combine to form a chemical compound is shown in *Figure 4.3*. Curves *AB* and *BC*

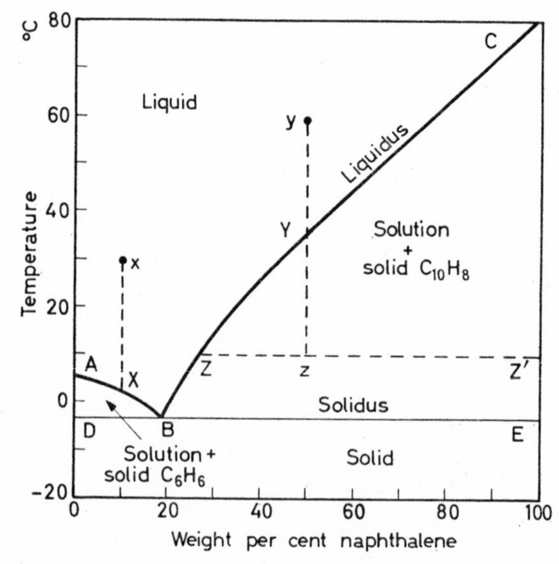

Figure 4.3. Phase diagram for the simple eutectic system naphthalene–benzene

represent the temperatures at which homogeneous liquid solutions of naphthalene in benzene begin to freeze or to crystallize. The curves also represent, therefore, the temperatures above which mixtures of these two components are completely liquid. The name 'liquidus' is generally given to this type of curve. In aqueous systems of this type one liquidus is the freezing point curve, the other the normal solubility curve. Line *DBE* represents the temperature at which solid mixtures of benzene and naphthalene begin to melt, or the temperature below which mixtures of these two components are completely solid. The name 'solidus' is generally given to this type of line. The melting or freezing points of pure benzene and naphthalene are given by points A (5·5° C) and B (80·2° C), respectively. The upper area enclosed by the liquidus, ABC, represents the homogeneous liquid phase, i.e. a solution of naphthalene in benzene, that enclosed by the solidus, DBE, indicates solid mixtures of benzene and naphthalene. The small and large 'triangular' areas ABD and BCE represent mixtures of solid benzene and solid naphthalene, respectively, and benzene–naphthalene solution.

If a solution represented by point x is cooled, pure solid benzene is deposited when the temperature of the solution reaches point X on curve AB. As solid benzene separates out, the solution becomes more concentrated in naphthalene and the equilibrium temperature of the system falls following curve AB. If a solution represented by point y is cooled, pure solid naphthalene is deposited when the temperature reaches point Y on the solubility curve; the solution becomes more concentrated in benzene and the equilibrium temperature follows curve CB. Point B, common to both curves, is the eutectic point (— 3·5° C and 0·189 mass fraction of naphthalene), and this is the lowest freezing point in the whole system. At this point, a completely solidified mixture of benzene and naphthalene of fixed composition is formed; it is important to note that the eutectic is a physical mixture, not a chemical compound. Below the eutectic temperature all mixtures are solid.

If the solution y is cooled below the temperature represented by point Y on curve BC to some temperature represented by point z, the composition of the system as a whole remains unchanged. The physical state of the system has been altered, however; it now consists of a solution of benzene and naphthalene containing solid naphthalene. The composition of the solution, or mother liquor, is given by point z on the solubility curve, and the proportions of solid naphthalene and solution are given by the ratio of the lengths zZ and zZ', i.e.

$$\frac{\text{mass of solid } C_{10}H_8}{\text{mass of solution}} = \frac{zZ}{zZ'}$$

A process involving both cooling and evaporation can be analysed in two steps. The first is as described above, i.e. the location of points z, Z and Z'; this represents the cooling operation. If benzene is evaporated from the system, z no longer represents the composition; so the new composition point z' (not shown in the diagram) is located along line ZZ' between points z and Z. Then the ratio $z'Z/z'Z'$ gives the proportions of solid and solution.

The systems $KCl–H_2O$ and $(NH_4)_2SO_4–H_2O$ are good examples of aqueous salt solutions which exhibit simple eutectic formation. In aqueous systems the eutectic mixture is usually called a *cryohydrate*, and the eutectic point, a 'cryohydric point'.

Compound Formation

The solute and solvent of a binary system may, and frequently do, combine to form one or more different compounds. In aqueous solutions these compounds are called 'hydrates', for non-aqueous systems the term 'solvate' is sometimes used. Two types of compound can be considered: one which can co-exist in stable equilibrium with a liquid of the same composition, and the other which cannot behave in this manner. In the former case, the compound is said to have a *congruent melting point*; in the latter, to have an *incongruent melting point.*

Figure 4.4 illustrates the phase reactions in the manganese nitrate–water system. Curve *AB* is the freezing point curve. The solubility curve *BCDE* for $Mn(NO_3)_2$ in water is not continuous, owing to the formation of two different hydrates. The area above curve *ABCDE* represents homogeneous

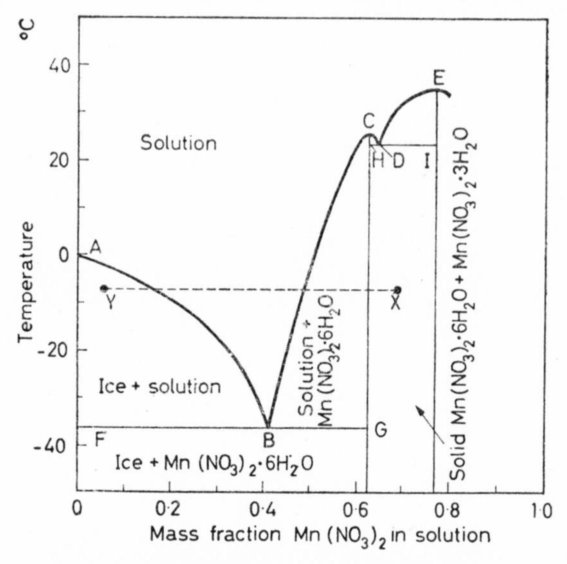

Figure 4.4. Phase diagram for the system $Mn(NO_3)_2$–H_2O

liquid solutions. Mixtures of the hexahydrate and solution exist in areas *BCD* and *CDH* of the trihydrate and solution in *DEI.* The rectangular areas under *FG* and *HI* both denote completely solidified systems, the former consisting of ice and hexahydrate, the latter of a solid mixture of the tri-hydrate and hexahydrate. Point *B* is a eutectic or cryohydric point with the co-ordinates – 36° C and 0·405 mass fraction of $Mn(NO_3)_2$.

Point *C* in *Figure 4.4* shows the melting point (25·8° C) and composition (0·624 mass fraction of $Mn(NO_3)_2$) of the hexahydrate. Thus when a solution of this composition is cooled to 25·8° C it solidifies to form the hexahydrate, i.e. no change in composition occurs. Point *C*, therefore, is a congruent point. Point *D* (23·5° C and 0·646 mass fraction) is the second cryohydric point of the system, and point *E* (35·5° C and 0·768 mass fraction) the congruent melting point of $Mn(NO_3)_2 \cdot 3H_2O$.

The behaviour of solutions of manganese nitrate in water on cooling can be traced in the same manner as that described above for simple eutectic systems. The solution concentrations and the proportions of solid and solution can similarly be deduced graphically. The process of isothermal evaporation in congruent melting systems presents an interesting phenomenon. For example, the mixture represented by point *X* in *Figure 4.4* is completely solid, being a mixture of the tri- and hexahydrates, but once sufficient water has been removed to reduce the $Mn(NO_3)_2$ content to 62·4 per cent, the system becomes partially liquefied. When more water is removed, so that curve *BC* is penetrated, the system becomes a homogeneous liquid solution, but solidifies again when curve *AB* is reached.

The formation of eutectics and solvates with congruent points is observed in many organic, aqueous inorganic and metallic systems. The case illustrated above is a rather simple example. Some systems form a large number of solvates and their phase diagrams can become rather complex. Ferric chloride, for example, forms four hydrates, and the $FeCl_3$–H_2O phase diagram exhibits five cryohydric points and four congruent points.

A solvate which is unstable in the presence of a liquid of the same composition is said to have an incongruent melting point. Such a solvate melts to form a solution and another compound which may or may not be a solvate. For instance, the hydrate $Na_2SO_4 \cdot 10H_2O$ melts at 32·4° C and immediately breaks down into the anhydrous salt and water, hence this temperature is the incongruent melting point of the decahydrate. The terms 'meritectic point' and 'transition point' are also used instead of the expression 'incongruent melting point'.

Figure 4.5 illustrates the behaviour of the system sodium chloride–water. The various areas are marked on the diagram. *AB* is the freezing point

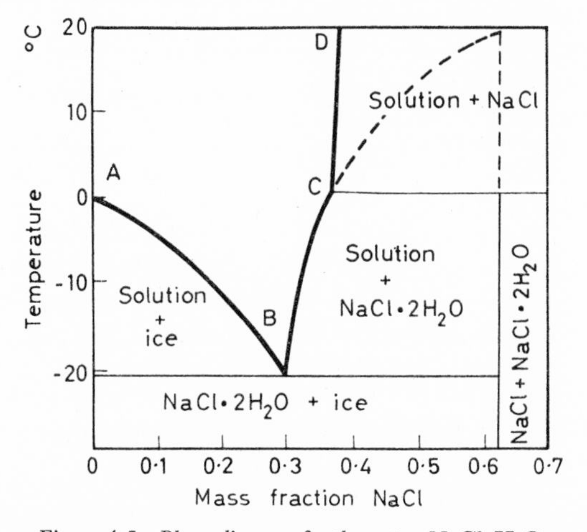

Figure 4.5. Phase diagram for the system NaCl–H_2O

curve, *BC* is the solubility curve for the dihydrate. Point *B* (− 21° C) is a eutectic or cryohydric point at which a solid mixture of ice and $NaCl \cdot 2H_2O$ of fixed composition (0·29 mass fraction of NaCl) is deposited. At point *C* (0·15° C) the dihydrate decomposes into the anhydrous salt and water; this is, therefore, the incongruent melting point, or transition point, of

$NaCl \cdot 2H_2O$. The vertical line commencing at 0·619 mass fraction of NaCl represents the composition of the dihydrate. If this system had a congruent melting point, which it does not have, this line would meet the peak of the extension of curve *BC* (e.g. see *Figure 4.4*).

Many aqueous and organic systems exhibit eutectic and incongruent points. Several cases are known of an inverted solubility effect after the transition point (see *Figure 2.1b*); the systems Na_2SO_4–H_2O and Na_2CO_3–H_2O are particularly well known examples of this phenomenon.

Solid Solutions

Many binary systems when submitted to a cooling operation do not at any stage deposit one of the components in the pure state: both components are deposited simultaneously. The deposited solid phase is, in fact, a solid solution. Only two phases can exist in such a system, a homogeneous liquid solution and a solid solution. Therefore, from the reduced phase rule, $F' = 1$, so an invariant system cannot result. One of three possible types of equilibrium diagram can be exhibited by systems of this kind. In the first type, illustrated in *Figure 4.6a*, all mixtures of the two components have

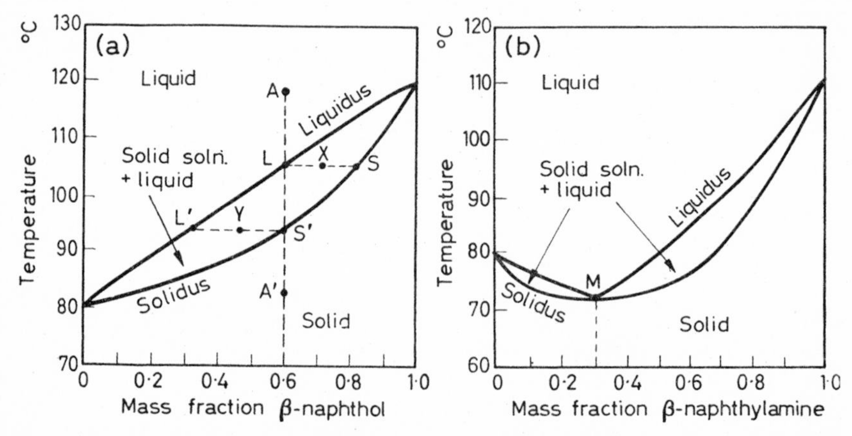

Figure 4.6. Solid solutions: (a) continuous series (naphthalene–β-naphthol; (b) minimum melting point (naphthalene–β-naphthylamine)

freezing or melting points intermediate between the melting points of the pure components. In the second type, shown in *Figure 4.6b*, a minimum is produced in the freezing and melting point curves. In the third, rare, type of diagram (not illustrated) a maximum is exhibited in the curves.

Figure 4.6a shows the temperature–concentration phase diagram for the system naphthalene–β-naphthol which forms a continuous series of solid solutions. The melting points of pure naphthalene and β-naphthol are 80° and 120° C, respectively. The upper curve is the liquidus or freezing point curve, the lower the solidus or melting point curve. Any system represented by a point above the liquidus is completely molten, and any point below the solidus represents a completely solidified mass. A point within the area enclosed by the liquidus and solidus curves indicates an equilibrium mixture of liquid and solid solution. Point *X*, for instance, denotes a liquid of composition *L* in equilibrium with a solid solution of composition *S*, and point *Y* a liquid *L′* in equilibrium with a solid *S′*.

The phase reactions occurring on the cooling of a given mixture can be traced as follows. If a homogeneous liquid represented by point *A* (60 per cent β-naphthol) is cooled slowly, it starts to crystallize when point *L* (105° C) is reached. The composition of the first crystals is given by point *S* (82 per cent β-naphthol). As the temperature is lowered further, more crystals are deposited but their composition changes successively along curve *SS′*, and the liquid composition changes along curve *LL′*. When the temperature is reduced to 94° C (points *L′* and *S′*), the system solidifies completely. The overall composition of the solid system at some temperature represented by say point *A′* is the same as that of the original homogeneous melt, assuming that no crystals have been removed during the cooling process, but the system is no longer homogeneous because of the successive depositions of crystals of varying composition. The changes occurring when a solid mixture *A′* is heated can be traced in a manner similar to the cooling operation.

Figure 4.6b shows the relatively uncommon, but not rare, type of binary system in which a common minimum temperature is reached by both the upper liquidus and lower solidus curves. These two curves approach and touch at point *M*. The example shown in *Figure 4.6b* is the system naphthalene–β-naphthylamine. Freezing and melting points of mixtures of this system do not necessarily lie between the melting points of the pure components. Three sharp melting points are observed: 80° C (pure naphthalene), 110° C (pure β-naphthylamine) and 72·5° C (mixture *M*, 0·3 mass fraction β-naphthylamine). Although the solid solution deposited at point *M* has a definite composition, it is not a chemical compound. The components of such a minimum melting point mixture are rarely, if ever, present in stoichiometric proportions. Point *M*, therefore, is not a eutectic point; the liquidus curve is completely continuous, it only approaches and touches the solidus at *M*. The phase reactions occurring when mixtures of this system are cooled can be traced in the same manner as that described for the continuous series solid solutions.

Thermal Analysis

Equilibrium in solid–liquid systems may be determined by the solubility methods discussed in Chapter 2. If these are not convenient or applicable another technique, known as thermal analysis, may be employed. A phase reaction is always accompanied by an enthalpy change, and this heat effect can readily be observed if a cooling curve is plotted for the system. In many cases a simple apparatus can be used; a 6 × 1 in. glass boiling tube, fitted with a stirrer and a thermometer graduated in $\frac{1}{10}$° C, suspended in a shielding vessel or refrigerant bath, will suffice. The temperature of the system is recorded at regular intervals of say 1 min.

A smooth cooling curve is followed until a phase reaction takes place, and then the accompanying heat effect causes an arrest or change in slope.

Figure 4.7a shows a typical example for a pure substance. *AB* is the cooling curve for the homogeneous liquid phase. At point *B* the substance starts to freeze and the system remains at constant temperature, the freezing point, until solidification is complete at point *C*. The solid then cools at a rate indicated by curve *CD*. It is possible, of course, for the liquid phase to cool below the freezing point, and some systems may exhibit appreciable degrees of supercooling. The dotted curve in *Figure 4.7a* denotes the sort of path followed if supercooling occurs. Seeding of the system will minimize these effects.

Figure 4.7b shows the type of cooling curve obtained for a binary system in which eutectic or compound formation occurs. The temperature of the

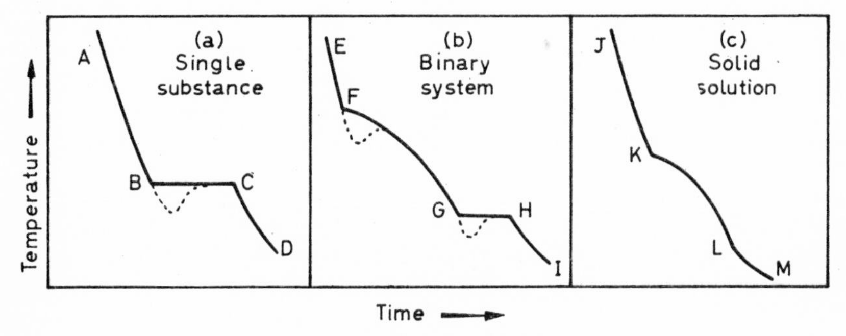

Figure 4.7. Some typical cooling curves

homogeneous liquid phase falls steadily along curve *EF* until, at point *F*, deposition of the solid phase commences. The rate of cooling changes along curve *FG* as more and more solid is deposited. The composition of the remaining solution changes until the composition of the eutectic is reached, then crystallization or freezing continues at constant temperature (line *GH*), i.e. the eutectic behaves as a single pure substance. The completely solidified system cools along curve *HI*. Supercooling, denoted by the dotted lines, may be encountered at both arrest points if the system is not seeded.

Figure 4.7c shows a typical cooling curve for a binary mixture that forms a series of solid solutions. The first arrest, *K*, in the curve corresponds to the onset of freezing, and this represents a point on the liquidus. The second arrest, *L*, occurs on the completion of freezing and represents a point on the solidus. It will be noted that no constant-temperature freezing point occurs in such a system.

Equilibria in solid solutions are better studied by a heating rather than a cooling process. This is the basis of the thaw-melt method first proposed by Rheinboldt (1925). An intimate mixture, of known composition, of the two pure components is prepared by melting, solidifying and then crushing to a fine powder. A small sample of the powder is placed in a melting-point tube, attached close to the bulb of a thermometer graduated in $\frac{1}{10}$° C, and immersed in a stirred bath. The temperature is raised slowly and regularly at a rate of about 1° in 5 min. The 'thaw point' is the temperature at which liquid first appears in the tube; this is a point on the solidus. The 'melt point' is the temperature at which the last solid particle melts; this is a point on the liquidus. Only pure substances and eutectic mixtures have sharp melting points. The thaw-melt method is particularly useful if the system is prone to supercooling, and it has the added advantage of requiring only small quantities of test material.

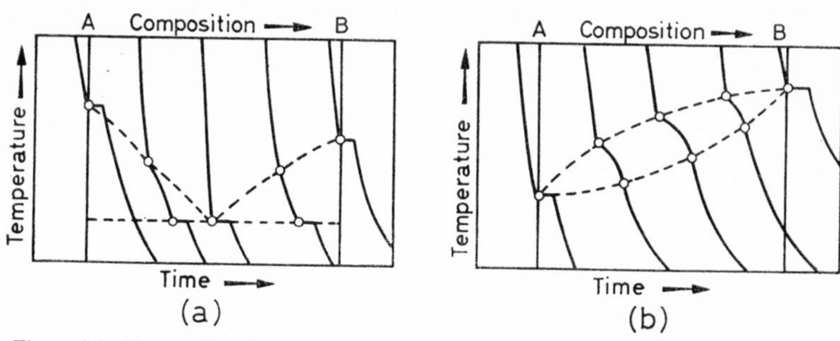

Figure 4.8. Construction of equilibrium diagrams from 'thaw–melt' data: (a) eutectic system; (b) solid solution

The construction of equilibrium diagrams from cooling or thaw-melt data is indicated in *Figure 4.8.* In practice, however, a large number of different mixtures of the two components A and B, covering the complete range from pure A to pure B, would be tested. The liquidus curves are drawn through the first-arrest points, the solidus curves through the second-arrest points. Only at 100 per cent A, 100 per cent B and at the eutectic point do the liquidus and solidus meet.

THREE-COMPONENT SYSTEMS

The phase equilibria in three-component, or ternary, systems can be affected by four variables, viz. temperature, pressure and the concentration of any two of the three components. This fact can be deduced from the phase rule:

$$P + F = 3 + 2$$

which indicates that a one-phase ternary system will have four degrees of freedom. It is impossible to represent the effects of the four possible variables in a ternary system on a two-dimensional graph. For solid–liquid systems, however, the pressure variable may be neglected, and the effect of temperature will be considered later.

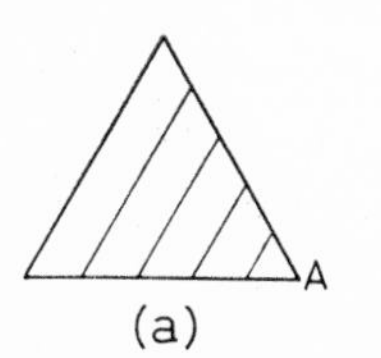

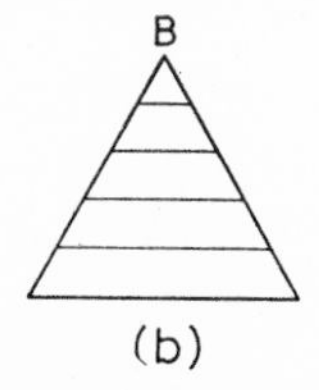

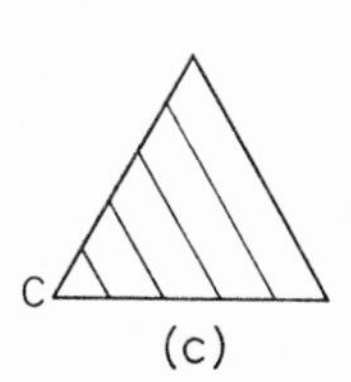

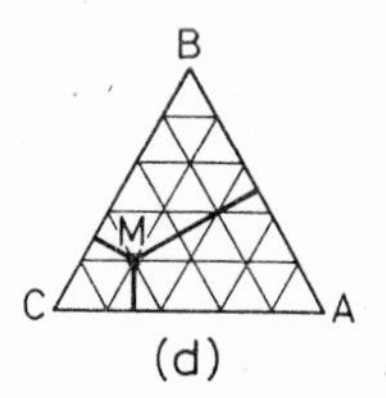

Figure 4.9. Construction of the equilateral triangular diagram

The composition of a ternary system can be represented graphically on a triangular diagram. Two methods are in common use. The first utilizes the equilateral triangle, and the method of construction is shown in *Figure 4.9.*

The apexes of the triangle represent the pure components A, B and C. A point on a side of the triangle stands for a binary system, AB, BC or AC, a point within the triangle represents a ternary system ABC. The scales may be constructed in any convenient units, e.g. weight or mole percent, weight or mole fraction, etc., and any point on the diagram must satisfy the equation $A + B + C = 1$ or 100. The quantities of the components A, B and C in a given mixture M (*Figure 4.9d*) are represented by the perpendicular distances from the sides of the triangle.

Special triangular graph paper is required if the equilateral diagram is to be used, and for this reason many workers prefer to employ the right-angled triangular diagram which can be drawn on ordinary linear graph paper. The construction of the right-angled isosceles triangle is shown in *Figure 4.10.*

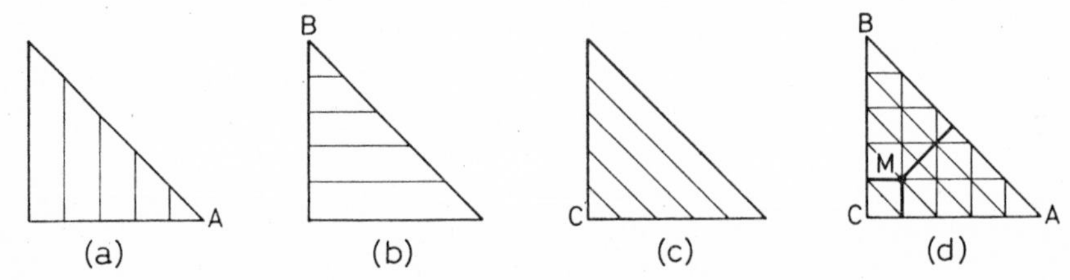

Figure 4.10. Construction of the right-angled triangular diagram

Again, as in the case of the equilateral triangle, each apex represents a pure component A, B or C, a point on a side a binary system, and a point within the triangle a ternary system; in all cases $A + B + C = 1$ or 100. The quantities of A, B and C in a given mixture M (*Figure 4.10d*) are represented by the perpendicular distances to the sides of the triangle. If two compositions, A and B, B and C, or A and C are known the composition of the third component is fixed on both triangular diagrams.

Two actual plots are shown in *Figure 4.11* to illustrate the interpretation of these diagrams. For clarity the C scale has been omitted from the right-angled diagram; the C values can be obtained from the expression $C = 1 - (A + B)$. The 'mixture rule' is also illustrated in *Figure 4.11.* When any

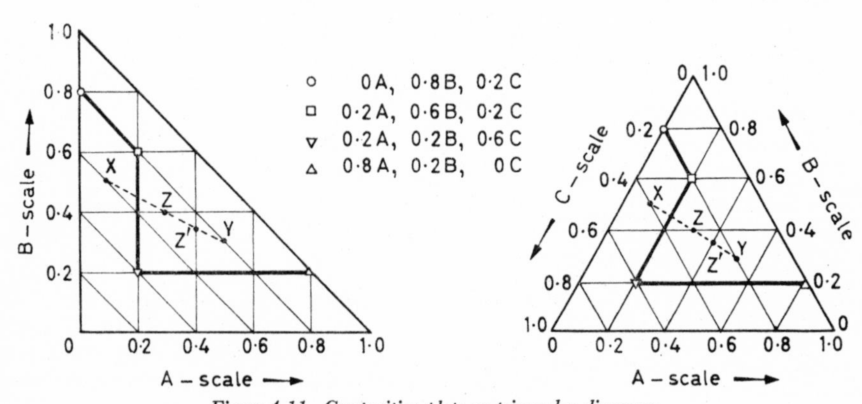

Figure 4.11. Composition plots on triangular diagrams

two mixtures X and Y are mixed together, the composition of the final mixture Z is represented by a point on the diagram located on a straight line drawn between the points representing the initial mixtures. The position of Z is located by the expression

$$\frac{\text{mass of mixture } X}{\text{mass of mixture } Y} = \frac{\text{distance } YZ}{\text{distance } XZ}$$

For example, if one part of a mixture X (0·1A, 0·5B, 0·4C) is mixed with one part of a mixture Y (0·5A, 0·3B, 0·2C), the composition of the final mixture Z (0·3A, 0·4B, 0·3C) is found on the line XY where $XZ = YZ$. Again, if 3 parts of Y are mixed with 1 part of X, the mixture composition Z' (0·4A, 0·35B, 0·25C) is found on the line XY where $XZ' = 3(YZ')$. The mixture rule also applies to the removal of one or more constituents from a system. Thus, one part of a mixture X removed from 2 parts of a mixture Z would yield one part of a mixture Y given by:

$$\frac{\text{mass of original } Z}{\text{mass of } X \text{ removed}} = \frac{YX}{YZ} = \frac{2}{1}$$

Similarly, one part of X removed from 4 parts of Z' would yield 3 parts of a mixture Y given by

$$\frac{\text{mass of original } Z'}{\text{mass of } X \text{ removed}} = \frac{YX}{YZ'} = \frac{4}{1}$$

The principle of the mixture rule is the same as that employed in the operation of lever-arm problems, i.e. $m_1 l_1 = m_2 l_2$ where m is a mass and l is the distance between the line of action of the mass and the fulcrum. For this reason, the mixture rule is often referred to as the lever-arm or centre of gravity principle.

Although ternary equilibrium data are most frequently plotted on equilateral diagrams, the use of the right-angled diagram has several advantages. Apart from the fact that special graph paper is not required, it is claimed that information may be plotted more rapidly on it, and some people find it easier to read. In this section, the conventional equilateral diagram will mostly be employed, but one or two illustrations of the use of the right-angled diagram will be given.

Eutectic Formation

Equilibrium relationships in three-component systems can be represented on a temperature–concentration space model as shown in *Figure 4.12a*. The ternary system *ortho-*, *meta-* and *para*-nitrophenol, in which no compound formation occurs, is chosen for illustration purposes. The three components will be referred to as O, M and P, respectively. Points O', M' and P' on the vertical edges of the model represent the melting points of the pure components *ortho-* (45° C), *meta-* (97° C) and *para-* (114° C). The vertical faces of the prism represent the temperature–concentration diagrams for the three binary systems O–M, O–P and M–P. These diagrams are each similar to that shown in *Figure 4.3* described in the section on binary eutectic systems. In this case, however, the solidus lines have been omitted for clarity.

The binary eutectics are represented by points A (31·5°C; 72·5 per cent O, 27·5 per cent M), B (33·5° C; 65·5 per cent O, 24·5 per cent M) and C (61·5° C; 54·8 per cent M, 45·2 per cent P). Curve AD within the prism represents the effect of the addition of the component P to the O–M binary eutectic A. Similarly, curves BD and CD denote the lowering of the freezing points of the binary eutectics B and C, respectively, on the addition of the third component. Point D, which indicates the lowest temperature at which solid and liquid phases can co-exist in equilibrium in this system, is a ternary eutectic point (21·5° C; 57·7 per cent O, 23·2 per cent M, 19·1 per cent P). At this temperature and concentration the liquid freezes invariantly to form a solid mixture of the three components. The section of the space model above the freezing point surfaces formed by the liquidus curves represents

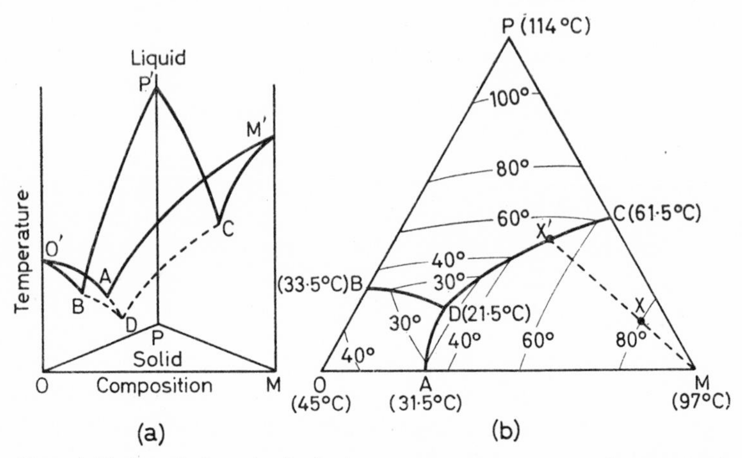

Figure 4.12. Eutectic formation in the three-component system o-, m- and p-nitrophenol: (a) temperature-concentration space model; (b) projection on a triangular diagram

the homogeneous liquid phase. The section below these surfaces down to a temperature represented by point D denotes solid and liquid phases in equilibrium. Below this temperature the section of the model represents a completely solidified system.

Figure 4.12b is the projection of the curves AD, BD and CD in *Figure 4.12a* on to the triangular base. The apexes of the triangle represent the pure components O, M and P and their melting points. Points A, B and C on the sides of the triangle indicate the three binary eutectic points, point D the ternary eutectic point. The projection diagram is divided by curves AD, BD and CD into three regions which denote the three liquidus surfaces in the space model. The temperature falls from the apexes and sides of the triangle towards the eutectic point D, and several isotherms showing points on the liquidus surfaces are drawn on the diagrams. The phase reactions occurring on cooling a given ternary mixture can now be traced.

A molten mixture with a composition as in point X starts to solidify when the temperature is reduced to 80° C. Point X lies in the region $ADCM$, so pure *meta-* is deposited on decreasing temperature. The composition of the remaining melt changes along line MXX' in the direction away from point M representing the deposited solid phase (the mixture rule). At X', where line MXX' meets curve CD, the temperature is about 50° C, and at this point a second component (*para-*) also starts to crystallize out. On further cooling, *meta-* and *para-* are deposited and the liquid phase composition changes in the direction $X'D$. When melt composition and temperature reach point D the third component (*ortho-*) crystallizes out, and the system solidifies without any further change in composition. A similar reasoning may be applied to the cooling, or melting, of systems represented by points in the other regions of the diagrams.

Aqueous Solutions

There are many different types of phase behaviour encountered in ternary systems consisting of water and two solid solutes. Only a few of the simpler cases will be considered here; attention will be devoted to a brief survey of systems in which there is (*a*) no chemical reaction, (*b*) formation of a double salt, and (*c*) formation of a solvate, e.g. a hydrate.

At one given temperature the composition of, and phase equilibria in, a ternary aqueous solution can be represented on an isothermal triangular diagram. The construction of these diagrams has already been described. Polythermal diagrams can also be constructed, but in the case of complex systems the charts tend to become congested and rather difficult to interpret.

No Compound Formed

This simplest case is illustrated in *Figure 4.13* for the system KNO_3–$NaNO_3$–H_2O at 50° C. Neither salt forms a hydrate, nor do they combine chemically.

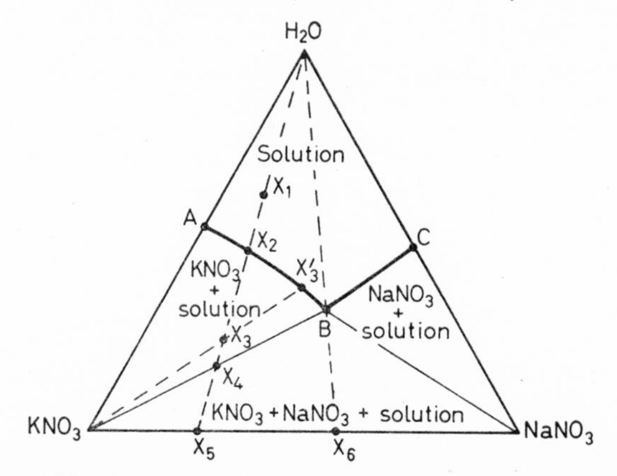

Figure 4.13. Phase diagram for the system KNO_3–$NaNO_3$–H_2O *at* 50°C

Point A represents the solubility of KNO_3 in water at the stated temperature (46·2 g/100 g of solution) and point C the solubility of $NaNO_3$ (53·2 g/100 g). Curve AB indicates the composition of saturated ternary solutions which are in equilibrium with solid KNO_3, curve BC those in equilibrium with solid $NaNO_3$. The upper area enclosed by ABC represents the region of unsaturated homogeneous solutions. The three 'triangular' areas are constructed by drawing straight lines from point B to the two apexes of the triangle; the compositions of the phases within these regions are marked on the diagram. At point B the solution is saturated with respect to both KNO_3 and $NaNO_3$, and from the reduced phase rule, $F' = 1$. This means that point B is univariant, or invariant when the temperature is fixed.

The effect of isothermal evaporation on such a system can be shown as follows. If water is evaporated from an unsaturated solution represented by point X_1 in the diagram, the solution concentration will increase, following line X_1X_2. Pure KNO_3 will be deposited when the concentration reaches point X_2. If more water is evaporated to give a system of composition X_3 the composition of the solution will be represented by point X_3' on the saturation curve AB, and when composition X_4 is reached, by point B; any further removal of water will cause the deposition of $NaNO_3$. All solutions in contact with solid will thereafter have a constant composition B which is referred to as the *drying-up point* of the system. After the complete evaporation of water the composition of solids in the residue is indicated by point X_5 on the base line.

Similarly, if an unsaturated solution, represented by a point located to the right of B in the diagram, were evaporated isothermally, only $NaNO_3$ would be deposited until the solution composition reached the drying-up point B when KNO_3 would also be deposited. The solution composition would thereafter remain constant until evaporation was completed. If water is removed isothermally from a solution of composition B the composition of the deposited solid is given by point X_6 on the base line, and it remains unchanged throughout the remainder of the evaporation process.

The effect of the addition of one of the salts to the system KNO_3–$NaNO_3$–H_2O at 50° C is shown in *Figure 4.14a*. This time the equilibria are plotted on a right-angled triangular diagram simply to demonstrate the use of this type of chart. Points A and C, as in *Figure 4.13*, refer to the solubilities at

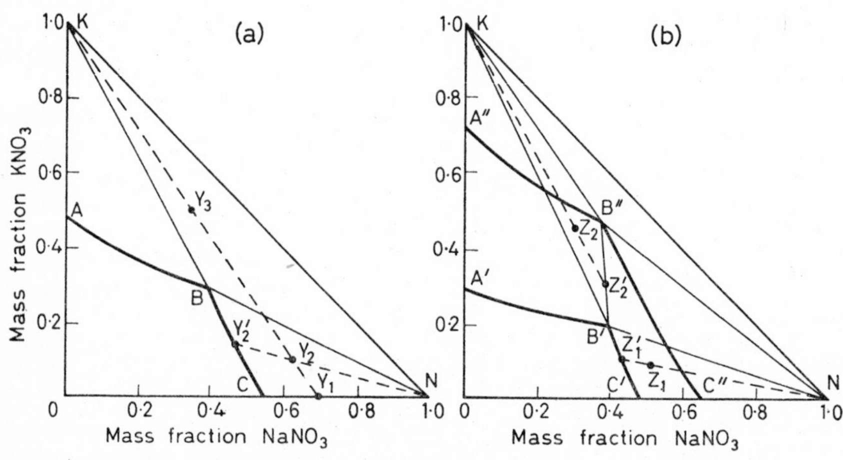

Figure 4.14. Phase diagrams for the system KNO_3–$NaNO_3$–H_2O*: (a) at* 50° C*;* *(b) at* 25 *and* 100° C

50° C of KNO_3 and $NaNO_3$, respectively. Curves *AB* and *BC* indicate the saturated ternary solutions in equilibrium with solid KNO_3 or $NaNO_3$, and show, for instance, that the solubility of KNO_3 in water is depressed when $NaNO_3$ is present in the system, and *vice versa*.

Take, for example, a binary system $NaNO_3$–H_2O represented by point Y_1 (0·7 mass fraction of $NaNO_3$ and 0·3 H_2O). As this point lies in the 'triangular' region to the right of curve *BC*, the system consists of a saturated solution of $NaNO_3$, with a composition given by point *C*, and excess solid $NaNO_3$. If a quantity of KNO_3 is added to this binary system, keeping the temperature constant at 50° C so that the new composition is represented by point Y_2 (0·64 $NaNO_3$, 0·1 KNO_3, 0·26 H_2O), the composition of the ternary saturated solution in contact with the excess solid $NaNO_3$ present is given by Y_2' (0·46 $NaNO_3$, 0·12 KNO_3, 0·39 H_2O) on the line drawn from the apex *N* through Y_2 to meet curve *BC*. As more KNO_3 is added, the solution concentration alters, following curve *CB*. At point *B* the solution becomes saturated with respect to both $NaNO_3$ and KNO_3; its concentration is 0·4 $NaNO_3$, 0·29 KNO_3, 0·31 H_2O. If after this point further quantities of KNO_3 are added to bring the system concentration up to some point Y_3, no more KNO_3 dissolves; the solution composition remains at point *B*.

The interpretation of these phase diagrams is aided by remembering the rule of mixtures, i.e. on the removal or addition of any component from or to a system, the composition of the system changes along a straight line drawn from the original composition point to the apex representing the pure given component. In *Figure 4.14a*, the right-angled apex represents pure water, the top apex *K* pure KNO_3 and the other acute apex *N* pure $NaNO_3$.

The effect of temperature on the system KNO_3–$NaNO_3$–H_2O is shown in *Figure 4.14b*. Two isotherms, *A'B'C'* and *A"B"C"*, for 25° and 100° C, respectively, are drawn on this diagram. The lower left-hand area enclosed by *A'B'C'* represents homogeneous unsaturated solutions at 25° C, the larger area enclosed by *A"B"C"* unsaturated solutions at 100° C. The line *B'B"* shows the locus of the drying-up points between 25° and 100° C. To illustrate the effect of temperature changes in the system, let point Z_1 refer to the composition (0·5 $NaNO_3$, 0·1 KNO_3, 0·4 H_2O) of a certain quantity of the ternary mixture. From the position of Z_1 in the diagram it can be seen that at 100° C the system would be a homogeneous unsaturated solution, but at 25° C it would consist of pure undissolved $NaNO_3$ in a saturated aqueous solution of $NaNO_3$ and KNO_3. Thus pure $NaNO_3$ would crystallize out of the solution Z_1 on cooling from say 100° to 25° C, in fact at about 50° C. Despite the phase changes, of course, the overall system composition remains at Z_1 until one or more components are removed. At 25° C the composition of the solution in contact with the crystals of $NaNO_3$ is given by the intersection of the line from *N* through Z_1 with curve *B'C'*, i.e. at point Z_1' (0·43 $NaNO_3$, 0·11 KNO_3, 0·46 H_2O). The quantity of $NaNO_3$ which would crystallize out at 25° C is given by the mixture rule

$$\frac{\text{mass of crystals deposited}}{\text{mass of saturated solution}} = \frac{\text{length } Z_1Z_1'}{\text{length } Z_1N}$$

where *N* represents the $NaNO_3$ apex of the triangle.

When a pure solute is to be crystallized from a ternary two-solute system by cooling, there is usually a temperature limit below which the desired solute becomes 'contaminated' with the other solute. This can be demonstrated by considering a system represented by point Z_2 in *Figure 4.14b*. The composition at Z_2 is 0·3 $NaNO_3$, 0·45 KNO_3, 0·15 H_2O; at 100° C the system is a homogeneous unsaturated solution. At 25° C, however, this point lies in the region where both solid $NaNO_3$ and KNO_3 are in equilibrium with a saturated solution of both salts, its composition being given by point B'. If it is desired to cool solution Z_2 in order to yield only KNO_3 crystals, then the temperature limitation is found by drawing a straight line from the KNO_3 apex K through point Z_2 and produce it to meet the drying-up line $B'B''$ at Z_2'. Point Z_2' occupies the position of an invariant point on an isotherm; by referring to *Figure 4.14a* it can be seen that it corresponds approximately to point B on the 50° C isotherm. Thus solution Z_2 must not be cooled below 50° C if only KNO_3 crystals are to be deposited.

Solvate Formation

When one of the solutes in a ternary system is capable of forming a compound with the solvent, the phase diagram will contain more regions to consider than in the simple case described above. A common example of solvate formation is the production of a hydrated salt in a ternary aqueous

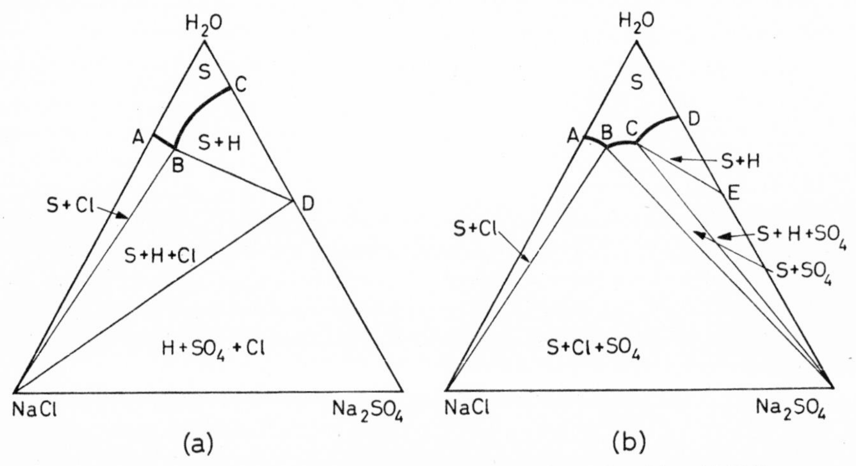

Figure 4.15. Phase diagrams for the system $NaCl$–Na_2SO_4–H_2O*: (a) at* 17·5° C; *(b) at* 25° C

system. *Figure 4.15* shows the isothermal diagrams for the system $NaCl$–Na_2SO_4–H_2O at two temperatures, 17·5° and 25° C, at which different phase equilibria are exhibited. Sodium sulphate combines with water, under certain conditions, to form $Na_2SO_4 \cdot 10H_2O$. Sodium chloride, however, does not form a hydrate at the temperatures being considered. *Figure 4.15a* shows the case where the decahydrate is stable in the presence of NaCl, and *Figure 4.15b* that of the decahydrate being dehydrated by the NaCl under certain conditions.

Points *A* and *C* in *Figure 4.15a* represent the solubilities of NaCl (26·5 per cent w/w) and Na_2SO_4 (13·8 per cent) in water at 17·5° C, curves *AB* and *BC* the ternary solutions in equilibrium with solid NaCl and $Na_2SO_4 \cdot 10H_2O$, respectively. Point *D* shows the composition of the hydrate $Na_2SO_4 \cdot 10H_2O$. For convenience, the following symbols are used on the diagram to mark the phase regions: S = solution; H = hydrate $Na_2SO_4 \cdot 10H_2O$; SO_4 = Na_2SO_4 and Cl = NaCl. The solution above curve *ABC* is unsaturated. The lowest triangular region represents a solid mixture of Na_2SO_4, $Na_2SO_4 \cdot 10H_2O$ and NaCl. Point *B* is the drying-up point of the system.

In *Figure 4.15b*, points *A* and *D* denote the solubilities of NaCl (26·6 per cent w/w) and Na_2SO_4 (21·6 per cent) in water at 25° C, point *E* the composition of $Na_2SO_4 \cdot 10H_2O$. In this diagram there are three curves, *AB*, *BC* and *CD*, which give the composition of the ternary solutions in equilibrium with NaCl, Na_2SO_4 and $Na_2SO_4 \cdot 10H_2O$. The various phase regions are indicated on the diagram. If NaCl is added to a system in the region *CDE*, i.e. to an equilibrium mixture of solid $Na_2SO_4 \cdot 10H_2O$ in a solution of NaCl and Na_2SO_4, the solution concentration will change along curve *DC*. When point *C* is reached, the NaCl can only dissolve by dehydrating the $Na_2SO_4 \cdot 10H_2O$, and anhydrous Na_2SO_4 is deposited. Further addition of NaCl will result in the complete removal of the decahydrate from the system, the solution concentration following curve *CB*; under these conditions the excess solid phase consists of anhydrous Na_2SO_4. At the drying-up point *B* the solution is saturated with respect to both NaCl and Na_2SO_4.

The effects of isothermal evaporation, salt additions and cooling can be traced from *Figures 4.15a* and *b* in a manner similar to that outlined for *Figures 4.13* and *4.14*.

Double Compound Formation

Cases are encountered in ternary systems where the two dissolved solutes combine in fixed proportions to form a definite double compound. *Figure 4.16* shows two possible cases for a hypothetical aqueous solution of two salts *A* and *B*. Point *C* on the *AB* side of each triangle represents the composition of the double salt, points *L* and *O* show the solubilities of salts *A* and *B* in water at the given temperature. Curves *LM* and *NO* denote ternary solutions saturated with salts *A* and *B*, respectively, curve *MN* ternary solutions in equilibrium with the double salt *C*. The significance of the various areas is marked on the diagrams.

The isothermal dehydration of solutions in *Figure 4.16a* can be traced in the manner described for *Figures 4.13* and *4.14*. Point *M* is the drying-up point for solutions located to the left of broken line *WR*, point *N* that for solutions to the right of this line. A solution on line *WM* behaves as a solution of a single salt in water; when its composition reaches point *M*, a mixture of salt *A* and double salt *C* crystallizes out in the fixed ratio of the lengths *PC*/*AP*. Similarly, a solution on line *WN* yields a mixture of *B* and *C*, in the ratio *CQ*/*QB*, when its composition reaches point *N*. A solution represented by a point on line *WR* also behaves as a solution of a single salt; when its composition reaches point *R*, the double compound *C* crystallizes out and neither salts A nor B are deposited at any stage. Point R, therefore, is the third drying-up point of the system.

The phase diagram in *Figure 4.16b* shows a different case. There are only two drying-up points, M and N, in this system, the first for solutions located to the left, the second for solutions to the right of line WN. Solutions on line

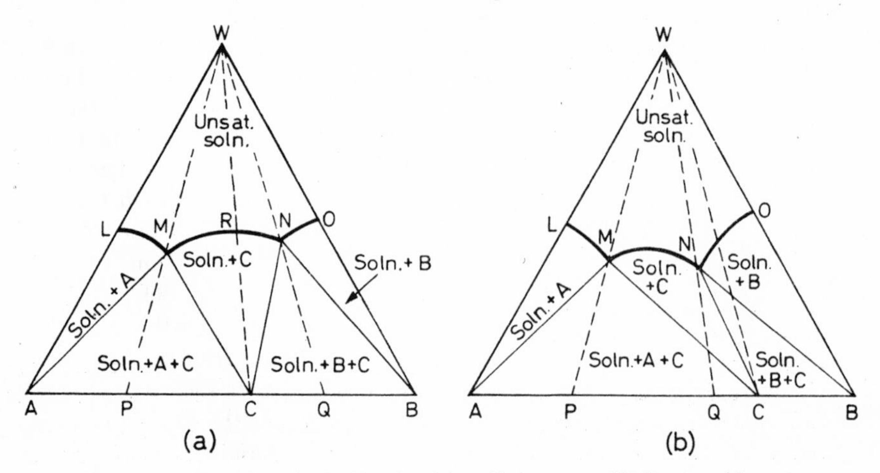

Figure 4.16. Formation of a double salt: (a) stable in water; (b) decomposed by water

WM and WN each behave as a solution of single salt in water. The line WC does not cross the saturation curve MN of the double salt but cuts the saturation curve for salt B, indicating that the double salt is not stable in water; it is decomposed and salt B is deposited.

Equilibrium Determinations

The solubility methods discussed in Chapter 2 are those most frequently used in the determination of equilibria in multicomponent systems, but for the complex cases the composition of the solid phase in equilibrium with the saturated solution is best analysed by the *wet-residue* method developed by Schreinemakers (1893). This method is illustrated below with reference to a ternary system.

A quantity of solvent containing excess solute is kept at a constant temperature until equilibrium is achieved. The clear supernatant solution is then analysed. Most of the solution is decanted off the remaining solute, and the composition of a sample of the wet solid determined. Ternary systems can, as described above, be represented on a triangular diagram, so the solution composition gives a point on the solubility curve and the wet-residue composition one within the triangle. By virtue of the properties of a triangular diagram and the mixture rule, the solubility and wet-residue points and the point representing the solid phase must lie on a straight line. Therefore, the point at which a line drawn through the solubility and wet-residue points meets the periphery of the triangle gives the composition of the solid phase.

An alternative procedure to the wet-residue method which is capable of yielding more rapid results is known as the *synthetic complex* method. Several mixtures of the solutes of known composition are prepared and a known quantity of solute is added to each sample. Thus a number of 'complex' points can be plotted within the triangular diagram. The samples are then allowed to achieve equilibrium at constant temperature by conventional methods, and the clear supernatant solution is analysed. Again, a line drawn through a solution point and its corresponding 'complex' point, extended to the periphery of the triangle, gives the composition of the solid phase.

FOUR-COMPONENT SYSTEMS

A one-phase, four-component or quaternary system has five degrees of freedom. Therefore, the phase equilibria in these systems may be affected by the five variables: pressure, temperature and the concentrations of any three of the four components. To represent quaternary systems graphically, one or more of the above variables must be excluded. The effect of pressure on solid–liquid systems may be ignored, and if only one temperature is considered an isothermal space model can be constructed. If the concentration of one of the components is excluded, usually the liquid solvent, a two-dimensional graph can be drawn, but this simplification will be described later.

Three Salts and Water

The first, simple, type of quaternary system to be considered here consists of three solid solutes, *A*, *B* and *C*, and a liquid solvent, *S*. No chemical reaction takes place between any of the components, e.g. water and three salts with a common ion. The isothermal space model for this type of system can be constructed in the form of a tetrahedron (*Figure 4.17a*) with

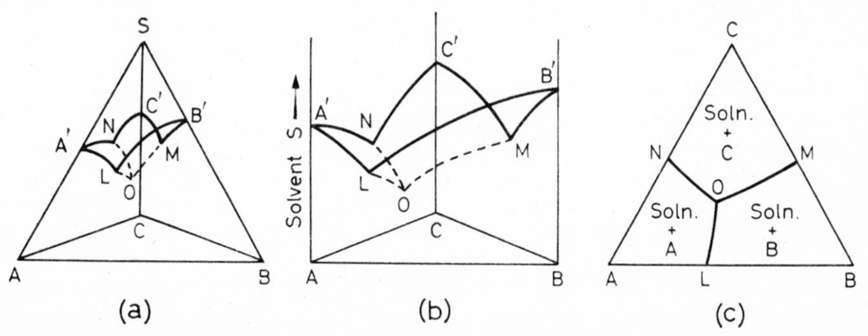

Figure 4.17. Isothermal representation of a quaternary system of the 'three salts in water' type: (a) tetrahedral space model; (b) triangular prism space model; (c) Jänecke projection

the solvent at the top apex and the three solid solutes on the base triangle. The four triangular faces of the tetrahedron represent the four ternary systems *A–B–C*, *A–B–S*, *A–C–S* and *B–C–S*. The three faces, excluding the base, have the appearance of the 'two salts and water' diagram shown in *Figure 4.15a*.

A point on an edge of the tetrahedron represents a binary system, a point within it, a quaternary. On the faces *ABS*, *BCS* and *ACS* the solubility curves meet at points *L*, *M* and *N*, respectively, which represent the solvent saturated with two solutes. They are the starting points for the three curves *LO*, *MO* and *NO* which denote solutions of three solutes in the solvent; point *O* represents the solution which, at the given temperature, is saturated with respect to all three solutes. All these curves form three curved surfaces within the space model. The section between these surfaces and the apex of the tetrahedron indicates unsaturated solution, that between the surfaces and the triangular base complex mixtures of liquid and solid.

Figure 4.17b shows another way in which systems of this type can be represented as a space model. Here it takes the form of a triangular prism where the apexes of the triangular base represent the three solid components and the vertical scale the liquid solvent. The interpretation of this model is similar to that just described for the tetrahedron; the same symbols have been used.

For a complete picture of the phase behaviour of quaternary systems a space model is essential, yet because of its time-consuming construction a two-dimensional 'projection' is frequently employed. Such a projection, named after E. Jänecke (1906), is shown in *Figure 4.17c*. In this type of isothermal diagram the solvent is excluded. The curved surfaces *A'LON*, *B'MOL* and *C'NOM* in *Figures 4.17a* and *4.17b*, which represent solutions in equilibrium with solutes *A*, *B* and *C*, respectively, are projected on to the triangular base and become areas *ALON*, *BMOL* and *CNOM* in *Figure 4.17c*. Curves *LO*, *MO* and *NO* denote solutions in equilibrium with two solutes, viz. *A* and *B*, *B* and *C*, *A* and *C*, respectively, while point *O* represents a solution in equilibrium with the three solutes. For this type of system the projection diagram can be plotted in terms of mass or mole fractions or percentages.

Reciprocal Salt Pairs

The second, and more important, type of quaternary system that will be considered is one consisting of two solutes and a liquid solvent where the two solutes inter-react and undergo double decomposition. This behaviour is frequently encountered in aqueous solutions of two salts which do not have a common ion. Typical examples of double decomposition reactions of commercial importance are

$$\begin{aligned}
&KCl + NaNO_3 &&\rightleftharpoons NaCl + KNO_3\\
&NaNO_3 + \tfrac{1}{2}(NH_4)_2SO_4 &&\rightleftharpoons NH_4NO_3 + \tfrac{1}{2}Na_2SO_4\\
&KCl + \tfrac{1}{2}Na_2SO_4 &&\rightleftharpoons NaCl + \tfrac{1}{2}K_2SO_4\\
&NaCl + \tfrac{1}{2}(NH_4)_2SO_4 &&\rightleftharpoons NH_4Cl + Na_2SO_4\\
&NaNO_3 + \tfrac{1}{2}K_2SO_4 &&\rightleftharpoons KNO_3 + \tfrac{1}{2}Na_2SO_4
\end{aligned}$$

The four salts in each of the above systems form what is known as a 'reciprocal salt pair'. Although all four may be present in aqueous solution, the composition of any mixture can be expressed in terms of three salts and water. Thus, from the phase rule point of view, an aqueous reciprocal salt pair system is considered to be a four-component system.

Reciprocal salt pair solutions may be represented on an isothermal space model, either in the form of a square-based pyramid or a square prism. *Figure 4.18a* indicates the pyramidal model; the four equilateral triangular faces stand for the four ternary systems AX–AY–W, AY–BY–W, BY–BX–W and AX–BX–W (W = water) for the salt pair represented by the equation

$$AX + BY \rightleftharpoons AY + BX$$

The apex of the pyramid denotes pure water, its base the anhydrous quaternary system AX–AY–BX–BY. Points L, M, N and O on the four triangular faces of the pyramid indicate the equilibria between two salts and water. Point P, which represents a solution of three salts AX, BX and BY in water saturated with all three salts, is a quaternary invariant point. So is Q, which

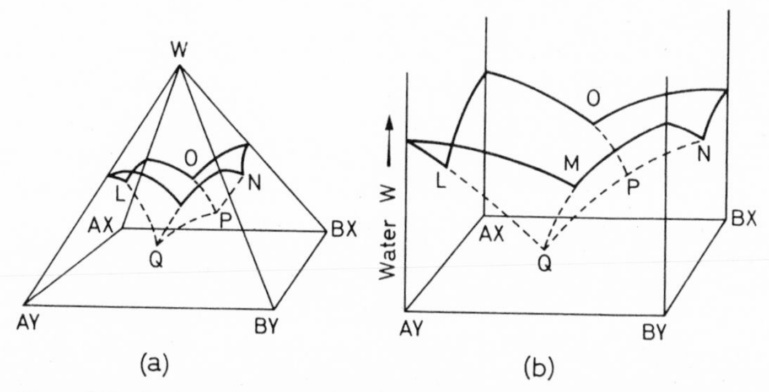

Figure 4.18. Isothermal representation of a quaternary system of the 'reciprocal salt pair' type: (a) square-based pyramid; (b) square prism

shows the equilibrium between salts AX, AY and BY and water. Curves OP, NP and LQ, MQ which join these quaternary invariant points P and Q to the corresponding ternary invariant points on the triangular faces of the pyramid, represent solutions of three salts in water saturated with two salts, and so does curve PQ, joining the two quaternary invariant points.

The square–prism space model (*Figure 4.18b*) illustrates another way in which a quaternary system of the reciprocal salt pair type may be represented. The vertical axis stands for the water content, and the points on the diagram are the same as those marked on *Figure 4.18a*. In both diagrams all surfaces formed between the internal curves represent solutions of three salts in water saturated with one salt, all internal curves solutions of three salts in water saturated with two salts, and the two points P and Q solutions of three salts in water saturated with the three salts. The section above the internal curved surfaces denotes unsaturated solutions, the section below them mixtures of liquid and solid.

Jänecke's Projection

In order to simplify the interpretation of the phase equilibria in reciprocal salt pair systems, the water content may be excluded. The curves of the space model can then be projected on to the square base to give a two-dimensional graph, called a Jänecke projection as described above. A typical projection is shown in *Figure 4.19a*; the lettering is that used in *Figures 4.18a* and *b*. The enclosed areas, which represent saturation surfaces, indicate solutions in equilibrium with one salt, the curves solutions in equilibrium with two salts, points *P* and *Q* solutions in equilibrium with three salts.

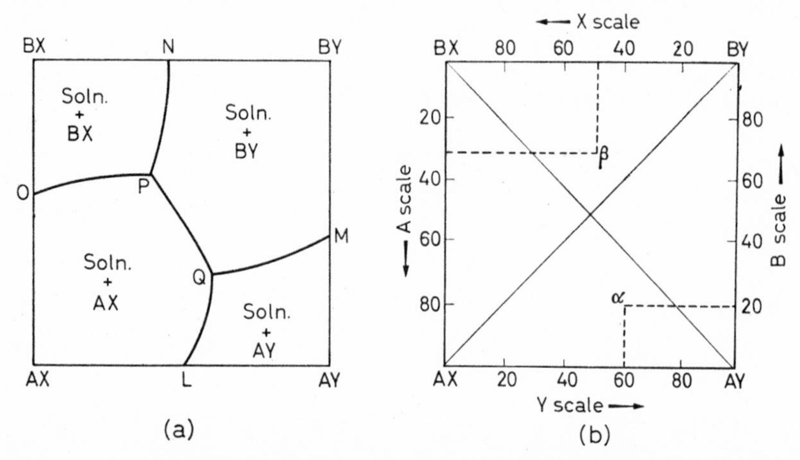

Figure 4.19. Interpretation of the Jänecke diagram for reciprocal salt pairs: (a) projection of the surfaces of saturation on to the base; (b) method of plotting

Molar or ionic bases must be used in this type of diagram for reciprocal salt pairs. The four corners of the square represent 100 moles of the pure salts *AX*, *BX*, *BY* and *AY*. Any point inside the square denotes 100 moles of a mixture of these salts; its composition can always be expressed in terms of three salts. The scales in *Figure 4.19b* are marked in ionic percentages, i.e. g-ions or lb.-ions of *A*, *B*, *X* and *Y*. Take, for example, 100 lb.-moles of a mixture expressed as

Salt	lb.-moles	lb.-ions			
		A	*B*	*X*	*Y*
AX . .	20	20		20	
AY . .	60	60			60
BX . .	20		20	20	
	100	80	20	40	60

The totals of the $A + B$ ions (e.g. the basic radicals) and the $X + Y$ ions (e.g. the acidic radicals) must always equal 100. Thus point α, indicating this mixture, can be plotted: the square is divided by the two diagonals into four right-angled triangles, and point α lies in triangles *AX.AY.BX* and *AX.AY.BY*. Therefore, the composition of the above mixture could also have been expressed in terms of salts *AX* (40 lb.-mole), *AY* (40 lb.-mole) and *BY* (20 lb.-mole). In a similar manner, it can be shown that point β which lies within the two triangles *AX.BX.BY* and *BX.BY.AY* represents 100 moles of a mixture with a composition expressed either by 50 *BY*, 30 *AX* and 20 *BX*, or by 50 *BX*, 30 *AY* and 20 *BY*.

Although it is generally more convenient to plot ionic percentages on the square, it is quite in order to plot mole percentages of the salts direct. The numerical scales marked on *Figure 4.19b* must now be ignored. If point α is considered to lie in triangle *AX.AY.BX*, representing a mixture 20 *AX*, 60 *AY* and 20 *BX*, the compositions of the two salts at opposite ends of the diagonal *AY* and *BX* are used for plotting purposes. Thus point α is located by 60 along the horizontal *AY* scale and 20 up the vertical *BX* scale. If α is taken to lie in triangle *AX.AY.BY*, the composition is represented by 40 *AX*, 20 *BY*, 40 *AY*, and the *AY* and *BX* compositions are used for plotting. A similar reasoning may be applied to the plotting of point β in triangles *AX.BX.BY* and *AY.BY.BX*.

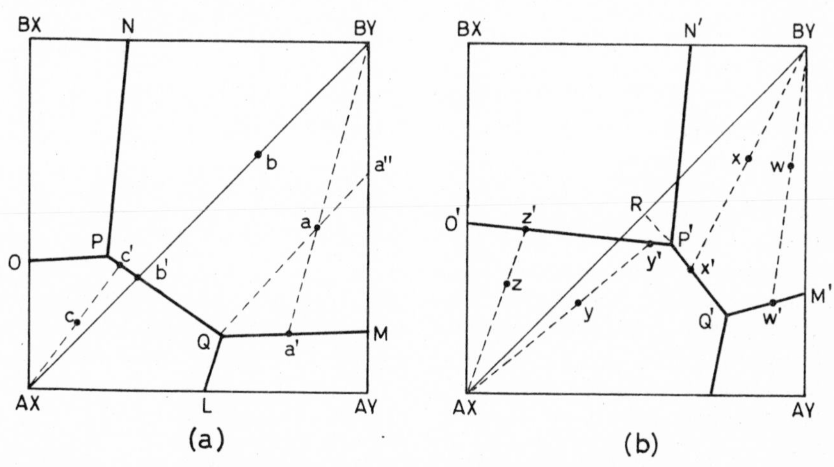

Figure 4.20. Jänecke projections for aqueous solutions of a reciprocal salt pair, showing (a) two congruent points; (b) congruent and incongruent points

Figures 4.20a and *b* show Jänecke diagrams for solutions of a given reciprocal salt pair at different temperatures. These two simple cases will be used to demonstrate some of the phase reactions that can be encountered in such systems. Both diagrams are curves by the saturation divided into four areas which are actually the projections of the surfaces of saturation (e.g. see *Figure 4.18b*). Salts *AX* and *BY* can co-exist in solution in stable equilibrium; the solutions are given by points along curve *PQ*. Salts *BX* and *AY*, however, cannot co-exist in solution because their saturation surfaces are separated from one another by curve *PQ*. Thus *AX* and *BY* are called the *stable salt pair*, or the *compatible salts*, *BX* and *AY* the *unstable salt pair*, or the *incompatible salts*. In *Figure 4.20a* the *AX–BY* diagonal cuts curve *PQ* which joins the two quaternary invariant points, while in *Figure 4.20b* curve *PQ* is not cut by either diagonal. These are two different cases to consider.

Point *P* represents a solution saturated with salts *AX*, *BX* and *BY*, *Q* one saturated with salts *AX*, *BY* and *AY*. In *Figure 4.20a* both *P* and *Q* lie in their 'correct' triangles, i.e. *AX.BX.BY* and *AX.BY.AY*, respectively, and solutions represented by P and Q are said to be congruently saturated. In *Figure 4.20b* point Q lies in its 'correct' triangle, $AX.BY.AY$, but P lies in the 'wrong' triangle, the same as Q. Point P, therefore, is *congruent* and point Q is *incongruent.*

Isothermal Evaporation

The phase reactions occurring on the removal of water from a reciprocal salt pair system will first be described with reference to *Figure 4.20a*. Point a which lies on the BY saturation surface represents a solution saturated with salt BY. When water is removed isothermally from this solution the pure salt BY is deposited and the solution composition (i.e. the composition of the salts in solution, the water content being ignored) moves from a towards a' along the straight line drawn from BY through a to meet curve QM. When a sufficient quantity of water has been removed, the solution composition reaches point a' and here the solution is saturated with two salts, BY and AY.

Further evaporation results in the deposition of AY as well as BY; the composition of the solid phase being deposited is given by point M. The overall composition of deposited solid therefore moves from BY towards a'' on the line $BY.AY$. The solution composition, being depleted in solid M, moves away from point M towards Q. On reaching point Q, three salts AX, AY and BY are deposited. The composition of the solid phase deposited is also given by point Q, the overall composition of the solid phase, assuming none has been removed from the system, by point a''. The solution composition, ignoring the water content, and the composition of the deposited solid phase remain constant at point Q for the rest of the evaporation process, and the overall solids content changes along line $a''a$, composition a representing the completely dry complex. Point Q is a quaternary drying-up point for all solutions represented by points within triangle $AX.AY.BY$.

The isothermal evaporation of solution b on the diagonal can be traced as follows. If point b lies on the saturation surface it represents a solution saturated with salt BY. While salt BY is being deposited the solution composition changes along the diagonal from b towards b'. At b' the solution becomes saturated with salts AX and BY. This ternary system (AX–BY–H_2O) thereafter dries up, without change in composition, at point b'. Point b', therefore, is a ternary drying-up point.

If point c lies on the saturation surface, it represents a solution saturated with salt AX. When this solution is evaporated isothermally, AX is deposited and the solution composition changes along line cc'. At c' salt BY also crystallizes out and the composition of the solid phase deposited is given by b', the point at which the diagonal crosses line PQ. The solution composition, therefore, changes along line $c'P$, and at P the three salts AX, BY and BX are co-deposited: point P is the quaternary drying-up point for all solutions represented by points within triangle $AX.BX.BY$.

The isothermal evaporation of a solution denoted by point w in *Figure 4.20b* can be traced in the same manner as that described for point a in *Figure 4.20a*. Q' is the drying-up point. The evaporation of solution x can be traced as follows. At x the solution is saturated with salt BY, and this salt is deposited until the solution composition reaches x' where the solution is saturated with the two salts AX and BY. The composition of the solid phase being deposited at this stage is given by point R on the diagonal. As evaporation proceeds, the solution composition changes from point x' along line $x'Q'$, i.e. in a direction away from point R, and at Q' the solution is saturated with the three salts AX, AY and BY. Both solution and deposited solids thereafter have a constant composition until evaporation is complete: Q' is the quaternary drying-up point.

Point Q' is also the drying-up point for a solution represented by point y. The solution composition changes along line yy' while salt AX crystallizes out, and then from y' towards P' while the two salts AX and BX of composition O' are deposited. At P' the solution is saturated with the three salts AX, BX and BY, the composition of the solid phase deposited at this point being given by R. On further evaporation, the solution composition remains constant at P' while salts AX and BY are deposited and salt BX is dissolved. When all BX has dissolved, the solution composition changes from P' towards Q', and the solution finally dries up at Q'.

Point P', therefore, is incongruent. It is not a true drying-up point except for the case where the original complex lies within the triangle representing the three salts of which it is the saturation point, i.e. AX, BY and BX. Point z may be taken as an example of this case. On evaporation, the solution composition changes from z to z' while salt AX is deposited, from z' towards P' while salts AX and BX are deposited. The composition of the solid phase at this latter stage is given by point O'. At P' this solution is saturated with salts AX, BX and BY. Further evaporation results in the deposition of AX and BY and the dissolution of BX. The solution dries up at point P'.

Representation of Water Content

So far in the discussion of Jänecke's projection for reciprocal salt pair systems the water content has been ignored. This is not a too serious omission because much information can be obtained from the projection before considering the quantity of water present. One way in which the water content can be represented is shown in *Figure 4.21a*; the plan shows the projection of the saturation surfaces, the elevation indicates the water contents. To avoid unnecessary complication, the elevation only shows the horizontal view of the particular saturation curve concerned in the problem.

The isothermal evaporation of water from a complex a was considered in *Figure 4.20a* where point a, representing the composition of the given complex, was taken to lie on the saturation surface. In *Figure 4.21a* the isothermal dehydration of an unsaturated solution S is considered, the dissolved salt having the same composition a as that in *Figure 4.20a*. Point S, therefore, is located on the elevation vertically above point a in the plan. The exact position of S is determined by the water content of the given solution, i.e. distance Sa_3 on the water scale denotes the moles of water per 100 moles of salt content. Line Sa_3, called the *water line*, represents the course of the isothermal dehydration. Points Q and M are similarly located on the elevation, according to their corresponding water contents, vertically above points Q and M on the plan. Point a' lies on curve QM vertically above a' in the plan. Point T on the elevation represents the water content of a saturated solution of pure salt BY, the salt to be deposited.

Three construction lines can now be drawn on the elevation. Line Ta' cuts the water line at point a. The BY saturation surface is assumed for simplicity to be plane, so Taa' is a line on this surface. The Y corner of the elevation represents pure salts AY and BY and all their mixtures. The line

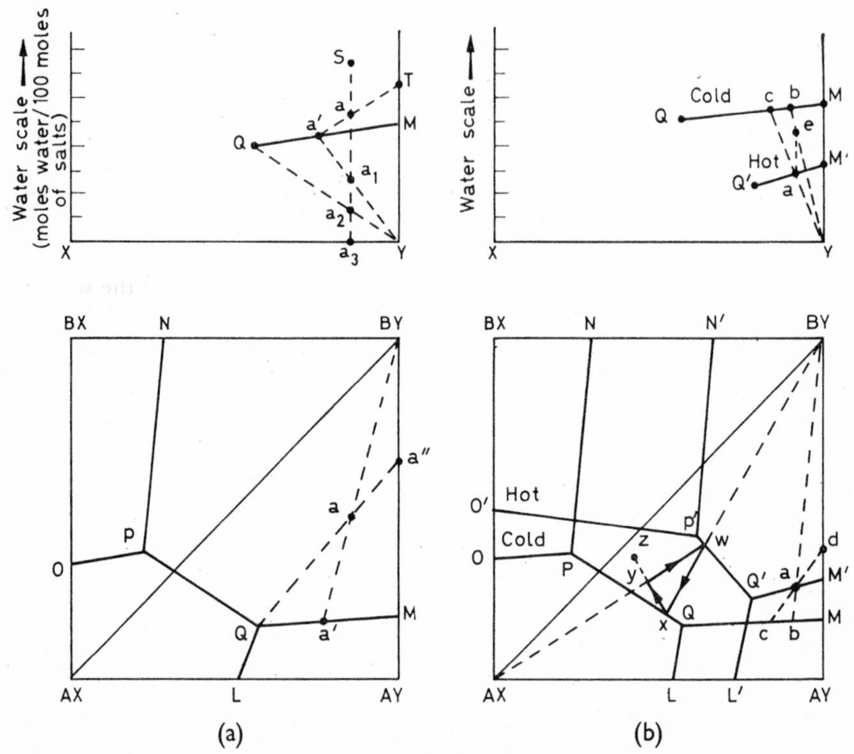

Figure 4.21. Representation of water content: (a) isothermal evaporation; (b) crystallization by cooling

drawn from a' to Y (BY on plan) cuts the water line at a_1, that from Q to Y (a'' on plan) at a_2.

When water is removed isothermally from the unsaturated solution S, the water content falls along the water line Sa_3. When point a is reached the solution is saturated with salt BY, and pure BY starts to crystallize out. The quantity of water to be removed to achieve this condition is determined from the water scale readings on the elevation diagram, i.e. Sa moles of water have to be removed from a system containing 100 moles of salts dissolved in Sa_3 moles of water. Salt BY is deposited while the water content falls from a to a_1, and at point a_1 the solution (of composition a') becomes saturated with salts BY and AY. Both salts are deposited while the water content falls from a_1 to a_2, and the overall deposited solids content changes along line BY/a'' on the plan. At point a_2 the solution (composition Q) is saturated with respect to the three salts AX, AY and BY, and further evaporation from a_2 to a_3 proceeds at constant solution composition Q. The solids composition changes along line $a''a$ on the plan.

Crystallization by Cooling

The graphical procedure described above, viz. the drawing of a plan and elevation, provides a simple pictorial representation of the phase reactions occurring in a given system at two different temperatures. *Figure 4.21b* shows two isotherms labelled 'hot' and 'cold', respectively; they are in fact the curves from *Figures 4.20a* and *b* plotted on one diagram, and the same lettering is used. By way of example, two different cooling operations will be considered.

Point a on curve QM represents a hot solution saturated with the two salts AY and BY. When it is cooled to the lower temperature, point a lies in the BY field of the projection. Line BY/a is drawn on the plan to meet curve QM at b, but point b represents the solution composition only if point a lies on the BY saturation surface in the 'cold' projection, i.e. if pure BY was crystallizing out. To find the actual solution composition and the composition of the deposited solid phase, point b is projected from the plan onto curve QM in the elevation.

Point Y on the elevation diagram represents salts AY or BY or any mixture of them. Line Ya is drawn on the elevation and then produced to meet curve QM at c. It can be seen that in this case points b and c do not coincide. This means that the deposited solid phase is not pure salt BY but some mixture of BY and AY. Point c is projected from the elevation onto the plan, and line cad is drawn. Thus the final solution composition is given by point c, and the overall solid phase composition by point d.

If pure salt BY was required to be produced during the cooling operation, the water content of the system would have to be adjusted accordingly. Solution point c has to move to become coincident with point d, and solid point d has to move to BY on the plan. In this case, therefore, water has to be added to the system, e.g. to the hot solution before cooling. The quantity of water required per 100 moles of salts is given by the vertical distance ae on the elevation.

A different sequence of operations is shown in another section of *Figure 4.21b*. Point w on curve $P'Q'$ represents a solution saturated with salts AX and BY at the higher temperature. At the lower temperature, however, point w lies in the BY field of the diagram. If the correct amount of water is present in the system, pure BY crystallizes out on cooling, and the solution composition is given by point x located on line BY/w produced to meet curve PQ. A cyclic process can now be planned.

The pure salt BY is filtered off and a quantity of a solid mixture, e.g. of composition z, is added to solution x. The quantity of solid z to be added, calculated by the mixture rule, must be the amount necessary to give complex y, the composition of which is chosen so that on heating to the higher temperature, it lies in the *AX* field, yields the original solution *w* and deposits the pure salt *AX*. Thus the sequence of operations is

(1) cool solution *w* to the lower temperature
(2) filter off solid *BY*
(3) add solid mixture *z* to the mother liquor *x* to give complex *y*
(4) heat the complex to the higher temperature
(5) filter off solid *AX*
(6) cool mother liquor *w*, and so on.

Of course, the water contents at each stage in the cycle must be adjusted so that the solutions deposit only one pure salt at a time. The quantities of water to be added or removed can be estimated graphically on the elevation diagram in the manner described above for solution *a*.

Only the simplest type of reciprocal salt pair diagram has been considered here. Many systems form hydrates or double salts, in others the stable salt pair at one temperature may become the unstable pair at another. For information on these more complicated systems reference should be made to specialized works on the Phase Rule. Purdon and Slater's publication (see Bibliography) is particularly noteworthy in this respect; many detailed graphical solutions of problems of commercial importance are given, and the analysis of five-component aqueous systems is discussed.

BIBLIOGRAPHY

BLASDALE, W. C., *Equilibria in Saturated Salt Solutions*, 1927. New York; Chemicals Catalog Co.
BOWDEN, S. T., *The Phase Rule and Phase Reactions*, 1950. London; Macmillan
FINDLAY, A. and CAMPBELL, A. N., *The Phase Rule and its Applications*, 8th Ed., 1938. London; Longmans
HILL, L. M., Phase Rule: Application to the Separation of Salt Solutions, in *Thorpe's Dictionary of Applied Chemistry*, 4th Ed., IX, 438, 1949. London; Longmans
PURDON, F. F. and SLATER, V. W., *Aqueous Solution and the Phase Diagram*, 1946. London; Arnold
RICCI, J. E., *The Phase Rule and Heterogeneous Equilibrium*, 1951. New York; Van Nostrand
WETMORE, F. E. W. and LEROY, D. J., *Principles of Phase Equilibria*, 1951. New York; McGraw-Hill

5

MECHANISM OF CRYSTALLIZATION

THE deposition of a solid crystalline phase from liquid and gaseous solutions, pure liquids and pure gases can only occur if some degree of supersaturation or supercooling has first been achieved in the system. The attainment of the supersaturated state is essential for any crystallization operation, and the degree of supersaturation, or deviation from the equilibrium saturated condition, is the prime factor controlling the deposition process. Any crystallization operation can be considered to comprise three basic steps:

(1) achievement of supersaturation or supercooling
(2) formation of crystal nuclei
(3) growth of the crystals

All three processes may be occurring simultaneously in different regions of a crystallization unit. The ideal crystallization, of course, would consist of a strictly controlled step-wise procedure, but the complete cessation of nucleation cannot normally be guaranteed in a growing mass of suspended and circulating crystals.

The supersaturation of a system may be achieved by cooling, evaporation, the addition of a precipitant or diluent or as a result of the chemical reaction between two homogeneous phases. Some of these operations have already been discussed in Chapter 2, and their practical utility will be described in Chapters 7 and 8; the other two factors, nucleation and growth, will be considered in this chapter. Modern theories and research trends in this field have been ably summarized in the many papers presented at three recent symposia[1, 2, 3].

NUCLEATION

The condition of supersaturation alone is not sufficient cause for a system to begin to crystallize. Before crystals can grow there must exist in the solution a number of minute solid bodies known as centres of crystallization, seeds, embryos or nuclei. Nucleation may occur spontaneously or it may be induced artificially; these two cases are frequently referred to as homogeneous and heterogeneous nucleation, respectively. It is not always possible, however, to decide whether a system has nucleated of its own accord or whether it has done so under the influence of some external stimulus.

Spontaneous Nucleation

Exactly how a crystal nucleus is formed within a homogeneous fluid system is not known with any degree of certainty. To take a simple example, the condensation of a supersaturated vapour to the liquid phase is only possible after the appearance of microscopic droplets, called condensation nuclei, on the condensing surface. However, as the vapour pressure at the surface of these minute droplets is exceedingly high, they evaporate rapidly even though the surrounding vapour is supersaturated. New nuclei form while old ones evaporate, until eventually stable droplets are formed either by coagulation or under conditions of very high vapour supersaturation.

The formation of crystal nuclei is an even more difficult process. Not only have the constituent molecules to coagulate, resisting the tendency to redissolve, but they also have to become orientated into a fixed lattice. The number of molecules in a stable crystal nucleus can vary from about 10 to several thousand; water (ice) nuclei, for instance, contain about 80–100 molecules. The actual formation of such a nucleus can hardly result from the simultaneous collision of the required number of molecules; this would constitute an extremely unlikely event.

Most probably, the mechanism of nucleation is as follows. Minute structures are formed, first from the collision of two molecules, then from that of a third with the pair, and so on. Short chains may be formed initially, or flat monolayers, and eventually the lattice structure is built up. The construction process, which occurs very rapidly, can only continue in local regions of very high supersaturation, and many of these 'sub-nuclei' fail to achieve maturity; they simply redissolve because they are extremely unstable. If, however, the nucleus grows beyond a certain critical size, as explained below, it becomes stable under the average conditions of supersaturation obtaining in the bulk of the fluid.

In Chapter 2, an account was given of Ostwald's concept of the stable, metastable and labile states of a solution. The Miers diagram (*Figure 2.4*) gives a pictorial representation of these regions in terms of temperature and solution concentration. The energy levels of these various states of stability can be represented by the simple analogy of a block or brick resting on a

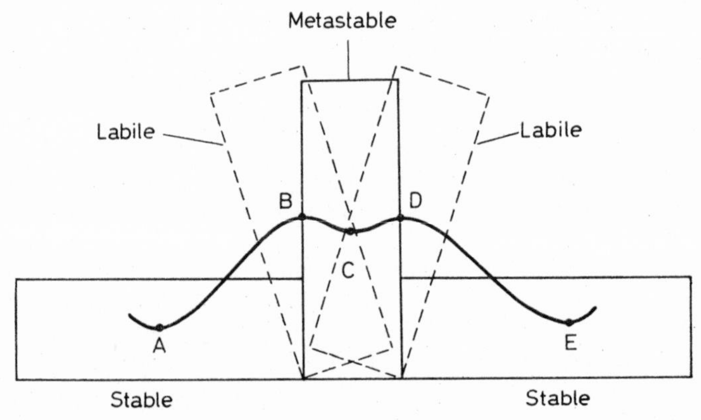

Figure 5.1. Demonstration of the stable, metastable and labile states

flat surface (*Figure 5.1*), although this picture probably oversimplifies the situation. The energy level (potential energy) in each of the various positions of the brick is denoted here by the height of the centre of gravity above the arbitrary datum of the flat surface. Cases A and E represent the lowest energy state, or state of maximum stability, which may be likened to the case of a saturated solution. Case C is also a stable state, but represents a higher energy level than A or E. A brick in this position cannot withstand any great displacement without reverting to the more stable positions A or E, so it may be considered to be metastable. This condition is similar, therefore, to the state of metastability in a supersaturated solution. Cases B and D represent unstable states, and also the highest energy level of the system. Any displacement would make the brick adjust itself into a more stable position. These cases may be taken to represent a labile supersaturated solution which would tend to nucleate spontaneously.

The phenonemon of spontaneous or homogeneous nucleation can be analysed by considering the various energy requirements. In this connection, the theoretical and practical contributions of such workers as GIBBS[4], OSTWALD[5], VOLMER[6, 7], BECKER and DÖRING[8, 9] have greatly assisted our understanding of the process. When a group of freely moving molecules becomes aggregated into a more condensed state, i.e. one in which the molecular movement is much more restricted, a quantity of energy is released. For example, when a vapour condenses to a liquid, the latent heat associated with the change of state is liberated. In a given system, therefore, the transitions from the gaseous to the liquid and then to the solid states represent a step-wise decrease in the degree of molecular mobility and likewise a decrease in the free energy of the system.

On the other hand, the formation of a liquid droplet or solid particle within a homogeneous fluid demands the expenditure of a certain quantity of energy in the creation of the liquid or solid surface. Therefore, the total quantity of work, W, required to form a stable crystal nucleus is equal to the sum of the work required to form the surface, W_s (a positive quantity), and the work required to form the bulk of the particle, W_v (a negative quantity)

$$W = W_s - W_v \tag{1}$$

For the formation of a spherical liquid droplet in a supersaturated vapour, for instance, equation 1 can be written as

$$W = a\sigma - v\Delta p \tag{2}$$

where σ is the surface energy of the droplet per unit area, Δp the difference in pressure between the vapour phase and the interior of the liquid droplet, and a and v are the surface area and volume, respectively, of the droplet. If the spherical droplet has radius r, then

$$a = 4\pi r^2$$

$$v = \tfrac{4}{3}\pi r^3$$

and

$$\Delta p = \frac{2\sigma}{r}$$

so equation 2 can be written

$$W = 4\pi r^2\sigma - \tfrac{4}{3}\pi r^3 \cdot \frac{2\sigma}{r} \tag{3}$$

$$= \tfrac{4}{3}\pi r^2\sigma \tag{4}$$

From equations 3 and 4 it can be seen that the work of formation of a droplet equals one-third of that required to form the surface of the droplet. This fact was first deduced by GIBBS[4].

The increase in the vapour pressure of a liquid droplet as its size decreases can be estimated from the Gibbs–Thomson formula, which may be written in the form

$$\ln \frac{p_r}{p^*} = \frac{2M\sigma}{\boldsymbol{R}T\rho r} \tag{5}$$

where p_r and p^* are the vapour pressures over a liquid droplet of radius r and a flat liquid surface, respectively; in other words, p^* is the equilibrium saturation vapour pressure of the liquid. M is the molecular weight, ρ the density of the droplet, T the absolute temperature and $\boldsymbol{R}$ the gas constant.

A formula similar to equation 5 has already been discussed (p. 33) with respect to the increase in solubility of a particle as its size is reduced; in that particular case the concentration ratio c_r/c^* was used in place of the term p_r/p^* in equation 5. Both these terms give a measure of the supersaturation, S, of the system, so equation 5 may be rewritten in a more general form as

$$\ln S = \frac{2M\sigma}{\boldsymbol{R}T\rho r} \tag{6}$$

or

$$r = \frac{2M\sigma}{\boldsymbol{R}T\rho \ln S} \tag{7}$$

If this value of r is substituted in equation 4 we get

$$W = \frac{16\pi\sigma^3 M^2}{3(\boldsymbol{R}T\rho \ln S)^2} \tag{8}$$

Equation 8 is an extremely important relationship: it gives a measure of the work of nucleation in terms of the degree of supersaturation of the system. It can be seen, for example, that when the system is only just saturated ($S = 1$, $\ln S = 0$), the amount of energy required for nucleation is infinite, so a saturated solution cannot nucleate spontaneously. However, equation 8 also suggests that any supersaturated solution can nucleate spontaneously because there is some finite work requirement associated with the process—it is merely a question of supplying the required amount of energy to the system. The necessary quantity of energy may be quite excessive, but it still remains a fact that spontaneous nucleation is theoretically possible at any degree of supersaturation.

The free energy changes associated with the process of homogeneous nucleation may be considered as follows. The overall excess free energy, ΔG, between a small solid particle of solute and the solute in solution is equal to the sum of the surface excess free energy, ΔG_s, i.e. the excess free energy between the surface of the particle and the bulk of the particle, and the volume excess free energy, ΔG_v, i.e. the excess free energy between a very large particle ($r = \infty$) and the solute in solution. ΔG_s is a positive quantity, the magnitude of which is proportional to r^2. In a supersaturated solution,

ΔG_v is a negative quantity proportional to r^3. These relationships are shown in *Figure 5.2*; as r increases from zero value, the overall excess free energy ΔG reaches a maximum value when the nucleus achieves a critical size, r_c, i.e.

$$\Delta G_{\text{crit}} = \frac{4\pi\sigma r_c^2}{3} \tag{9}$$

The behaviour of a newly created crystalline lattice structure in a supersaturated solution depends on its size; it can either grow or redissolve, but the process which it undergoes should result in the decrease in the free energy of the particle. The critical size r_c, therefore, represents the minimum size of a stable nucleus. Particles smaller than r_c will dissolve, or evaporate if the

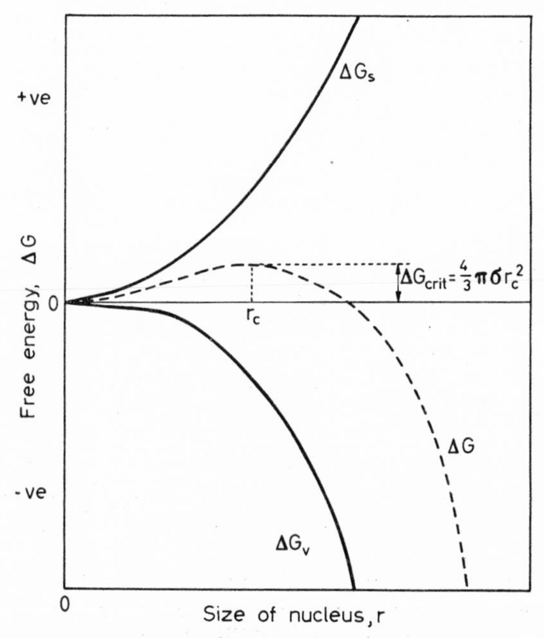

Figure 5.2. Free energy diagram for nucleation explaining the existence of a 'critical' nucleus

particle is a liquid in a supersaturated vapour, because only in this way can the particle achieve a reduction in its free energy. Similarly, particles larger than r_c will continue to grow.

Although it can be seen from the free energy diagram why a particle of size greater than the critical size is stable, it does not explain how the amount of energy, ΔG_{crit}, necessary to form a stable nucleus is produced. This may be explained as follows. The energy of a fluid system at constant temperature and pressure is constant, but this does not mean that the energy level is the same in all parts of the fluid. There will be fluctuations in the energy about the constant mean value, i.e. there will be a statistical distribution of energy, or molecular velocity, in the molecules constituting the system, and in those supersaturated regions where the energy level rises temporarily to a high value nucleation will be favoured.

The rate of nucleation, N, e.g. the number of nuclei formed per unit time per unit volume, can be expressed in the form of the Arrhenius reaction velocity equation

$$N = A \,.\, \exp\,(-\,\Delta G/\boldsymbol{R}T) \tag{10}$$

where A is a constant of proportionality, and ΔG the overall excess free energy of the particle, i.e. the work of nucleation, W. From equations 8 and 11 we get

$$N = A \,.\, \exp\left[-\,\frac{16\pi\sigma^3 M^2}{3\boldsymbol{R}^3 T^3 \rho^2 (\ln S)^2}\right] \tag{11}$$

This equation indicates that three main variables govern the rate of nucleation: temperature, T, degree of supersaturation, S, and interfacial tension, σ. The dominant effect of the degree of supersaturation can be shown by calculating the time required for the spontaneous appearance of nuclei in supercooled water vapour[7]

Supersaturation, S	*Time*
1·0	∞
2·0	10^{62} years
3·0	10^{3} years
4·0	0·1 sec
5·0	10^{-13} sec

Therefore, in this case, a 'critical' degree of supersaturation exists in the region of $S \sim 4{\cdot}0$, but it is also clear that nucleation would have occurred at any value of $S \sim 1{\cdot}0$ if sufficient time had elapsed.

LaMer[10] and Pound[11] have discussed the implications of these theoretical equations, and Turnbull *et al.*[12] have shown how a relationship similar to equation 11 can be modified and utilized to calculate either values of the interfacial tension, σ, or the minimum temperature to which a liquid can be supercooled before nucleating spontaneously.

Tamman[13] studied the formation of nuclei in a number of supercooled organic melts by a simple but rather ingenious method. A small volume of the substance was melted in a closed glass tube, then cooled rapidly by immersing it in a bath at a given temperature for a fixed period of time. Any nuclei formed during this time were invisible to the naked eye, so they were 'developed' by heating the contents of the tube to some given temperature at which the nuclei would grow. The tiny crystals were then counted. Results obtained with supercooled piperine (m.p. 129° C) are shown in *Figure 5.3*. In this case, the melts were maintained at a given temperature for 10 minutes, after which they were heated to 100° C for 4 minutes. The temperature of maximum nucleation was about 40° C (89° supercooling) and the temperature range over which an appreciable number of nuclei were formed was approximately 10°–70° C (59°–119° supercooling).

Nucleation curves similar in shape to the one shown in *Figure 5.3* were obtained by Tamman for many other systems. For all the substances studied, the optimum temperature for nucleation was found to be much lower than that required for maximum crystal growth. Below the optimum nucleation temperature, the viscosity of the liquid increased to such a value as to prohibit nucleation, whereas above this temperature the vigorous molecular motion in the liquid prevented the formation of a fixed crystal lattice.

The initial rapid increase in the nucleation rate with a decrease in the temperature of the system, i.e. an increase in the degree of supercooling, shown in *Figure 5.3* is to be expected from the nucleation rate equation 11.

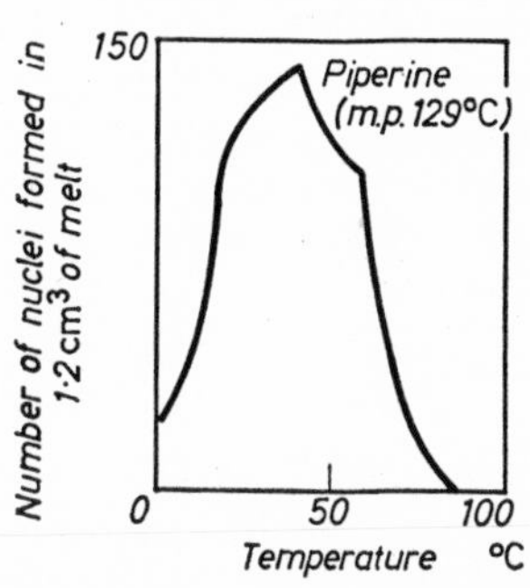

Figure 5.3. Relationship between nucleation and degree of supercooling. (*After* G. TAMMAN[13])

However, this also predicts that the rate of nucleation should continue to increase rapidly with an increase in the degree of supersaturation, but this in point of fact does not happen; after reaching a maximum value the rate of nucleation falls off in the regions of high supersaturation. The most probable explanation is that as the system cools it becomes more viscous, the mobility of the solute molecules is restricted and nucleation inhibited. It has been suggested[8] that the nucleation rate equation should also contain an exponential term to account for the viscosity effect.

Different systems require different degrees of supersolubility or supercooling before nucleation will occur. Many inorganic aqueous solutions allow only a few degrees of supercooling, whereas organic melts and most metals require a considerable amount of supercooling before crystallizing. In fact some 'solid' substances, such as glass, are simply liquids supercooled many hundreds of degrees below their melting points. The formation of the glassy state is by no means uncommon; TAMMAN[13] reported that out of some 150 selected organic compounds, all capable of being crystallized fairly easily, over 30 per cent yielded the glassy state on cooling their melts slowly. In all probability, rapid cooling would have increased this percentage.

Induced Crystallization

To the organic chemist, the reluctance to crystallize exhibited by many substances is a well known and often exasperating situation. Many techniques and rituals can be performed on the unwilling system, and occasionally stories are told, some of them apocryphal, of individuals who seemed to possess the uncanny power of inducing crystallization. The late W. H. Perkin was one of these 'gifted' persons; his beard was reputed to be the secret of his success—apparently it contained hidden supplies of crystal seeds of wide variety! Despite the aura of witchcraft that has surrounded this fascinating subject, there are many good reasons why some of the time-honoured methods should be successful; there are also some good reasons why several other methods often tried will not produce the required result.

Excessive cooling, for instance, does not aid nucleation; the results of Tamman (*Figure 5.3*) show this. There is an optimum temperature for nucleation, and any reduction below this value is prohibitive. As indicated by equation 11, nucleation can theoretically occur at any temperature, provided the system is supersaturated, or supercooled in the case of a one-component system, but under normal conditions the temperature range over which nucleation can occur readily and vigorously may be quite restricted. Therefore, if a system has set to a clear, highly viscous mass, further cooling is pointless. The temperature would have to be increased to some value in the region of the optimum nucleation temperature before crystallization could commence.

Agitation or the bubbling of a gas through a solution can often induce nucleation. Stirred water, for example, will only allow about $\frac{1}{2}$° C of supercooling before spontaneous nucleation occurs, whereas undisturbed water will allow over 5° C of supercooling. Actually, very pure water, free from all extraneous matter, can be supercooled some 39° C[12]. Most agitated solutions nucleate spontaneously at lower degrees of supersaturation than quiescent ones; in other words, the supersolubility curve (*Figure 2.4*) tends to approach the solubility curve more closely in agitated solutions, and the width of the metastable band is reduced. Van Hook[14] reported that the speed of a stirrer has little effect on the rate of nucleation in dust-free sugar solutions, especially at degrees of supersaturation less than 1·4. At supersaturation greater than this value increased stirrer speeds increased the rate of nucleation. In all cases, however, any accidental contact of the glass stirrer with the sides of the vessel promptly induced nucleation, and the presence of glass beads within the solution did likewise.

The vigorous agitation of a solution invariably results in the production of small crystals; prolific nucleation is induced, and each stable nucleus grows into a crystal. If the agitation is violent, the crystals may rub together, and the tiny fragments or nodules which break off will act as additional nuclei. The tendency of crystals to suffer attrition in solution varies; some crystals are very hard whereas others are so friable that even mild degrees of agitation can cause severe breakdown. Pure sodium carbonate, for instance, cannot resist agitation during crystallization, but the presence of small quantities of sodium sulphate in the solution allows the production of quite large crystals under conditions of moderate agitation.

Mechanical shock, friction and extreme pressures within melts and solutions can readily induce nucleation, as shown by the experiments of Young[15,16] and Berkeley[17]. The effects of external influences such as electric and magnetic fields, ultra-violet light, X-rays, β-rays, sonic and ultrasonic radiation have been studied in recent years. Tipson[18] has summarized the findings of a large number of workers on this aspect of the subject. None of these methods has yet found application in large-scale crystallization practice, but many of them show a great deal of promise for use in the laboratory. VAN HOOK[14] reported work carried out into the effect of sonic irradiation on the graining of sugar solutions. It was found that the extent of nucleation depends on time and frequency of irradiation, power input and degree of supersaturation. Sonic irradiation has also been found to assist the crystallization of such substances as *l*-arabinose, sorbitol, *dl*-glyceraldehyde and *d*-gluconic acid.

Probably the best method for inducing crystallization is to inoculate or seed the supersaturated solution with small particles of the material to be crystallized. The seeds should be dispersed uniformly throughout the solution by means of gentle agitation, and if the temperature is carefully regulated, considerable control is possible over the final product size. Deliberate seeding is frequently employed in industrial crystallizations; the actual weight of seed material added depends on the required amount of solute deposition and on the size of the seeds (see *Figure 7.2*). The use of seeds as small as 5 microns has been reported in sugar boiling practice[14]; these tiny particles are produced by prolonged ball-milling in an inert medium, e.g. *iso*propyl alcohol or mineral oil, and 1 lb. of such seeds may be quite sufficient for 12,000 gal. of massecuite.

The seeds do not necessarily have to consist of the material being crystallized, unless absolute purity of the final product is required. A few tiny crystals of some isomorphous substance may be used to induce crystallization. For example, phosphates will often nucleate solutions of arsenates. Small quantities of sodium tetraborate decahydrate can induce the crystallization of sodium sulphate decahydrate[19]. Crystalline organic homologues, derivatives and isomers are frequently used for inducing crystallization; phenol can nucleate *m*-cresol, and ethyl acetanilide can nucleate methyl acetanilide[18]. PRECKSHOT and BROWN[20] reported the effects of some ionic crystals on the nucleation of quiescent potassium chloride solutions; the substances used were crystallographically similar to KCl but virtually insoluble in water. It was found that lead selenide was the best nucleation agent, i.e. the required subcooling of the KCl solution for nucleation was least. Lead telluride was also quite effective, and lead sulphide least of all.

Heterogeneous Nucleation

The rate of nucleation of a solution or melt is affected considerably by the presence of traces of impurities in the system, even though the temperature at which the maximum rate occurs is not appreciably altered. For a melt of β-naphthyl salicylate (m.p. 91° C) TAMMAN[13] showed that the maximum nucleation rate occurred at about 16° C irrespective of the presence of impurities. Traces of anisic acid, benzamide, fine emery or quartz particles increased the nucleation rate, traces of salicin and cane sugar decreased it, and traces of powdered glass prevented nucleation altogether. However, an impurity which acts as a nucleation inhibitor in one case may act as an accelerator in another. No general rule applies and each case must be investigated separately.

Many reported cases of spontaneous nucleation are found on careful examination to have been induced in some way. Indeed, some authors suggest that true examples of spontaneous nucleation are rarely encountered. For example, a supercooled system can be seeded unknowingly by the presence of atmospheric dust. Systems which apparently nucleate spontaneously can often be rendered immune to nucleation after very careful filtration. Large bulk samples frequently nucleate spontaneously at smaller degrees of supercooling than small samples; the most probable reason is that they stand a greater chance of being contaminated with dust.

Atmospheric dust frequently contains particles of the product itself, especially in industrial plants or in laboratories where samples of the crystalline material have been handled. Quite often some inert amorphous material in the dust acts as a seed; deliberate addition of small quantities of fine particles such as kieselguhr, silica and powdered glass has been found helpful for inducing crystallization in specific cases.

In laboratory and large-scale crystallizations, the first sign of nucleation often appears in one given region of the vessel, usually in regions where there is a local high degree of supersaturation, such as near a cooling surface or at the surface of the liquid. It is not uncommon to find some particular spot on the vessel wall or on the stirrer acting as a crystallization centre. The most reasonable explanation of this phenomenon is that minute cracks and crevices in the surface retain tiny crystals from a previous batch which seed the system when it becomes supercooled. It is possible, of course, for a metal or glass surface to be in a condition where it acts as a catalyst for nucleation.

As the presence of a suitable foreign body or 'sympathetic' surface can induce nucleation at degrees of supercooling lower than those required for spontaneous nucleation, the overall free energy change associated with the formation of a critical nucleus under heterogeneous conditions, $\Delta G'_{crit}$, must be less than the corresponding free energy change, ΔG_{crit}, associated with homogeneous nucleation, i.e.

$$\Delta G'_{crit} = \phi \Delta G_{crit} \tag{12}$$

where the factor ϕ is less than unity.

It has been indicated above that the interfacial energy, σ, is one of the important factors controlling the nucleation process. *Figure 5.4* shows a typical interfacial energy diagram for three phases in contact; in this case,

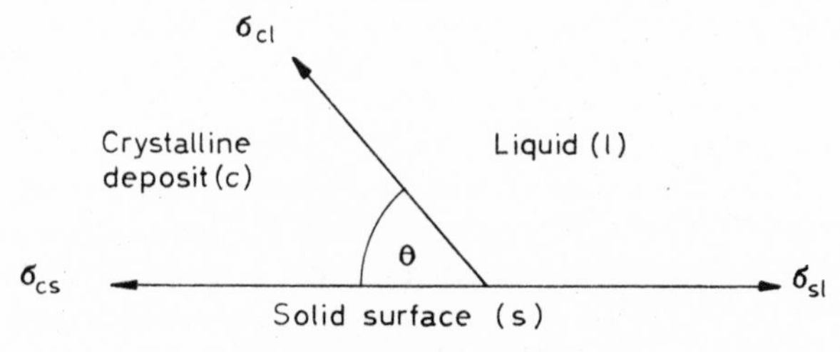

Figure 5.4. Surface energies at the boundaries between three phases (two solids, one liquid)

however, the three phases are not the more familiar solid, liquid and gas, but two solids and a liquid. The three interfacial energies are denoted by σ_{cl} (between the solid crystalline phase, c, and the liquid, l), σ_{sl} (between another foreign solid surface, s, and the liquid) and σ_{cs} (between the solid crystalline phase and the foreign solid surface). Resolving these forces in a horizontal direction

$$\sigma_{sl} = \sigma_{cs} + \sigma_{cl} \cos \theta \tag{13}$$

or

$$\cos \theta = \frac{\sigma_{sl} - \sigma_{cs}}{\sigma_{cl}} \tag{14}$$

The angle θ, the angle of contact between the crystalline deposit and the foreign solid surface, corresponds to the angle of wetting in liquid–solid systems.

VOLMER[6, 7] has suggested that the factor ϕ in equation 12 can be expressed as

$$\phi = \frac{(2 + \cos \theta)(1 - \cos \theta)^2}{4} \tag{15}$$

Thus, when $\theta = 180°$, $\cos \theta = -1$ and $\phi = 1$, equation 12 becomes

$$\Delta G'_{crit} = \Delta G_{crit} \tag{16}$$

When θ lies between 0 and 180°, $\phi < 1$, therefore

$$\Delta G'_{crit} < \Delta G_{crit} \tag{17}$$

When $\theta = 0$, $\phi = 0$, and

$$\Delta G'_{crit} = 0 \tag{18}$$

The three cases represented by equations 16 to 18 can be interpreted as follows. For the case of complete non-affinity between the crystalline solid and the foreign solid surface (corresponding to that of complete non-wetting in liquid–solid systems), $\theta = 180°$, and equation 16 applies, i.e. the overall free energy of nucleation is the same as that required for homogeneous or spontaneous nucleation. For the case of partial affinity (cf. partial wetting), $0 < \theta < 180°$, and equation 17 applies, indicating that nucleation is easier to achieve because the overall excess free energy required is less than that for homogeneous nucleation. For the case of complete affinity (cf. complete wetting) $\theta = 0$, and the free energy of nucleation is zero. This case corresponds to the seeding of a supersaturated solution with tiny crystals of the required solute, i.e. no nuclei have to be formed in the solution. Recent work on catalyzed nucleation has been reviewed by TURNBULL and VONNEGUT[21].

CRYSTAL GROWTH

As soon as stable nuclei, i.e. particles larger than the critical size, have been formed in a supersaturated or supercooled system they begin to grow into crystals of visible size. Many attempts have been made to explain the mechanism and rate of crystal growth, and these may be broadly classified under the three general headings of 'surface energy', 'diffusion' and 'adsorption-layer' theories.

The surface energy theories are based on the postulation of Gibbs (1878) and Curie (1885) that the shape a growing crystal assumes is that which has a minimum surface energy. This approach, although not completely abandoned, has largely fallen into disuse. The diffusion theories originated by Noyes and Whitney (1897) and Nernst (1904) presume that matter is deposited continuously on a crystal face at a rate proportional to the difference in concentration between the point of deposition and the bulk of the solution. The mathematical analysis of the operation is similar to that used for other diffusional and mass transfer processes. In 1922 Volmer suggested that crystal growth was a discontinuous process, taking place by adsorption, layer by layer, on the crystal surfaces. Several notable modifications of this adsorption-layer theory have been proposed in recent years.

For a comprehensive account of the historical development of the many crystal growth theories, reference should be made to the critical reviews given by WELLS[22] and BUCKLEY[23]. Over 80 papers dealing with recent advances in this field were presented at two symposia on crystal growth in 1949[2] and 1958[3].

Surface Energy Theories

An isolated droplet of a fluid is most stable when its surface free energy, and hence its area, is a minimum. GIBBS[4] suggested that the growth of a crystal could be considered as a special case of this principle: the total free energy of a crystal in equilibrium with its surroundings at constant temperature and pressure would be a minimum for a given volume. If the volume free energy per unit volume is assumed to be constant throughout the crystal, then

$$\sum_{1}^{n} a_i g_i = \text{minimum} \tag{19}$$

where a_i is the area of the ith face of a crystal bounded by n faces, and g_i the surface free energy per unit area of the ith face. Therefore, if a crystal is allowed to grow in a supersaturated medium it should develop into an 'equilibrium' shape, i.e. the development of the various faces should be in such a manner as to ensure that the whole crystal has a minimum total surface free energy for a given volume.

Of course, a liquid droplet is very different from a crystalline particle; in the former the constituent atoms or molecules are randomly dispersed, whereas in the latter they are regularly located in a lattice structure. Gibbs was fully aware of the limitations of his simple analogy, but CURIE[24] found it a useful starting point for an attempt to evolve a general theory of crystal growth.

WULFF[25] showed that the equilibrium shape of a crystal is related to the free energies of the faces; he suggested that the crystal faces would grow at rates proportional to their respective surface energies. LAUE[26] has modified Wulff's theory, pointing out that all possible combinations of faces must be considered to determine which of the overall surface free energies represents a minimum. Wulff had also indicated, however, that the rate of growth of a face would be inversely proportional to the reticular or lattice density of the respective lattice plane, so that faces having low reticular densities would grow rapidly and eventually disappear.

The velocity of growth of a crystal face is measured by the outward rate of movement in a direction perpendicular to that face. In order to maintain constant interfacial angles in the crystal (Haüy's law) the successive displacements of a face during growth or dissolution must be parallel to each other. Except for the special case of a geometrically regular crystal the velocity of growth will vary from face to face. *Figure 5.5a* shows the ideal

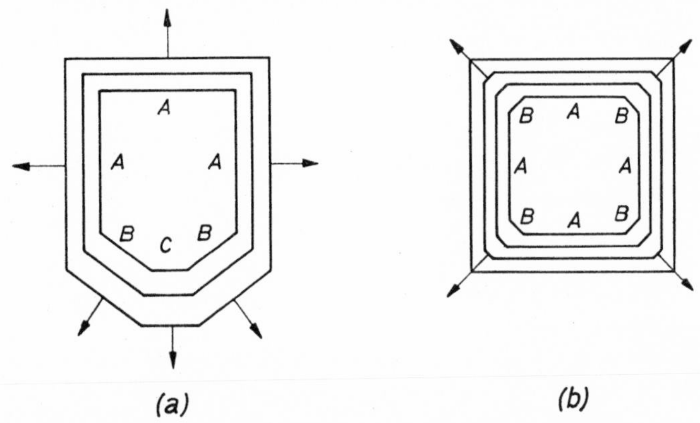

Figure 5.5. Velocities of crystal growth faces: (a) invariant crystal; (b) overlapping

case of a crystal which maintains its geometric pattern as it grows; such a crystal is called 'invariant'. The three equal *A* faces grow at an equal rate, the smaller *B* faces grow faster, while the smallest face, *C*, grows fastest of all. A similar, but reverse, behaviour may be observed when a crystal of this type dissolves in a solvent; the *C* face dissolves at a faster rate than the other faces.

In practice, a crystal does not always maintain geometric similarity during growth; the smaller, faster-growing faces are often eliminated, and this mode of crystal growth is known as 'overlapping'. *Figure 5.5b* shows the various stages of growth or dissolution of such a crystal. The smaller *B* faces, which grow much faster than the *A* faces, gradually disappear from the pattern.

So far there is no general acceptance of the surface energy theories of crystal growth; there is little quantitative evidence available to support them. In recent years, more attention has been devoted to studies of the connection between the initial shape of a small crystal (*Ausgangskörper*) and its final form (*Endkörper*) after it has been allowed to grow under fixed conditions of temperature and supersaturation[23].

Adsorption Layer Theories

The concept of a mechanism of crystal growth based on the existence of an adsorbed layer of solute atoms or molecules on a crystal face was first suggested by VOLMER[6, 7]. Other workers who have contributed to, and modified, this theory include BRANDES[27], STRANSKI[28] and KOSSEL[29], although Kossel's theory, based on a sequence of repeatable steps, deviates to a large degree from Volmer's original postulation. BUCKLEY[23] has made a comprehensive survey of these and other modern theories of crystal growth. The brief account given here will serve merely to indicate how a crystal face can be built up and how imperfections in crystals can be initiated.

Atoms or molecules in the vicinity of a crystal face will tend to attach themselves onto the surface in positions where the attractive forces are greatest, i.e. they will migrate towards positions where a maximum number of like elements are located (*Figure 5.6a*). This step-wise build up will

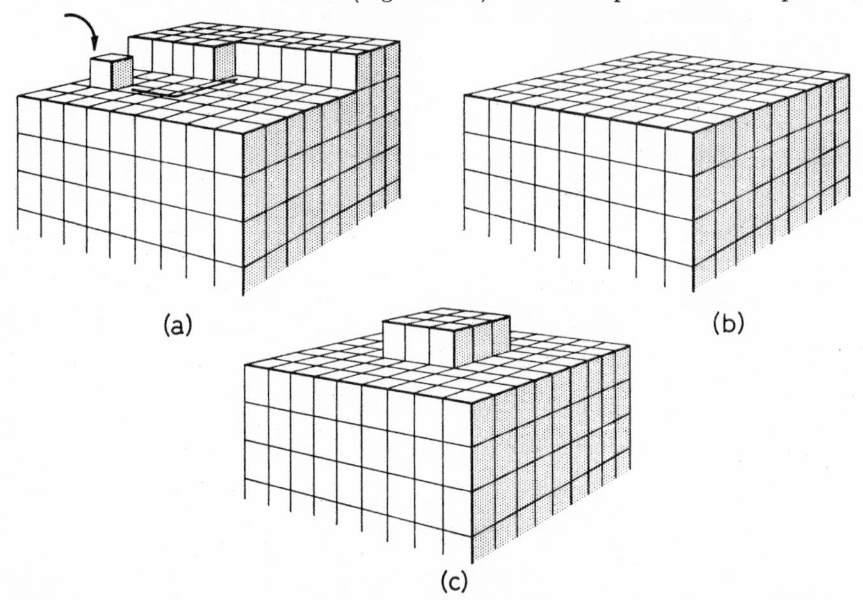

Figure 5.6. A mode of crystal growth without dislocations: (a) migration towards desired location; (b) completed layer; (c) surface nucleation

continue until the whole plane surface is completed (*Figure 5.6b*). Before the crystal face can continue to grow, i.e. before a further layer can be built up again, another 'centre of crystallization' must come into existence on the plane surface, and it is suggested that a monolayer island nucleus, usually called a two-dimensional nucleus, is created (*Figure 5.6c*).

Expressions for the critical radius of a two-dimensional nucleus, assuming for simplicity that the island is circular, and the free energy change associated with the nucleation process can be derived in a manner similar to that described above for a three-dimensional nucleus. In general, it may be assumed that a high degree of local supersaturation is necessary for this type of nucleation to occur, but lower than that required for the formation of three-dimensional nuclei under equivalent conditions.

Dislocations

Few crystals ever grow in the ideal layer-by-layer fashion without some imperfection occurring in the pattern. With regard to the lattice structure itself, foreign atoms or ions may replace certain lattice elements, or they may be included interstitially. Some of the lattice elements may be misplaced, leaving vacant sites in the lattice; the misplaced elements may be included interstitially. Another type of flaw, called a dislocation, consists of some geometrical irregularity in the crystal lattice. Two recent books[30, 31] have been devoted solely to the effect of dislocations on crystal growth.

One type of crystal defect that is of particular interest is the screw dislocation (*Figure 5.7a*)—during the solute deposition process a step may develop on the crystal face, extending over only part of the surface. FRANK[32] has suggested that the presence of such a screw dislocation on a crystal face obviates the need for the succesive formation of two-dimensional nuclei after

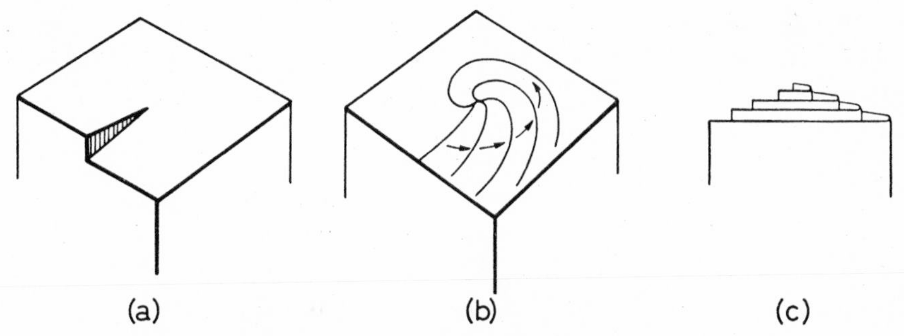

Figure 5.7. Development of a growth spiral starting from a screw dislocation

each layer has been built up. Once a screw dislocation has been formed, the crystal face can grow perpetually 'up a spiral staircase'. As a plane face can never appear under these conditions there is no need for nucleation to occur, and growth can continue uninterrupted at a very low degree of supersaturation, much lower than that required for spontaneous (two-dimensional) nucleation.

Figure 5.7a to *c* indicates the successive stages in the development of a growth spiral starting from a screw dislocation. The curvature of the spiral cannot exceed a certain maximum value, determined by the critical radius for a two-dimensional nucleus under the conditions of supersaturation obtaining in the medium where the crystal is growing.

Screw dislocations and growth spirals are discussed in some detail by BURTON, CABRERA and FRANK[32, 33], and many workers in the past decade have studied these growth phenomena. Quite often very complicated spirals are formed, especially when several screw dislocations grow together. VERMA[30] gave a detailed account of the optical techniques by means of which growth spirals may be observed and published several quite remarkable photographs of this phenomenon.

Diffusion Theories

NOYES and WHITNEY[34] considered that deposition of solid on the face of a growing crystal was essentially a diffusional process. They also assumed that crystallization was the reverse of dissolution, and that the rates of both processes were governed by the difference between concentration at the solid surface and in the bulk of the solution. An equation for crystallization was proposed in the form

$$\frac{dm}{dt} = k_m A(c - c^*) \tag{20}$$

where m = mass of solid deposited in time t
A = surface area of the crystal
c = solute concentration in the solution (supersaturated)
c^* = equilibrium saturation concentration
k_m = coefficient of mass transfer

On the assumption that there would be a thin laminar film of liquid adjacent to the growing crystal face, through which molecules of the solute would have to diffuse, NERNST[35] modified equation 20 to the form

$$\frac{dm}{dt} = \frac{D}{x} A(c - c^*) \tag{21}$$

where D = coefficient of diffusion of the solute
x = length of the diffusion path

The thickness x of the laminar film would obviously depend on the relative solid–liquid velocity, i.e. on the degree of agitation of the system. Film thicknesses varying from 20 to 150 μ have been measured on stationary crystals in stagnant aqueous solution, but MARC[36] found that the film thickness was virtually zero in vigorously stirred solutions. As this would imply an almost infinite rate of growth in agitated systems, it is obvious that the concept of film diffusion alone is not sufficient to explain the mechanism of crystal growth.

It was also shown by Marc that crystallization is not necessarily the reverse of dissolution. A substance generally dissolves at a faster rate than it crystallizes, under the same conditions of temperature and concentration. Another important finding was made by MIERS[37] who determined, by refractive index measurements, the solution concentrations near the faces of crystals of sodium chlorate growing in aqueous solution; he showed that the solution in contact with a growing crystal face is not saturated but supersaturated.

In the light of these facts, a considerable modification was made to the diffusion theory of crystal growth by BERTHOUD[38] and VALETON[39] who suggested that there were two steps in the mass deposition, viz. a diffusion process, whereby solute molecules are transported from the bulk of the fluid phase to the solid surface, followed by a first-order 'reaction' when the solute molecules arrange themselves into the crystal lattice. These two stages, occurring under the influence of different concentration driving forces, can be represented by the equations

$$\frac{dm}{dt} = k_d A(c - c_i) \qquad \text{(diffusion)} \tag{22}$$

and

$$\frac{\mathrm{d}m}{\mathrm{d}t} = k_r A(c_i - c^*) \qquad \text{(reaction)} \tag{23}$$

where k_d = a coefficient of mass transfer by diffusion
k_r = a rate constant for the surface reaction
c_i = solute concentration in the solution at the crystal–solution interface

A pictorial representation of these two processes is shown in *Figure 5.8* where the various concentration driving forces can be seen. It must be clearly understood, however, that this is only diagrammatic: the driving forces will rarely be of equal magnitude, and the concentration drop across the laminar film is not necessarily linear.

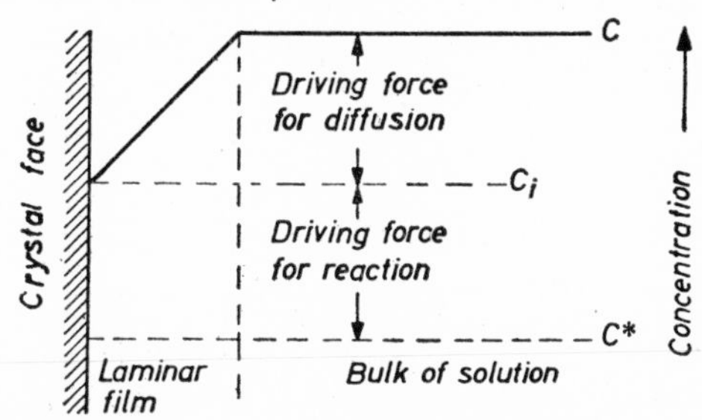

Figure 5.8. Concentration driving forces in crystallization from solution

The validity of the assumption of a first-order surface reaction is questionable. Many inorganic salts crystallizing from aqueous solution show rates slightly greater than first-order, while others indicate a second-order reaction. Furthermore, the value of the rate constant k_r varies from face to face on a crystal. The individual face constants are difficult, if not impossible, to determine, but average values for whole crystals have been measured.

Values of k_d can also differ for different faces of the same crystal, but average values of the quantity are most frequently measured. It is even possible for k_d to vary over one given face: whilst it is true, as Miers found, that the solution in contact with a growing crystal face is always supersaturated, the degree of supersaturation can vary at different points over the face[40]. From refractive index measurements Berg[41] showed, surprisingly, that the concentration was highest at the corners and lowest at the centre of the face, but whether this condition arose from the mode of crystal growth or whether it actually caused a particular mode of growth would be extremely difficult to decide.

Equations 22 and 23 are not easy to apply in practice because they involve interfacial concentrations which are difficult to measure. It is usually more convenient to eliminate the term c_i by considering an 'overall' concentration driving force, $c - c^*$, which is quite easily measured. A general equation for crystallization based on this overall driving force can be written as

$$\frac{\mathrm{d}m}{\mathrm{d}t} = KA(c - c^*) \tag{24}$$

where K is an overall mass transfer coefficient. From equations 22 and 23

$$\frac{dm}{dt} = \frac{A(c - c^*)}{1/k_d + 1/k_r} \tag{25}$$

i.e.

$$\frac{1}{K} = \frac{1}{k_d} + \frac{1}{k_r} \tag{26}$$

For cases of extremely rapid surface reaction, i.e. large k_r, $K \sim k_d$, the crystallization process is controlled by the diffusional operation. Similarly, if the value of k_d is large, i.e. if the diffusional resistance is low, $K \sim k_r$, and the process is then controlled by the surface reaction.

Theories of Mass Transfer

The two-film theory due to WHITMAN[42] has proved to be useful for the analysis of mass transfer between two immiscible phases. The assumptions are essentially similar to those of Nernst, i.e. stagnant or laminar films of fluid exist adjacent to the interface between the two phases, even though turbulence occurs in the rest of the system. The transfer of matter across the interfacial region is considered to take place by molecular diffusion, and as this represents a much slower process than the transport of matter from the bulk of the fluids to the interfacial films by convection currents or eddy diffusion, the major resistance to mass transfer is assumed to lie in the interfacial films. The two-film theory, originally proposed for gas absorption, has been applied successfully to a wide variety of unit operations of chemical engineering.

It is generally agreed, however, that the film concept, which forms the basis of the mathematical theory of mass transfer, presents an oversimplified picture. There has long been a considerable amount of doubt as to the possibility of stagnant or laminar layers of fluid existing at the interface between two phases, one or both of which may be flowing in a turbulent manner. With regard to dissolution and crystallization processes, for example, there is evidence[43] that turbulent eddies can extend to the surface of a solid located in an agitated liquid. In view of these objections, several alternative theories for the mechanism of mass transfer have been developed which do not depend on the existence of films of fluid at an interface.

HIGBIE[44] suggested that the mechanism of mass transfer in gas absorption processes could be explained by assuming that a systematic renewal of the liquid surface occurred. The surface in contact with the gas phase was envisaged as being composed of a large number of tiny elements of liquid; each element remains at the interface for a short fixed period of time, after which it is replaced by a fresh element of liquid from the bulk of the liquid phase. Mass transfer occurs between the gas and the liquid during this small interval of time. DANCKWERTS[45, 46] later modified this theory by suggesting that the surface renewal was random, i.e. the elements of liquid remained at the interface for random, not fixed, intervals of time.

The surface renewal theory, like the film theory, can be applied to mass transfer operations other than gas absorption. In both cases the driving force for mass transfer is considered to be the difference in concentration between the bulk of the fluid and the interface. For crystallization processes, therefore, both theories can be represented in mathematical form by equation 22 where

$$k_d = \frac{D}{x} \quad \text{(stagnant film)} \tag{27}$$

$$= \sqrt{D.f} \quad \text{(random surface renewal)} \tag{28}$$

where D = diffusivity of the solute molecules in the liquid, x = equivalent film thickness, and f = fractional rate of surface renewal.

There is, as yet, no conclusive evidence available to decide between the film and surface renewal theories. It might seem at first sight that this would be a rather simple matter to decide because the film theory suggests that $k_d \propto D$, whereas the surface renewal theory assumes that $k_d \propto \sqrt{D}$. Unfortunately, equation 27 contains a hypothetical film thickness, x, whereas equation 28 contains a surface renewal rate term, f, neither of which is capable of being measured directly. However, as both theories are based on the same driving force $(c - c_i)$, the mass transfer coefficient, k_d, can still be used as a constant of proportionality in the mass transfer equation despite the doubts regarding its true significance.

Crystallization and Dissolution

If both crystallization and dissolution processes were purely diffusional in nature, they should exhibit a true reciprocity; the rate of crystallization should equal the rate of dissolution at a given temperature and under equal concentration driving forces, i.e. at equal displacements away from the equilibrium saturated conditions. In addition, all faces of a crystal should grow and dissolve at the same rate. These conditions rarely obtain in practice. Crystals usually dissolve faster than they grow, and different faces usually grow or dissolve at different rates, although this may not always be so with crystals belonging to the cubic system.

These facts have led most workers to support the view that in a crystallization process diffusion is followed by a surface 'reaction' at the growing crystal face. There have, however, been other suggestions put forward. Some authors hold that a crystal dissolves faster than it grows because the exposed solid surfaces are not the same in each case. When a crystal grows the faces are plane, whereas when it dissolves the faces are usually pitted, resulting in a much greater area of solid–liquid contact. GRUT[47] has shown that a crystal previously pitted by partial dissolution grows initially at a very fast rate; then, as time proceeds, the rate falls and eventually becomes fairly constant (*Figure 5.9*).

It is generally agreed that surface pitting can increase the rate of mass transfer in a dissolution process, but this does not exclude the possibility of a time-consuming surface reaction taking place in the reverse process of crystallization. As VAN HOOK[48] pointed out, even an over-generous allowance of an extra surface area due to pitting cannot possibly explain the greater rate of solution compared with the rate of crystallization of sucrose under comparable conditions. Other workers have expressed similar views,

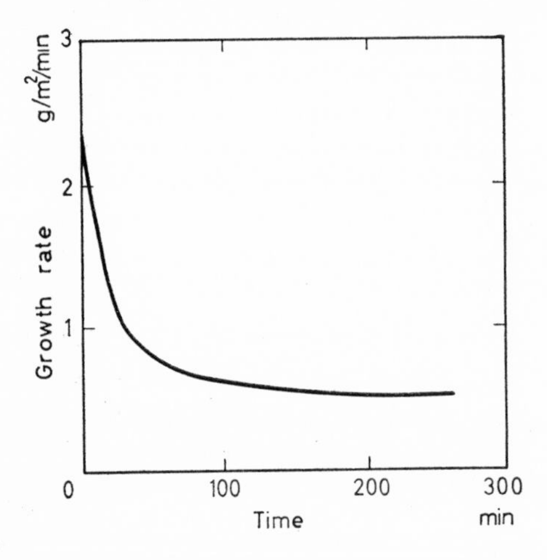

Figure 5.9. Growth rate of a sucrose crystal initially etched by partial dissolution. (After data of E. GRUT[47])

and it has even been shown that some dissolution processes can also involve a slow chemical step at the surface of the crystal lattice[49].

Effect of Temperature

The relationship between a reaction rate constant, k, and the absolute temperature, T, is given by the ARRHENIUS[50] equation

$$\frac{\mathrm{d} \ln k}{\mathrm{d}T} = \frac{E}{\boldsymbol{R}T^2} \tag{29}$$

The symbol E, which represents a quantity of energy associated with the particular reaction, is usually called the energy of activation of the process. On integration, equation 29 gives

$$k = A \,.\, \exp\,(-E/\boldsymbol{R}T) \tag{30}$$

or, taking logarithms

$$\ln k = \ln A - \frac{E}{\boldsymbol{R}T} \tag{31}$$

Therefore, a plot of $\ln k$ against $1/T$ should give a straight line of slope $-E/\boldsymbol{R}T$ and intercept $\ln A$ if the Arrhenius equation applies. This method of plotting is most frequently employed for representing temperature–reaction velocity data, measuring values of E and predicting values of k at any given temperature.

Alternatively, if only two measurements of the rate constant are available, k_1 at temperature T_1 and k_2 at T_2, the following equation may be used

$$E = \frac{\boldsymbol{R}T_1T_2}{T_2 - T_1} \ln \frac{k_1}{k_2} \tag{32}$$

Equation 32 is obtained by integrating equation 29 between the limits T_1 and T_2, assuming that E remains constant over this temperature range.

Applying the above equations to a crystallization process, k can be taken as the rate constant associated with the diffusion process or the observed growth. Plots of $\log K$ and $\log D$, where K is the overall mass transfer coefficient as defined by equation 24 and D is the diffusivity, can be used to calculate the corresponding activation energies E_{diff} and E_{cryst}. As the diffusivity is a function of the reciprocal of the absolute viscosity, η, a plot of $1/\eta$ against $1/T$ would yield a value of E_{visc}. These three plots can be made for one given system over a range of temperature; the process which demands the highest activation energy at a given temperature will be the controlling one.

Figure 5.10 was prepared by VAN HOOK[48] for the case of the crystallization of sucrose from its aqueous solution. The diagram shows that the energy of

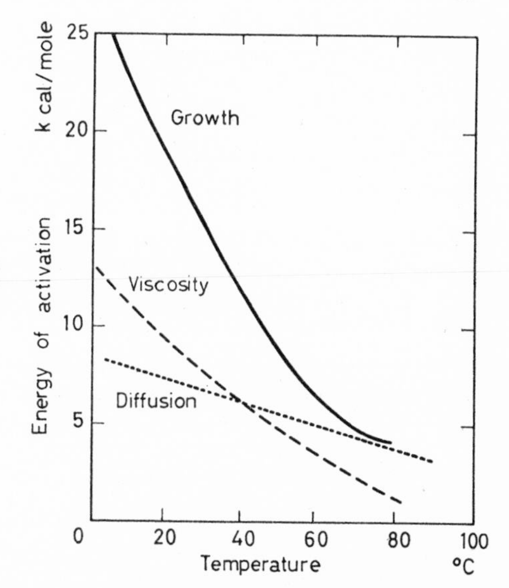

Figure 5.10. Activation energies of growth diffusion and viscosity for sucrose at constant supersaturation. (After A. VAN HOOK[48])

activation for growth is much greater than the activation energies for diffusion and viscosity at low temperatures, suggesting that diffusion is not an important factor in the growth process. At higher temperatures, however, the activation energies for diffusion and viscosity are comparable with the activation energy for crystallization, suggesting that the process is diffusion controlled under these conditions. The results of ENGLISH and DOLE[51] tend to support Van Hook's deductions.

DEDEK[52] came to a similar conclusion when he compared the rate at which sucrose crystallizes from aqueous solution to the theoretical rate at which sucrose becomes available at a growing crystal face by diffusion through the solution. The data shown in *Table 5.1*, compiled for a fixed degree of supersaturation of 1·05 g/cm^3, indicate that excess sucrose is available at the growing face at temperatures lower than about 45° C, but above this temperature there is a deficiency of sucrose. These results would suggest, therefore, that above about 45° C the crystallization process is diffusion-controlled. Below about 45° C the rate of diffusion is not important, and the rate of the surface 'reaction' probably controls the crystallization rate.

Table 5.1. Observed Crystallization Rates of Sucrose Compared with Theoretical Diffusion Rates at Different Temperatures
(*After* DEDEK[52], *reported by* VAN HOOK[48])

Temperature °C	*Growth rate*, r_g g/m² . mm	*Diffusion rate*, r_d g/m² . mm	r_d/r_g
20	0·42	10·8	25·7
30	1·34	5·9	4·5
40	2·1	5·2	2·1
50	5·4	2·8	0·5
60	13·5	1·6	0·2
70	30	1	0·03

In a similar manner RUMFORD and BAIN[53] showed that the crystallization of sodium chloride from aqueous solution is diffusion-controlled above about 50° C and reaction rate controlled below this temperature. COOKE[54], however, is of the opinion that the crystallization of sodium chloride is controlled by the diffusional mechanism at all temperatures, even though the contribution of the surface reaction increases at low temperatures and low degrees of supersaturation.

Effect of Agitation

The rate at which a crystal grows at a given temperature under constant conditions of supersaturation can be altered appreciably by agitating the liquid or by rotating the crystal in the liquid. The rate of growth increases

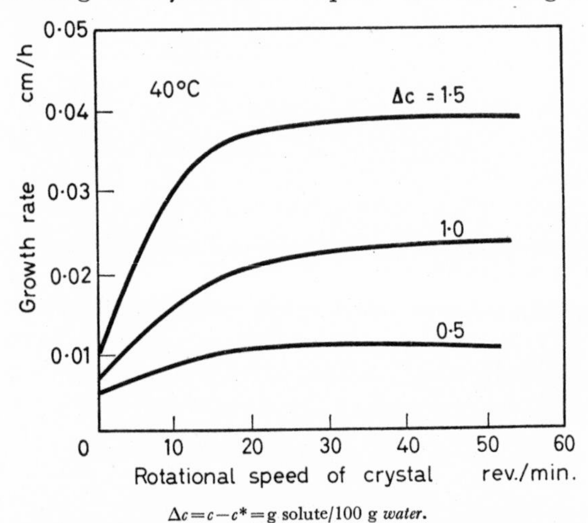

$\Delta c = c - c^* =$ g solute/100 g *water*.

Figure 5.11. Effect of the speed of rotation of the growth rate of a crystal of sodium thiosulphate. (After J. M. COULSON *and* J. F. RICHARDSON[55])

considerably in the initial stages as the relative velocity between crystal and liquid increases, but conditions are soon reached when further agitation has no effect. The results[55] shown in *Figure 5.11* illustrate the behaviour of single crystals of sodium thiosulphate in aqueous solutions of different degrees of supersaturation. In all cases, rotational speeds of the crystals in excess of about 20 rev/min had little effect on the growth rate. Similar results have been reported for the growth of sucrose[48] and copper pentahydrate[56] crystals from their supersaturated aqueous solutions.

As an increase in the relative velocity between the crystal and the liquid would be expected to decrease the thickness of any laminar film of liquid at the solid surface, the above results indicate that diffusion is not the only, nor indeed the most important, factor to be considered in a crystallization process. McCabe and Stevens[56] correlated their growth measurements on crystals of $CuSO_4 \cdot 5H_2O$ at constant supersaturation by means of the equation

$$\frac{1}{r_g} = \frac{1}{r_0 + \beta v} + \frac{1}{r_i} \tag{33}$$

where r_g = mean rate of growth (length/unit time)
r_0 = rate of growth at zero solid–solution reactive velocity, v
r_i = interfacial growth rate
β = dimensionless constant

Thus, the total resistance to crystallization, $1/r_g$, is equal to the sum of the interfacial resistance, $1/r_i$, and a resistance, $1/r_0 + \beta v$, which incorporates two parallel processes, a diffusion effect independent of velocity and a flow effect dependent on velocity. When $v \rightarrow \infty$, $r_g = r_i$, and the rate of growth depends solely on the rate of the surface reaction. When $v = 0$ and $r_i = \infty$, $r_g = r_0$, and the growth is diffusion-controlled.

The growth rate was found to be relatively unaffected by the size of the crystal, but it was pointed out that there would be an apparent size effect if (*a*) the relative velocities of crystals of different size differed, and (*b*) the velocities were in the low range where the velocity effect was marked. It was concluded that in a crystallizer in which crystals are maintained in suspension by agitation, the large crystals will grow faster than the small ones because the former will benefit from a higher relative solid–solution velocity.

For the case of a crystal growing in a supersaturated solution under a constant concentration driving force, the rate of mass transfer, k_d, in a diffusion-controlled system can be considered to depend on the following variables: the viscosity, η, and density, ρ, of the solution, the relative velocity, v, between the solid surface and the solution, the diffusivity, D, of the solute molecules in the liquid, and some chosen length parameter, or diameter, d, of the crystal, i.e.

$$k_d = f(\eta, \rho, v, D, d) \tag{34}$$

The relationship between these quantities can be written in the form of a dimensionless equation containing three dimensionless groups

$$\frac{k_d d}{D} = \text{const} \left(\frac{\rho v d}{\eta}\right)^a \left(\frac{\eta}{\rho D}\right)^b \tag{35}$$

where the groups $\rho v d/\eta$, $\eta/\rho D$ and $k_d d/D$ are the dimensionless Reynolds

(Re), Schmidt (Sc) and Sherwood (Sh) numbers, respectively. The Reynolds number indicates the effect of agitation; it expresses the ratio of the fluid momentum and the viscous drag force. The Schmidt number gives the ratio of the fluid properties which govern the rate of transfer of momentum under the influence of a velocity gradient and those governing the rate of mass transfer by molecular diffusion under the influence of a concentration gradient. The Sherwood number denotes the ratio of the actual to the theoretical rate of mass transfer by diffusion in a stationary fluid.

HIXSON and his co-workers[57–60] have carried out an extensive research programme on the effect of agitation on mass transfer in liquid–solid systems. The major part of this work was devoted to the process of dissolution, but in the paper with KNOX[60] the growth rates of single crystals of $CuSO_4 \cdot 5H_2O$ and $MgSO_4 \cdot 7H_2O$ were considered. The growth rates were found to depend on the diffusional mass transfer coefficient, k_d, which varied with the fluid velocity, and on the rate constant of the surface reaction, k_r, which varied with temperature. The mass transfer coefficients, k_d, in lb.-mol/h·ft.², were correlated by means of a relationship similar in form to equation 35

$$\frac{k_d d}{D_m} = \beta \left(\frac{\rho v d}{\eta}\right)^{0 \cdot 6} \left(\frac{\eta}{\bar{M} D_m}\right)^{0 \cdot 3} \tag{36}$$

where D_m = molal diffusivity (lb.-mole/ft.·h)
d = equivalent diameter of the crystal (ft.)
v = relative solution–crystal velocity (ft./h)
ρ = density of solution (lb./ft.³)
η = viscosity of solution (lb./ft.·h)
$\bar{M}$ = mean molecular weight of solution

The equivalent diameter was taken to be the diameter of a sphere of the same surface area as the crystal. Values of the constant β were reported as 0·48 and 0·29 for $CuSO_4 \cdot 5H_2O$ and $MgSO_4 \cdot 7H_2O$, respectively.

The rate coefficient, r_i, for the surface reaction for copper sulphate crystals had a value of 1,600 lb.-mole/h·ft.² at 53·5° C. The activation energy was 13·6 kcal/mole. For magnesium sulphate, k_r was 0·331 lb.-mole/h·ft.² at 30·5° C, and the corresponding activation energy was 2·43 kcal/mole.

Apart from the above-mentioned work, a large number of publications have appeared in recent years concerning the effects of variables on mass transfer in liquid–solid systems[61–70]. Most of this work is concerned primarily with dissolution rather than crystallization, but it is still of direct interest to the present problem. Comprehensive accounts of the development of the theoretical aspects of the subject have been given in several of these papers.

Heat and Mass Transfer

The rate of crystallization from a melt depends on the rate of heat transfer from the crystal face to the bulk of the liquid. As the process is generally accompanied by the liberation of heat of crystallization, the surface of the crystal will have a slightly higher temperature than the supercooled melt.

These conditions are shown in *Figure 5.12* where the melting point of the substance is denoted by T^* and the temperature of the bulk of the supercooled melt by T. The overall degree of supercooling, therefore, is $T^* - T$. The temperature at the surface of the crystal, the solid–liquid interface, is

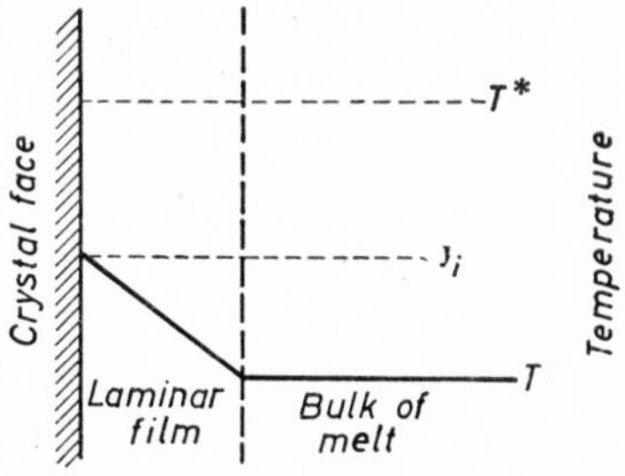

Figure 5.12. Temperature gradients near the face of a crystal growing in a melt

denoted by T_i, so the driving force for heat transfer across the 'effective' film of liquid close to the crystal face is $T_i - T$. The rate of heat transfer, $\mathrm{d}q/\mathrm{d}t$, can be expressed in the form of the equation

$$\frac{\mathrm{d}q}{\mathrm{d}t} = hA(T_i - T) \tag{37}$$

where A is the area of the growing solid surface, and h is a film coefficient of heat transfer defined by

$$h = \frac{\kappa}{x'} \tag{38}$$

where κ is the thermal conductivity, x' the effective film thickness for heat transfer. There is a distinct similarity between the form of equation 37 for heat transfer and equation 22 for mass transfer by diffusion. Agitation of the system will reduce the effective film thickness, increase the film coefficient of heat transfer and tend to increase the interfacial temperature, T_i, to a value near to that of the melting point, T^*.

TAMMAN[13] showed that the rate of crystallization of a supercooled melt achieves a maximum value at a lower degree of supercooling, i.e. at a temperature higher than that required for maximum nucleation. The nucleation and crystallization rate curves are dissimilar: the former has a relatively sharp peak (*Figure 5.3*), the latter usually a rather flat one. Tamman suggested that the maximum rate of crystallization would occur at a melt temperature, T, given by

$$T = T^* - \left(\frac{\Delta H_{\mathrm{cryst}}}{c_m}\right) \tag{39}$$

where $\Delta H_{\mathrm{cryst}}$ is the heat of crystallization, c_m the mean heat capacity of the melt.

The analogy between heat and mass transfer was demonstrated by CHILTON and COLBURN[71] who proposed the use of the semi-theoretical relationships now known as 'j factors'; j_h is a factor for heat transfer, j_m for mass transfer. These are usually expressed in the form

$$j_h = \frac{h}{Gc_p}\left(\frac{\eta c_p}{\kappa}\right)^{\frac{2}{3}} \tag{40}$$

and

$$j_m = \frac{k\rho}{G}\left(\frac{\eta}{\rho D}\right)^{\frac{2}{3}} \tag{41}$$

where h and k are coefficients of heat and mass transfer, respectively, G is the mass flow rate of the fluid per unit area, e.g. in a conduit. η, ρ, D, κ and c_p are viscosity, density, diffusivity, thermal conductivity and heat capacity, respectively. These relationships have proved invaluable for the correlation of heat and mass transfer data but, as HIXSON and BAUM[59] pointed out, they cannot be applied to agitated liquid–solid systems, as they involve a mass velocity term, G, which can neither be measured nor defined in these cases.

Heat transfer data can also be correlated by means of the Nusselt equation which can be derived by dimensional analysis in the form

$$\frac{hd}{\kappa} = \text{const}\left(\frac{\rho vd}{\eta}\right)^x\left(\frac{\eta c_p}{\kappa}\right)^y \tag{42}$$

where d is a chosen length parameter, or diameter. The similarity between equations 42 and 35 is very close. Both incorporate the Reynolds number, but the Schmidt and Sherwood groups in equation 35 are replaced in equation 42 by the dimensionless Prandtl number, Pr $(= \eta c_p/\kappa)$, and Nusselt number, Nu $(= hd/\kappa)$, respectively.

Equations 35 and 42 have been used by a large number of workers to demonstrate the heat and mass transfer analogy in liquid–gas systems. HIXSON and BAUM[59] have also demonstrated this analogy for liquid–solid systems in agitated vessels. For the case of mass transfer they studied the dissolution of benzoic acid, benzene and several other liquids, sodium and barium chlorides in water, and of naphthalene in methanol. For the case of heat transfer they studied melting rates in the one-component systems water, benzene, nitrobenzene and acetic acid. The Reynolds number used in this work was defined by the equation

$$\text{Re} = \frac{nd^2\rho}{\eta} \tag{43}$$

where d was the diameter of the vessel, n the rotational speed of the stirrer (revolutions per unit time). The following correlations were made

(*a*) *mass transfer*

$$\frac{kd}{D} = 0{\cdot}16(\text{Re})^{0{\cdot}62}(\text{Sc})^{0{\cdot}5} \tag{44}$$

(*b*) *heat transfer*

$$\frac{hd}{\kappa} = 0{\cdot}207(\text{Re})^{0{\cdot}63}(\text{Pr})^{0{\cdot}5} \tag{45}$$

FORSYTH and WOOD[72] attempted the more difficult task of trying to apply the heat and mass transfer analogy to the case of the crystallization of organic binary melts. They used a modified mass transfer equation, with constants that had previously been deduced[64, 73] for liquid–gas systems, in the form

$$\frac{k_l}{v}\left(\frac{c_{BM}}{c_A + c_B}\right)(\mathrm{Sc})^{\frac{2}{3}} = 0{\cdot}023(\mathrm{Re})^{-0{\cdot}17} \tag{46}$$

where k_l = a mass transfer coefficient for the liquid phase, v = fluid velocity, c_A and c_B = concentrations of components A and B, respectively, in the liquid, and c_{BM} = the logarithmic mean concentration of component B between the phase boundary and the bulk liquid. The application of this equation to the data obtained from the system naphthalene–β-naphthol showed that the effect of the Reynolds number was similar to that in gas–liquid systems, but in other respects the application of the proposed equation was less successful.

Surface Reaction

The second resistance to crystallization, which is usually referred to as the surface reaction, has received relatively little attention compared with the first, viz. the transport of matter to the solid surface. In this second step the molecules of solute arrange themselves into the crystal lattice. A theoretical approach to this problem was made by AMELINCKX[74] who proposed an equation for the particle integration rate of homopolar-type crystals. He derived the equation statistically by considering the rate of particle attachment at the crystal surface and the rate of detachment; the particle integration rate was defined as the difference between these two rates. His final equation was of the form

$$g = (c_f - c)\left[\frac{v}{4} + \phi m \nu_0 n \,.\, \exp\left(-U/\boldsymbol{k}T\right)\right] \tag{47}$$

where g = growth rate when the surface reaction controls
c_f = solution concentration at crystal face
c = solution concentration in bulk liquid
v = mean relative velocity between crystal and liquid
m = mass of the crystal
ν_0 = average frequency of the lattice vibration
n = reticulate density of the *hkl* plane of the crystal face
U = attachment energy of the crystallizing particles
$\boldsymbol{k}$ = Boltzmann's constant
T = absolute temperature
$\phi = -(\mathrm{d}U/\mathrm{d}c)/\boldsymbol{k}T$

As Amelinckx's equation was not in a form which readily allowed experimental verification, CARTIER *et al.*[75] derived a modified relationship, which satisfactorily correlated the particle integration rate data of citric and itaconic acids

$$g = K\{\exp\left[\phi(c_f - c)\right] - 1\} \tag{48}$$

where K, called a particle integration factor, was defined by

$$K = \frac{v}{4\phi} - mn\nu_0 \exp\left(-U/\boldsymbol{k}T\right) \tag{49}$$

It was found that K and ϕ could be correlated simply from experimental data in terms of the absolute temperature

$$K = \alpha T^{\frac{3}{2}} + \beta \tag{50}$$

and

$$\phi = -\frac{1}{\rho}\left(\frac{A}{T} - B\right) \tag{51}$$

where α, β, A and B were constants for the particular system, and ρ = solution density.

Apart from indicating the successful applications of these equations to the growth rate of crystals under conditions where the effect of mass transfer by diffusion is negligible, i.e. in a fast-moving stream of solution, the paper by Cartier *et al.* is of particular interest because it describes a neat and versatile apparatus for measuring crystal growth. The growth cell is shown in *Figure 5.13.* A small crystal, 0·5 to 1 mm in length, was mounted with

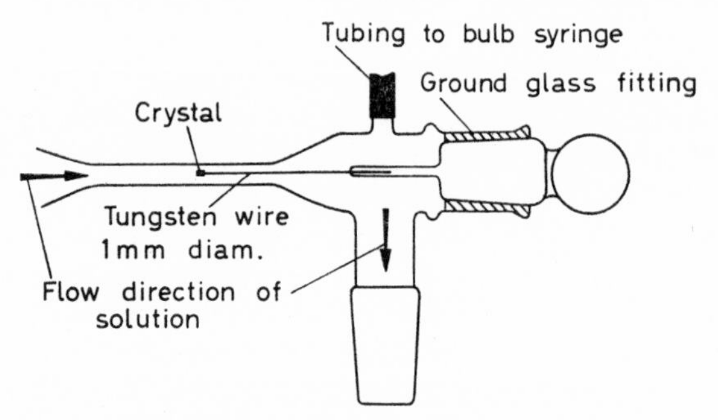

Figure 5.13. Crystal growth cell. (*After* R. CARTIER, D. PINDZOLA *and* P. F. BRUINS[75])

Canadian balsam cement on a tungsten wire and placed in the centre of the glass tube. Measurements of linear growth were made by observing the particular crystal face through a microscope (100 × magnification) fitted with a screw micrometer eyepiece. The unit could also be used for the determination of solubility data.

HABIT MODIFICATION

The habit of a crystal may be defined as the type of external shape which results from the different rates of growth of the various faces. Under certain conditions of crystallization one set of faces may be induced to grow faster than others, or the growth of another set of faces may be retarded. Crystals of one given substance produced by different methods may be completely dissimilar in appearance, even though still belonging to the same crystal system. For example, one method of crystallization may favour an acicular (needle) habit, while another may give a tabular habit (plates or flakes). A large number of factors can affect the habit of a crystal, e.g. the type of solvent, the pH of the solution, the presence of impurities, the degree of supersaturation or supercooling, the rate of cooling, the temperature of crystallization, the degree of agitation, and so on.

Rapid cooling of a solution or melt will often cause the preferential growth of a crystal in one particular direction, leading to the formation of needles. Under these conditions there is a need for a fast rate of heat dissipation from the solid phase, and an elongated crystal is better suited for this purpose than say a granular or tabular crystal. Crystallization from the vapour phase, as in sublimation processes, generally leads to the formation of needle crystals. Here again a habit which permits a rapid heat dissipation is required; the heat of crystallization is usually fairly high (heat of fusion plus heat of vaporization), and the gaseous phase is a poor conductor of heat.

Dendritic growth, in which tree-like formations are produced, may also result from vapour phase crystallization; the thin acicular branches provide the necessary exposed solid surface to dissipate the heat quickly. This type of growth is most usually associated with crystallization from quiescent media, especially in thin layers of solution. The dendritic frosting of window panes is a typical example of this phenomenon. Many metals and organic melts crystallize initially in the form of dendrites, and several inorganic salts can be induced to crystallize in this form under the influence of traces of certain dyestuffs. Dendritic crystals of potassium chlorate are readily deposited from aqueous solution in the presence of Congo Red[23].

A change of solvent will often result in a change of the habit of the crystallizing solute. Naphthalene, for instance, crystallizes in the form of needles from *cyclo*hexane, as thin plates from methanol. Iodoform crystallizes as hexagonal prisms from *cyclo*hexane, as hexagonal pyramids from aniline. Pentaerythritol crystallizes in the form of tetragonal bipyramids from water, of tetragonal plates from acetone. The possibility of a habit change is an important factor to be borne in mind when making a choice of a solvent for crystallization. Occasionally the pH of an aqueous solution has an influence on the type of crystal which grows. Copper sulphate, for example, which normally crystallizes in the form of large granules, can be crystallized in the form of thin plates from acid solution. In general, however, pH is not regarded as one of the major factors affecting crystal habit.

The most common cause of habit modification is probably the presence of impurities in the crystallizing solution. One of the most frequently quoted examples is sodium chloride which crystallizes in the form of cubes from water and as octahedra when urea is present in the solution. The habit of naphthalene crystallizing from methanol can be changed from plates to needles when traces of collodion are added[76]. In the beet-sugar industry the effect of raffinose on the crystal form of sucrose is well known—peculiar flat crystals are induced[77]. Often minute traces of a third component can cause a startling change in the crystal habit, but this is not always the case; many known examples of habit modification have resulted from the presence of large quantities of impurity in the crystallizing system, and this fact is not always made clear. Tervalent ions such as ferric iron, aluminium and chromium can act as active habit modifiers, and small quantities of salts of these metals are often added to crystallizing systems to effect a desired habit change.

Many complex dyes act as habit modifiers for inorganic salts. WHETSTONE and BUTCHART[78, 79] studied the habit-modifying powers of a number of dyestuffs and showed that the anionic and cationic polar substituent groups, and the nature of their substitution in the dyestuff molecule, exert an important influence on their effectiveness. BUCKLEY[23] has summarized a large number of case histories of habit modification due to dyestuffs and gives an indication of the concentrations necessary to induce the required change. From about 400 known active habit-modifying dyes, Buckley selects about 30 which are particularly effective for the four salts $KClO_3$, K_2SO_4, K_2CrO_4 and NH_4ClO_4, and from these again five which he considers to be the best all-rounders: Brilliant Azurine B (511), Tryptan Red (438), Naphthol Black B (315), Brilliant Congo R (456) and Sky Blue FF (518); the numbers in parentheses are the Colour Index numbers.

Habit modification by impurities is essentially a surface phenomenon; impurity molecules or ions are attracted to various faces of the crystal and physically adsorbed or chemisorbed on the surface. In this way available sites for surface nucleation or solute deposition are reduced, and growth on that particular face is retarded. If the active impurity is present in the solution only in small traces, it may be adsorbed only on to the most receptive faces. However, if large quantities of the active impurity are present, adsorption may occur on all faces, not necessarily to the same degree but sufficient to produce a habit quite different from that effected by trace quantities. In some extreme cases the impurity can 'block' all the faces and prevent crystallization even at very high degrees of supersaturation.

If the impurity is adsorbed or chemisorbed very strongly on to the crystal surfaces, a considerable amount of contamination of the crystal is unavoidable. When dyestuffs or coloured ions, for example, have been used as habit modifiers, hour-glass patterns or striations may be formed in the grown crystal. Some habit modifiers, however, are only weakly adsorbed on the surfaces; they retard crystal growth temporarily, but soon become desorbed. Further growth on the particular face occurs, and then the impurity becomes adsorbed again. This sequence of adsorption and desorption continues, with the result that the crystal habit is modified due to the intermittent retardation effect, but the grown crystal is relatively free from impurity.

Many surface-active agents have been used in habit modification studies, but as in most of this work, the majority of published results are qualitative only. The recent work of MICHAELS and COLVILLE[80], therefore, is of particular interest because the growth rates of different faces of crystals of adipic acid in aqueous solution were actually measured. Controlled degrees of supersaturation in the temperature range 25–45° C were used, and the individual growth rates of the (001), (010) and (110) faces were determined with and without the presence of surface-active agents in the solution. The two agents, used in concentrations of 50–100 p.p.m., were sodium dodecylbenzenesulphonate (SDBS) (anionic) and trimethyl dodecylammonium chloride (TMDAC) (cationic). Some of the results of this work are shown in *Figure 5.14*.

The growth rates of the normal faces of adipic acid crystals were found to increase in the order (110) > (010) > (001), and the rate constants were correlated with the hydroxyl densities on the faces. Trace quantities of the anionic agent caused a much greater reduction in the growth rate of the (010) and (110) faces than of the (001) face, leading to the formation of

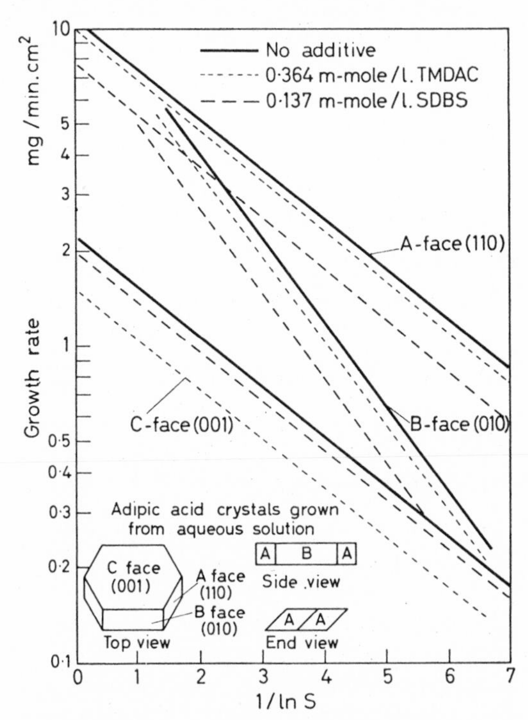

Figure 5.14. Specific growth rates of the faces of adipic acid crystals as a function of the degree of supersaturation, S, with and without the presence of surface-active agents. (*After* A. S. MICHAELS *and* A. R. COLVILLE[79])

prisms or needles. The cationic agent had the opposite effect, favouring the formation of plates or flakes. Anionic agents appeared to be physically adsorbed on to the crystal faces, whereas the cationic agents appeared to be chemisorbed. It was also observed that surface-active agents in general exhibited a far greater retarding influence on the growth of very small crystals than of large ones, and it was suggested that growth from dislocations is less sensitive to the influence of adsorbed contaminants than is the growth by two-dimensional nucleation.

Industrial Importance

The majority of reported cases of habit modification have been concerned with laboratory investigations, but the phenomenon is of the utmost importance in industrial crystallization and by no means a mere laboratory curiosity. Certain crystal habits are disliked in commercial crystals because they give the crystalline mass a poor appearance, others make the product prone to caking (see Chapter 7), impart poor flow characteristics or give rise to difficulties in the handling or packaging of the material. For most commercial purposes, a granular or prismatic habit is usually desired, but there are specific occasions when plates or needles may be wanted.

In nearly every industrial crystallization some form of habit modification is necessary to control the type of crystal produced. This may be done by controlling the rate of crystallization, e.g. the rate of cooling or evaporation, the degree of supersaturation or the temperature at which crystallization occurs, by choosing the correct type of solvent, adjusting the pH of the solution, or deliberately adding some 'impurity' which acts as a habit modifier to the system. A combination of several of these methods may have to be used. It is also worthwhile remembering that the results of small-scale laboratory trials on habit modification may not always prove of value for large-scale application; they may even be quite misleading. Pilot-plant trials, however, on batches greater than about 50 gal. will usually yield very useful information.

GARRETT[80] recently reviewed many of the reported cases of industrial applications of habit modification, and also discussed the factors that must be considered when selecting a suitable habit modifier. A few of his examples may be quoted. Borax crystals can be modified to a tabular or flaky habit by the addition of gelatin or casein to the crystallizing solution; these substances are selectively adsorbed on to the basal pinacoid faces and reduce their growth rate. Large crystals of sodium, potassium and ammonium chlorides, normally difficult to produce, can be grown when Pb^{++} ions are present in the solution. Large granular crystals of sodium sulphate decahydrate (Glauber salt) can be produced from by-product rayon liquors when certain surface-active agents, e.g. alkyl aryl sulphonates, are added; otherwise the salt usually tends to crystallize in the form of small needles from these liquors.

Other examples of the use of trace additives in the control of crystal habit have been given by several authors. For instance, the production of large prism-shaped crystals of ammonium dihydrogen phosphate (ADP), used for their piezo-electric properties, is facilitated by incorporating traces of iron in the crystallizing solution[82]. A trace of ferric ion impurity in a slightly acid solution allows ammonium sulphate to be produced as stout pseudo-hexagonal prisms[83], a desirable form in this material for use as a fertilizer. The crystal habit of sodium chloride can be altered to a dendritic form by adding small quantities of sodium ferrocyanide to a vacuum crystallizer; the crystals have low bulk density ($< 0{\cdot}7$ g/cm^3) and are less prone to caking than the small cubic crystals of normal 'vacuum' salt[84].

The use of temperature and concentration control in habit modification is illustrated by work carried out by EDWARDS[85] on paraffin wax, where the type of habit formed is a very important factor in its efficient processing. When crystallization occurs at temperatures near the melting point of paraffin wax, or from concentrated solutions, needle crystals are usually formed. Thin plates grow when the crystallizing temperature is much lower than the melting point, or on crystallization from dilute solutions. The cooling rate was also shown to exert a considerable influence on the crystal habit.

SVANOE[86] discussed many of the factors affecting the habit of industrial crystals, i.e. the effects of supersaturation, the type of solvent, pH, the presence of inorganic and organic impurities and surface-active agents on the type of crystal produced. Several interesting photographs of the 'before and after' type are given showing crystals of K_2CO_3, KCl, $CuSO_4 \cdot 5H_2O$, $Na_2CO_3 \cdot 10H_2O$ and $(NH_4)_2SO_4$, produced under different conditions.

REFERENCES

1 Symposium on Nucleation Phenomena, *Industr. Engng Chem.* 44 (1952) 1269 (17 papers)
2 Symposium on Crystal Growth, *Disc. Faraday Soc.* No. 5, 1949 (44 papers)
3 *Growth and Perfection of Crystals* (Proceedings of an International Conference held in New York, 1958, 42 papers). R. H. Doremus, B. W. Roberts and D. Turnbull (Eds.). 1958. New York; Wiley
4 GIBBS, J. W., *Collected Works*, 1928. London; Longmans Green
5 OSTWALD, Wilhelm, Studien über die Bildung und Umwandlung fester Körper, *Z. phys. Chem.* 22 (1897) 289
6 VOLMER, M., Über Keimbildung und Keimwirkung als Spezialfälle der heterogenen Katalyse, *Z. Electrochem.* 35 (1929) 555
7 VOLMER, M., *Kinetik der Phasenbildung*, 1939. Dresden and Leipzig; Steinkopff
8 BECKER, R. VON and DÖRING, W., Kinetische Behandlung der Keimbildung in übersättigten Dämpfen, *Ann. Phys., Lpz.* 24 (1935) 719
9 BECKER, R. VON, Die Keimbildung bei der Ausscheidung in metallischen Mischkristallen, *Ann. Phys., Lpz.* 32 (1938) 128
10 LAMER, V. K., Nucleation in phase transitions, p. 1270 in reference 1
11 POUND, G. M., Liquid and crystal nucleations, p. 1278 in reference 1
12 FISHER, J. C., HOLLOMAN, J. H. and TURNBULL, D., Rate of nucleation of solid particles in a sub-cooled liquid, *Science* 109 (1949) 168
13 TAMMAN, G., *States of Aggregation*, transl. R. F. Mehl, 1925. New York; van Nostrand. 1926. London; Constable
14 VAN HOOK, A., Nucleation in supersaturated sucrose solutions, in *Principles of Sugar Technology*, Vol. 2, P. Honig (Ed.), 1959. Amsterdam; Elsevier
15 YOUNG, S. W., Mechanical stimulus to crystallization in supercooled liquids, *J. Amer. chem. Soc.* 33 (1911) 148
16 YOUNG, S. W. and VAN SICKLEN, W. J., Mechanical stimulus to crystallization, *J. Amer. chem. Soc.* 35 (1913) 1067
17 BERKELEY, The Earl of, Solubility and supersolubility from the osmotic standpoint, *Phil. Mag.* 24 (1912) 254
18 TIPSON, R. S., Crystallization and recrystallization, in *Techniques of Organic Chemistry*, Vol. 3, A. Weissberger (Ed.), 1950. New York; Interscience
19 TELKES, M., Nucleation of supersaturated inorganic salt solutions, p. 1308 in reference 1
20 PRECKSHOT, G. W. and BROWN, G. G., Nucleation of quiet supersaturated potassium chloride solutions, p. 1314 in reference 1
21 TURNBULL, D. and VONNEGUT, B., Nucleation catalysts, p. 1292 in reference 1
22 WELLS, A. F., Crystal growth, *Ann. Rep. Chem. Soc., Lond.* 43 (1946) 62
23 BUCKLEY, H. E., *Crystal Growth*, 1952. London; Chapman & Hall
24 CURIE, P., *Bull. Soc. franç. Mineral* 8 (1885) 145
25 WULFF, G., *Z. Kristallogr.* 34 (1901) 449
26 LAUE, M. VON, Der Wulffsche Satz für die Gleichgewichtsform von Kristallen, *Z. Kristallogr.* 105 (1943) 124
27 BRANDES, H., Zur Theorie des Kristallwachstums, *Z. phys. Chem.* 126 (1927) 196
28 STRANSKI, I. N., Zur Theorie des Kristallwachstums, *Z. phys. Chem.* 136 (1928) 259

[29] KOSSEL, W., Zur Energetik von Oberflächenvorgängen, *Ann. Phys., Lpz.* 21 (1934) 457
[30] VERMA, A. R., *Crystal Growth and Dislocations*, 1953. London; Butterworths
[31] READ, W. T., *Dislocations in Crystals*, 1953. New York; McGraw-Hill
[32] FRANK, F. C., The influence of dislocations on crystal growth, p. 48 in reference 2
[33] BURTON, W. K., CABRERA, N. and FRANK, F. C., The growth of crystals and the equilibrium structure of their surfaces, *Phil. Trans.* A243 (1951) 299
[34] NOYES, A. A. and WHITNEY, W. R., Rate of solution of solid substances in their own solution, *J. Amer. chem. Soc.* 19 (1897) 930 and *Z. phys. Chem.* 23 (1897) 689
[35] NERNST, W., Theorie der Reaktionsgeschwindigkeit in heterogenen Systemen, *Z. phys. Chem.* 47 (1904) 52
[36] MARC, R., Über die Kristallisation aus wässerigen Lösungen, *Z. phys. Chem.* 61 (1908) 385; 67 (1909) 470; 68 (1909) 104; 73 (1910) 685
[37] MIERS, H. A., The concentration of the solution in contact with a growing crystal, *Phil. Trans.* A202 (1904) 492
[38] BERTHOUD, A., Théorie de la formation des faces d'un crystal, *J. chim. Phys.* 10 (1912) 624
[39] VALETON, J. J. P., Wachstum und Auflösung der Kristalle, *Z. Kristallogr.* 59 (1923) 135, 335; 60 (1924) 1
[40] BUNN, C. W., Concentration gradients and the rates of growth of crystals, p. 132 in reference 2
[41] BERG, W. F., Crystal growth from solutions, *Proc. roy. Soc.* A164 (1938) 79
[42] WHITMAN, W. G., The two-film theory of absorption, *Chem. metall. Engng* 29 (1923) 146
[43] HANRATTY, T. J., Turbulent exchange of mass and momentum with a boundary, *Amer. Instn chem. Engrs J.* 2 (1956) 359
[44] HIGBIE, R., The rate of absorption of pure gas into a still liquid during short periods of exposure, *Trans. Amer. Instn chem. Engrs* 31 (1935) 365
[45] DANCKWERTS, P. V., Significance of liquid–film coefficients in gas absorption, *Industr. Engng Chem.* 43 (1951) 1460
[46] DANCKWERTS, P. V. and KENNEDY, A. M., Kinetics of liquid–film processes in gas absorption, *Trans. Instn chem. Engrs, Lond.* 32 (1954) S49
[47] GRUT, E., Nachprodukt-Maischenarbeit, *Zucker* 6 (1953) 411
[48] VAN HOOK, A., Kinetics of Crystallization (in reference 14); relevant papers on the same subject in *Industr. Engng Chem.* 36 (1944) 1042, 1048; 37 (1945) 782; 38 (1946) 50; 40 (1948) 85
[49] BENNETT, J. A. R. and LEWIS, J. B., Dissolution of solids in mercury and aqueous liquids: development of a new type of rotating dissolution cell, *Amer. Instn chem. Engrs J.* 4 (1958) 418
[50] ARRHENIUS, S., Über die Reaktionsgeschwindigkeit bei der Inversion von Rohrzucker durch Säuren, *Z. phys. Chem.* 4 (1889) 226
[51] ENGLISH, A. C. and DOLE, M., Diffusion of sucrose [and glucose] in supersaturated solutions, *J. Amer. chem. Soc.* 72 (1950) 3261; 75 (1953) 3900
[52] DEDEK, J. (reported in reference 48)
[53] RUMFORD, F. and BAIN, J., The controlled crystallization of sodium chloride, *Trans. Instn chem. Engrs, Lond.* 38 (1960) 10
[54] COOKE, E. G. (in discussion to reference 53)
[55] COULSON, J. M. and RICHARDSON, J. F., *Chemical Engineering* Vol. 2, p. 817, 1955. London; Pergamon Press
[56] MCCABE, W. L. and STEVENS, R. P., Rate of growth of crystals in aqueous solution, *Chem. Engng Progr.* 47 (1951) 168
[57] HIXSON, A. W. and CROWELL, J. H., Dependence of reaction velocity upon surface and agitation, *Industr. Engng Chem.* 23 (1931) 923
[58] HIXSON, A. W. and WILKENS, G. A., Performance of agitators in liquid–solid chemical systems, *Industr. Engng Chem.* 25 (1933) 1196; 34 (1942) 120, 194
[59] HIXSON, A. W. and BAUM, S. J., Agitation: heat and mass transfer coefficients in liquid–solid systems, *Industr. Engng Chem.* 33 (1941) 478, 1433
[60] HIXSON, A. W. and KNOX, K. L., Effect of agitation on the rate of growth of single crystals, *Industr. Engng Chem.* 43 (1951) 2144

[61] WILHELM, R. H., CONKLIN, L. H. and SAUER, T. C., Rate of solution of crystals, *Industr. Engng Chem.* 33 (1941) 453
[62] WAGNER, C., The dissolution rate of sodium chloride with diffusion and natural convection as rate determining factors, *J. phys. Chem.* 53 (1949) 1030
[63] AMELINCKX, S., Phénomènes de diffusion pendant la croissance et la dissolution de cristaux dans une solution, *J. chim. Phys.* 47 (1950) 208
[64] LINTON, W. H. and SHERWOOD, T. K., Mass transfer from solid shapes to water in streamline and turbulent flow, *Chem. Engng Progr.* 46 (1950) 258
[65] BIRCUMSHAW, L. L. and RIDDIFORD, A. C., Transport control in heterogeneous reactions, *Quart. Rev. Chem. Soc., Lond.* 6 (1952) 157
[66] RIDDIFORD, A. C., The temperature coefficient of heterogeneous reactions, *J. phys. Chem.* 56 (1952) 745
[67] DAVION, M., Étude sur la vitesse de dissolution des sels cristallisés, *Ann. Chim.* 12 (8) (1953) 259
[68] JOHNSON, A. I. and HUANG, C. J., Mass transfer studies in an agitated vessel, *Amer. Instn chem. Engrs J.* 2 (1956) 412
[69] HUMPHREY, D. W. and VAN NESS, H. C., Mass transfer in a continuous-flow mixing vessel, *Amer. Instn chem. Engrs J.* 3 (1957) 283
[70] GARNER, F. H. and SUCKLING, R. D., Mass transfer from a soluble solid sphere, *Chem. Engng Sci.* 9 (1958) 119
[71] CHILTON, T. H. and COLBURN, A. P., Mass transfer (absorption) coefficients—predictions from data on heat transfer and fluid friction, *Industr. Engng Chem.* 26 (1934) 1183
[72] FORSYTH, J. S. and WOOD, J. T., The separation of organic mixtures by crystallization from the melt, *Trans. Instn chem. Engrs, Lond.* 33 (1955) 122
[73] GILLILAND, E. R. and SHERWOOD, T. K., Diffusion of vapours into air streams, *Industr. Engng Chem.* 26 (1934) 516
[74] AMELINCKX, S., Sur la vitesse de croissance des faces d'un cristal et la loi de Bravais–Donnay–Harker, *J. chim. Phys.* 47 (1950) 213
[75] CARTIER, R., PINDZOLA, D. and BRUINS, P. F., Particle integration rate in crystal growth, *Industr. Engng Chem.* 51 (1959) 1409
[76] JENKINS, J. D., The effect of various factors upon the velocity of crystallization of substances from solution, *J. Amer. chem. Soc.* 47 (1925) 903
[77] HUNGERFORD, E. H. and NEES, A. R., Raffinose: preparation and properties, *Industr. Engng Chem.* 26 (1934) 462
[78] WHETSTONE, J., The crystal habit modification of inorganic salts with dyes, *Trans. Faraday Soc.* 51 (1955) 973, 1142
[79] BUTCHART, A. and WHETSTONE, J., The effect of dyes on the crystal habits of some oxy-salts, p. 254 in reference 2
[80] MICHAELS, A. S. and COLVILLE, A. R., The effect of surface active agents on crystal growth rate and crystal habit, *J. phys. Chem.* 64 (1960) 13
[81] GARRETT, D. E., Industrial crystallization: influence of chemical environment, *Brit. chem. Engng* 4 (1959) 673
[82] JAFFE, H. and KJELLGREN, B. R. F., Controlled growth inhibition in large-scale crystal growth, p. 319 in reference 2
[83] ADAM, W. G., MURDOCH, D. G. and The Gas Light and Coke Co., Ltd., *Brit. Pat.* 330, 945 and 330, 947 (1929)
[84] MAY, W. E., SCOTT, T. R. and Imperial Chemicals Industries, Ltd., *Brit. Pat.* 667, 101 (1949) and *U.S. Pat.* 2, 642, 335 (1953)
[85] EDWARDS, R. T. Crystal habit of paraffin wax, *Industr. Engng Chem.* 49 (1957) 750
[86] SVANOE, H., Solids recovery by crystallization, *Chem. Engng Progr.* 55 (5) (1959) 47

6

RECRYSTALLIZATION

A SINGLE crystallization operation performed on a solution or melt often fails to produce a pure crystalline product. It may be that the impurity has solubility characteristics similar to those of the desired pure component, and both substances are co-precipitated. On the other hand, the impurity may be present in large amounts and the deposited crystals become contaminated. A pure substance cannot be produced from a melt in one crystallization stage if the impurity and the required substance form a solid solution. The word 'pure' is used rather indiscriminately; no substance can be prepared absolutely pure, so the meaning of the word is somewhat arbitrary. For example, a chemist who requires a sample of a compound for the determination of its physical characteristics may demand a purity of 99·999 per cent, while an industrial chemist may reckon that he had produced a pure chemical, i.e. pure enough for his purpose, if a purity of 99 per cent had been achieved.

Recrystallization, i.e. repeated crystallization steps, from solutions or melts is an extremely useful and widely employed separation technique; if the operation is repeated often enough, very high degrees of purity can be attained. Recrystallization often offers several advantages over the processes of extraction and distillation: a less complicated apparatus is usually required, and one crystallization step can, in many cases, effect a much larger removal of impurity than one extraction or distillation step. Unfortunately, the inclusion of impure mother liquor by a crystalline mass can reduce the efficiency of a crystallization process quite considerably. Many systems, such as azeotropes, close-boiling mixtures and heat-sensitive products which cannot be separated by distillation, can often be processed with ease by crystallization. Another point to be borne in mind is that less heat energy is required for crystallization than for distillation because latent heats of fusion, solution and crystallization are very much lower than latent heats of vaporization.

Recrystallization from Solutions

It is often possible to remove the impurities from a crystalline mass by dissolving the crystals in a small amount of fresh hot solvent and cooling the solution to produce a fresh crop of purer crystals, provided the impurities

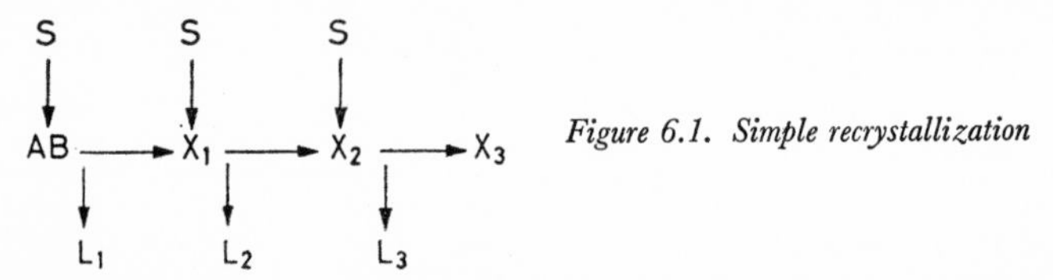

Figure 6.1. Simple recrystallization

are more soluble in the solvent than is the main product. This step may have to be repeated many times before a yield of crystals of the desired purity is obtained. Such an operation is called a simple recrystallization. A typical scheme is shown in *Figure 6.1.* An impure crystalline mass *AB*

(A is the less soluble pure component, B the more soluble impurity) is dissolved in the minimum amount of hot solvent S and then cooled. The crop of crystals X_1 will contain less impurity B than the original mixture, but if the desired degree of purity has not been achieved the procedure can be repeated; crystals X_1 are dissolved in more fresh solvent S and recrystallized to give a crop X_2, and so on.

In a sequence of operations of the above kind the losses of the desired component A can be considerable, and the final amount of 'pure' crystals may easily be a minute fraction of the starting mixture AB. This question of yield from recrystallization processes is of paramount importance, and many schemes have been designed with the object of increasing both yield and separation efficiency. The choice of solvent depends on the nature of the required substance A and the impurity B. Ideally, B should be very soluble in the solvent at the lowest temperature employed, and A should have a high temperature coefficient of solubility so that high yields of A can be obtained while operating within a small temperature range. Some of the factors affecting the choice of a solvent are discussed in Chapter 2.

Figure 6.2 indicates a modification of the simple recrystallization scheme; additions of the original impure mixture AB are made to the system at

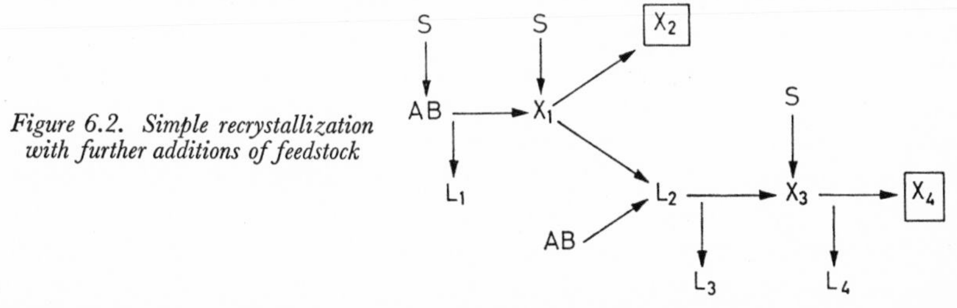

Figure 6.2. Simple recrystallization with further additions of feedstock

certain intervals. The first two stages are identical with those illustrated in *Figure 6.1*, but the crystals X_2 are set aside and fresh feedstock AB dissolved by heating in mother liquor L_2. On cooling, a crop of crystals X_3 are obtained and a mother liquor L_3 which is discarded. Crystals X_3 are recrystallized from fresh solvent to give a crop X_4 which is set aside; the mother liquor L_4 can, if required, be used as a solvent from which more feedstock can be crystallized. If the crops X_2, X_4, X_6, etc. are still not pure enough they can be bulked together and recrystallized from fresh solvent.

The triangular fractional recrystallization scheme shown in *Figure 6.3* makes better use of the successive mother liquor fractions than that of

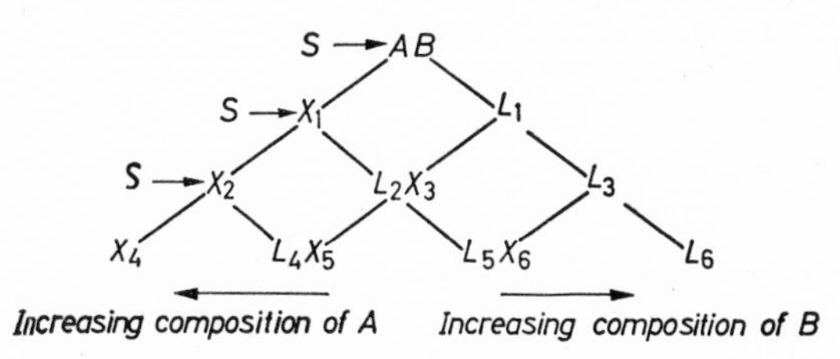

Figure 6.3. Fractional recrystallization of a solution

Figure 6.2. Again, A and B are taken to be the less and more soluble constituents, respectively. The mixture AB is dissolved in the minimum amount of hot solvent S, and then cooled. The crop of crystals X_1 which is deposited is separated from the mother liquor L_1 and then dissolved in fresh solvent. The cooling and separating operations are repeated, giving a further crop of crystals X_2 and a mother liquor L_2. The first mother liquor L_1 is concentrated to yield a crop of crystals X_3 and a mother liquor L_3. Crystals X_3 are dissolved by warming in mother liquor L_2 and then cooled to yield crystals X_5. Crystals X_2 are dissolved in fresh hot solvent and cooled to yield crystals X_4 and liquor L_4. Liquor L_3 is concentrated giving crystals X_6 and liquor L_6. The scheme can be continued until the required degree of separation is effected. The less soluble substance A is concentrated in the fractions on the left-hand side of the diagram, the more soluble constituent B on the right-hand side. If any other substances with intermediate solubilities were present, they would be concentrated in the fractions in the centre of the diagram.

If the starting material contained a unit quantity of component A, and each crystallization step resulted in the deposition of a proportion P of this component, the proportions of A which would appear at any given point in

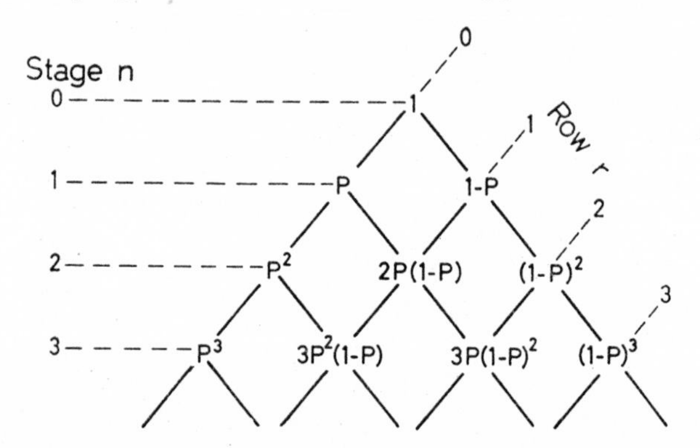

Figure 6.4. Analysis of the triangular fractional crystallization scheme

the triangular scheme (see *Figure 6.4*) would be given by a term in the binomial expansion

$$[P + (1 - P)]^n = 1 \tag{1}$$

$$p_{r,n} = \frac{n!}{r!\,(n-r)!} \cdot P^{n-r}(1 - P)^r \tag{2}$$

where $p_{r,n}$ = proportion of A at a point represented by row r and stage n. Thus, for example

$$p_{2,3} = \frac{3!}{2!} \cdot P^{(3-2)}(1 - P)^2$$

$$= 3P(1 - P)^2$$

A modification of the triangular scheme is shown in *Figure 6.5*. In this case, further quantities of the feedstock AB are added to the system by dissolving it in successive mother liquors on the right-hand side of the diagram.

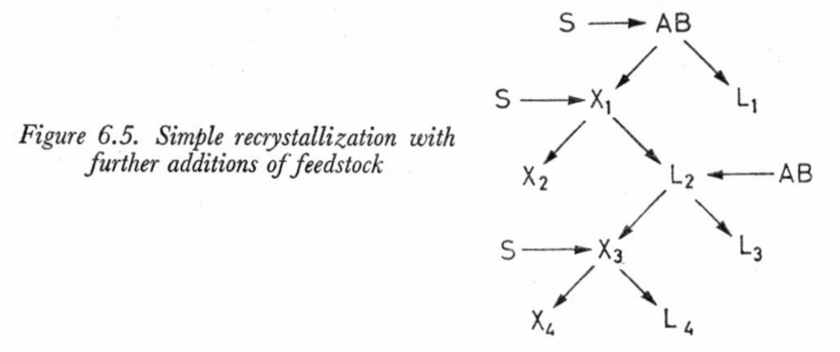

Figure 6.5. Simple recrystallization with further additions of feedstock

This scheme is particularly useful if component A has a high temperature coefficient of solubility.

Several other much more complex schemes for fractional recrystallization can be used, their aim being to increase the yield of the desired constituent by further re-use of the mother liquors. A detailed account of these methods has been given by TIPSON[1, 2]; *Figure 6.6* illustrates two of them. In the 'diamond' scheme (*Figure 6.6a*) the outermost fractions are set aside when they have reached a pre-determined degree of purity, and fractionation is

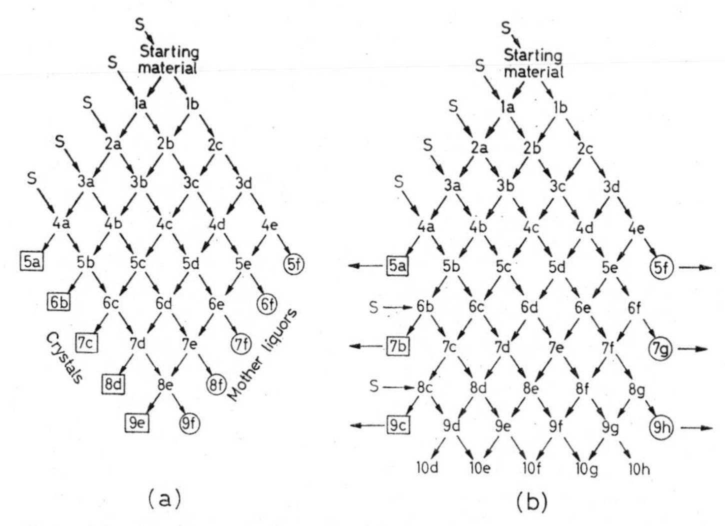

Figure 6.6. Fractional recrystallization schemes: (a) diamond; (b) double withdrawal. (After R. S. TIPSON[1,2])

continued until all the material is obtained either in a crystalline form or in solution in a final mother liquor. If necessary, various crystal fractions can be bulked together and recrystallized, and in a similar manner the mother liquors can receive further treatment.

Unfortunately, the various fractions obtained by the above method will differ in composition, and relatively pure crystals will be mixed with relatively impure ones. This difficulty can be overcome by the use of the 'double-withdrawal' scheme shown in *Figure 6.6b*. The procedure is the same as that used in the diamond scheme up to the point where no further fresh solvent is used (line 5 in *Figure 6.6a*). At this point it is assumed that crystals 5*a* and mother liquor 5*f* have reached the desired degree of purity, and they are both set aside from the system. Fresh solvent is then added to the crystal crop 6*b* to yield a purer crop 7*b* which is arranged to have a purity similar to that of crop 5*a*. Crop 7*b* is set aside, crop 8*c* crystallized from pure solvent, and so on.

Theoretical analyses and surveys of the factors affecting the choice of many fractional crystallization schemes have been made by several authors[1, 3, 4]. GARBER and GOODMAN[5] presented a rigorous mathematical analysis of the fractional crystallization of multicomponent systems. JOY and PAYNE[6] investigated the fractional crystallization of similar substances when one substance is present in very small amounts, e.g. radium–barium chromate mixtures. The separation efficiency for a particular fractionation step was considered to be given by the ratio of the entropy change for the step to the maximum obtainable entropy change for the system. For many precipitation and crystallization operations the optimum cut at each step for maximum entropy of fractionation was considered to be about 0·5.

The concentration of heat-sensitive substances in aqueous solution by freezing, followed by removal of the ice, has been employed for a considerable number of years. OLIVE[7] discussed an industrial approach to this technique whereby the growth of ice crystals, e.g. in citrus fruit juices, is controlled in countercurrent stages. In this manner, the largest crystals are in contact with the least viscous and most dilute juice, and a clean separation of ice with a minimum loss of valuable product can be effected. GANE[8] has also described the freeze concentration of fruit juices, and his results show that this method produces a concentrate with a very high vitamin C content.

Recrystallization from Melts

The fractional recrystallization of melts can be carried out by schemes similar to those described for the fractional recrystallization of solutions (*Figures 6.1–6.6*). Of course, in most cases no solvent need be added to a molten system; the usual procedure consists simply of sequences of melting, partial freezing and separation. Selected fractions may be mixed at intervals according to the type of scheme employed, and fresh additions of feedstock may be introduced at certain stages if necessary.

Solid Solutions

When two substances constitute a system which forms a simple series of solid solutions, it should be possible to separate them almost completely by repeated melting, cooling and separating. The process is similar in many respects to fractional distillation. *Figure 6.7* shows the phase diagram for a solid solution of two components, A and B, which do not form a eutectic, or maximum or minimum melting point mixtures. In this particular example the melting point of A is lower than that of B. The region above the liquidus represents the homogeneous, that below the solidus the solid phase. Liquid and solid co-exist in equilibrium in the region between the liquidus and solidus, and it is within this region that phase separation is effected. The recrystallization process can be illustrated as follows. If the solid solution is first melted by heating to some temperature T_1, giving a liquid of composition L_1, then cooled to some temperature T_2 within the region bounded by the liquidus and solidus curves, it will partially solidify giving a liquid of composition L_2 and a solid (solid solution) of composition S_2. If this solid S_2 is separated from the liquid, melted by heating to some temperature T_3

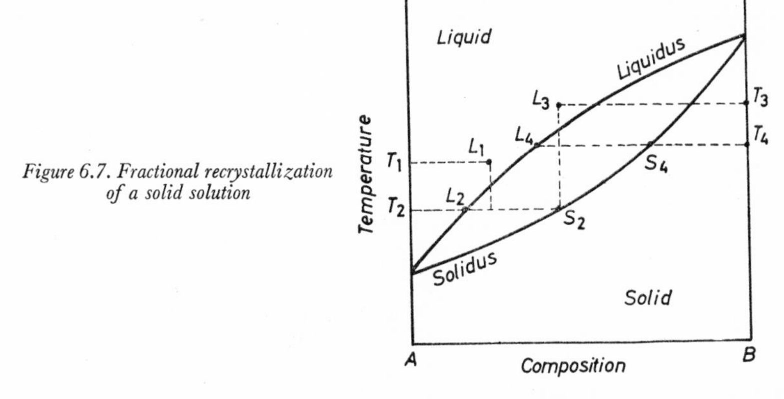

Figure 6.7. Fractional recrystallization of a solid solution

and then cooled to some temperature T_4, a liquid L_4 and solid S_4 will be obtained. It can be seen, therefore, that the solid fractions become richer and richer in component B, and a repetition of this process will result in an almost complete separation of the two components.

This simple recrystallization method is also applicable to solid solutions containing more than two components, but many recrystallization steps will be required to effect any appreciable degree of separation. When two substances form a solid solution with a maximum or minimum melting point (see Chapter 4), a complete separation of both components by fractional recrystallization will be impossible; these cases are analogous to the formation of azeotropes in distillation processes.

FORSYTH and WOOD[9] investigated the possibilities of the separation of organic mixtures by crystallization from the melt. A molten binary organic system, e.g. naphthalene–β-naphthol, was pumped in turbulent flow down the inside of a cooled tube. The solid which was deposited on the wall differed in composition from the melt which continued down the tube, and such an arrangement acted as a combined crystallization and phase separation unit in a stagewise countercurrent cascade. For systems which formed solid solutions, stage efficiencies of over 50 per cent at a rate of deposition of solid of 10 lb./h · ft.^{2} were reported.

Impure systems which do not form solid solutions can be purified by partial melting or partial freezing, followed by separation of the liquid and solid phases. Liquids, too, can be submitted to these operations. Commercial grade benzene, for example, can be partially frozen or completely frozen and then partially melted, depending on the nature of the impurity which has to be removed. ASTON and MASTRANGELO[10] discussed the method of fractional melting as an alternative to fractional freezing, or crystallization from solution, for certain organic compounds. They described an apparatus, later improved[11], for the stepwise melting of many impure hydrocarbons, including cyclohexane, iso-octane, n-heptane and *cis*-2-butene.

Eutectic Systems

As in the case of solid solutions which form maximum and minimum melting mixtures, the components of a eutectic system cannot be isolated in the pure state by normal fractional crystallization techniques. The best that can be done in the case of a binary eutectic system, for example, is the isolation of one pure component and a eutectic mixture containing a fixed proportion of both components. The phase reactions that can occur in eutectic systems have been described in Chapter 4. *Figure 6.8* shows the

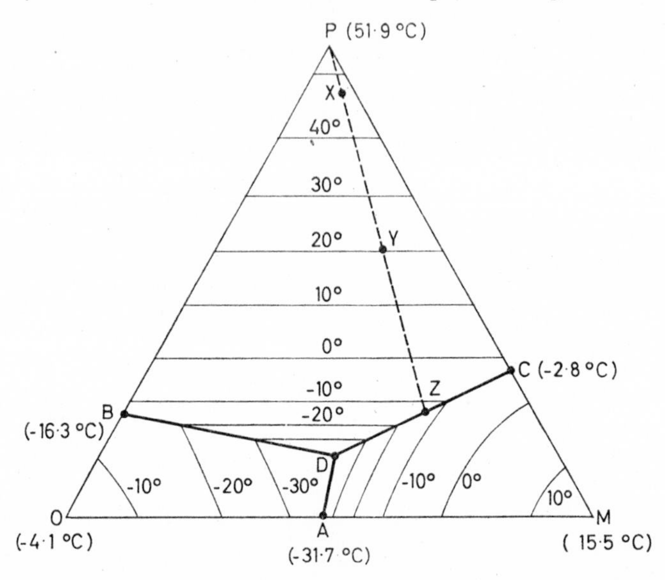

Figure 6.8. Phase diagram for the ternary system o-, m- and p-nitrotoluene (point D = − 40·0° C)

phase equilibria for the ternary system, *ortho*-, *meta*- and *para*-nitrotoluene; the three pure components are represented by the letters *O*, *M* and *P*, respectively, at the apexes of the triangle. Four different eutectics can exist in this system, three binaries and one ternary:

Eutectic points		*Per cent by weight*			*Temperature* °C
Symbol	*Components*	*O*	*M*	*P*	
A	*O–M*	52	48	—	− 31·7
B	*O–P*	76	—	24	− 16·3
C	*M–P*	—	67	33	− 2·8
D	*O–M–P*	42	44	14	− 40·0

COULSON and WARNER[12] used a diagram similar to *Figure 6.8* to demonstrate the analysis of a crystallization problem encountered in the commercial manufacture of mononitrotoluene; the following example is taken from this publication.

Example—The bottom product from a continuous distillation column has the composition 3·0 per cent *ortho*-, 8·5 per cent *meta*- and 88·5 per cent *para*-nitrotoluene. Suggest the operating conditions of a cooling crystallizer to recover pure *para*-nitrotoluene from this material and estimate the yield.

Solution—Point X in *Figure 6.8* represents the composition of feedstock; it is located in between the 40° and 50° C isotherms. By interpolation, the temperature at which this system starts to freeze can be estimated as about 46° C. As point X lies in the region $PBDC$, pure *para*- will crystallize out once the temperature falls below 46° C, and the composition of the mother liquor will follow line XYZ (i.e. away from point P) as cooling proceeds. At point Z (about − 15° C) on curve DC, *meta*- also starts crystallizing out. It is not necessary, however, to cool to near − 15° C in order to get a high yield of *para*-, as *Table 6.1*, based on 100 lb. of feedstock, shows.

Table 6.1. Yield of para-Nitrotoluene from an O–P–M Mixture by Cooling Crystallization
(*After* COULSON and WARNER[12])

Temperature ° C	*Para-deposited* (lb.)	*Mother liquor* (lb.)	*Composition of mother liquor (per cent by weight)*		
			O	*M*	*P*
46	0	100	3·0	8·5	88·5
40	39·6	60·4	5·0	14·0	81·0
30	66·7	33·3	8·7	24·8	66·5
20	75·0	25·0	12·0	34·0	54·0
10	79·6	20·4	14·7	41·6	43·7
0	82·3	17·7	16·7	48·0	35·3
− 10	84·8	15·2	17·0	53·9	27·1

The above mother liquor compositions are read off *Figure 6.8* at the point which the line $P \rightarrow Z$ cuts the particular isotherm. The total weight of *para*- crystallized out is calculated by the mixture rule. For example, for 100 lb. of original mixture X at 20° C (point Y)

$$\textit{para-}\text{ deposited} = 100\left(\frac{\text{distance } XY}{\text{distance } PY}\right) = 75 \text{ lb.}$$

In commercial practice 20° C is a reasonable temperature which can be attained by the use of normal cooling water. If this is adopted as the operating temperature, the yield of pure *para*- crystals per 100 lb. of feedstock is 75 lb., equivalent to a recovery of about 85 per cent.

Countercurrent Fractional Crystallization

FINDLAY and WEEDMAN[13] described the first successful application of the technique of fractionation to a crystallization process. Basically, the operation comprises a simple cooling crystallization carried out in a column where the pure and impure streams flow countercurrently, just as the reflux flows countercurrently to vapours in a distillation column. The method is best described with reference to an actual process, the separation of *p*-xylene from a mixture of xylenes, developed by the Phillips Petroleum Co. in the United States. Distillation cannot be used for the separation of the xylenes because the atmospheric boiling points of the *ortho*-, *meta*- and *para*-isomers are 282·4, 291·9 and 281·0° F, respectively. The corresponding freezing points, however, are − 54·2, − 13·3 and + 55·9° F, so crystallization is the logical method of separation to employ.

A schematic diagram of the Phillips process for the production of *p*-xylene is shown in *Figure 6.9*. The feedstock containing about 15–20 per cent of

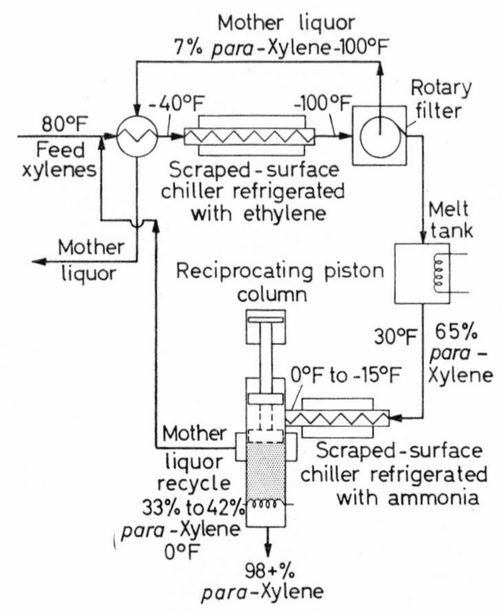

Figure 6.9. Continuous countercurrent fractional crystallization—the Phillips p-xylene process. (*From* R. A. Findlay *and* J. A. Weedman[13], *by courtesy of* Interscience)

p-xylene is pre-cooled to about − 40° F before being passed to a scraped-surface chiller (see Chapter 8) cooled with ethylene. The emergent stream, now containing a mass of tiny crystals of *p*-xylene, passes at about − 100° F to a rotary filter where a filter cake containing about 65 per cent *p*-xylene is recovered. The mother liquor containing about 7 per cent *p*-xylene is rejected after being used as a medium for pre-cooling the incoming feedstock.

The filter cake, of high *p*-xylene content, is melted and passed to a second scraper-chiller, this time cooled with ammonia, from which it emerges as a crystalline slurry at about 0 to − 15° F; the crystals, of course, are minute. The slurry is then pulsed through a column containing a filter at the top end, which removes the mother liquor (33–42 per cent *p*-xylene), and a melting section at the lower end. The column itself is quite small, rarely exceeding

3 ft. in height, and its diameter is very small compared with its capacity; production rates from 50–150 gal./h · ft.^{2} have been reported. When the high-purity crystals (> 98 per cent *p*-xylene) fall to the bottom of the tower, they are melted; part of the melt is removed from the system as the finished product, while the rest is forced back up the column countercurrently to the falling crystals. The countercurrent contacting of the warm, pure melt with the cold, impure crystals results in the partial freezing of the melt and the melting of the crystals. All the *p*-xylene in the upward flowing stream is re-frozen and the crystals are returned to the melting section. The overall result is the almost perfect separation of crystals and mother liquor. Careful control of the temperature gradient across the column (– 100 to 56° F) can produce a product containing 99·5 per cent *p*-xylene.

This crystallization technique could well be applied to many other commercial separations, and with this in mind McKay *et al.*[14] have developed the laboratory-scale model of a column crystallizer which can be used for

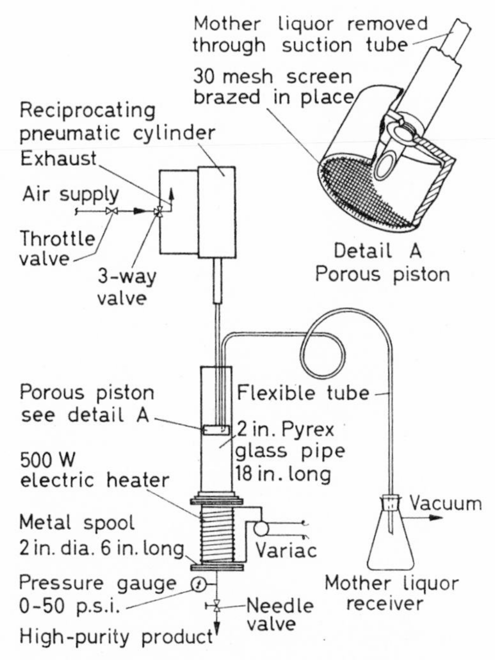

Figure 6.10. Laboratory-scale column crystallization. (*From* D. L. McKay, G. H. Dale *and* J. A. Weedman[14])
By courtesy of *Industrial Engineering Chemistry*

test runs on only a few gallons of feedstock. The equipment, shown in *Figure 6.10*, consists of a column, a reciprocating porous piston, a vacuum source to remove mother liquor, and a Variac to control the heat input. The 2 in. diam. borosilicate glass column must have perfectly smooth inner walls; no surface scratches can be permitted. The piston is faced with a 30-mesh screen to permit the passage of liquid as the piston pushes crystals down the column; the liquid is removed from the column through a suction tube. The clearance between column wall and piston should be < $\frac{1}{16}$ in. to prevent excessive quantities of crystals by-passing. No insulation is considered necessary for products melting in the range 20–150° F.

Before using the column on the desired feedstock it is recommended that a standardization test be made with the system n-heptane–benzene. A test mixture consisting of 70 per cent by weight of benzene is used, cooled by addition of solid CO_2 until it becomes a 'stiff' slurry at about − 20° F. The slurry is scooped into the column, the piston moved downwards. As the slurry is compressed, the mother liquor passes through the screen. This procedure is repeated until the whole working space in the column is filled with a firm crystal bed. The heat input to the base of the column is then adjusted so that the bed descends at a rate of about 1 in./min. When this condition is obtained, the bottom outlet (for high-purity liquor) is partially closed so that some of the liquid is forced back up the column through the crystal bed, thus effecting a purifying action.

A calibration test is necessary to fix the operating conditions of the column, because its performance cannot be predicted from the physical characteristics of the system. When the column is operating properly, the results on the benzene–n-heptane system will be as shown in *Table 6.2*. Apart from the

Table 6.2. Purification Tests for n-*Heptane–Benzene in a 2-inch Diameter Crystallization Column to Develop Standard Operating Procedures*

(*After* McKay *et al.*[14])

Drive cylinder pressure 50–100 p.s.i.
Column base pressure 2–10 p.s.i.
Piston stroke length 6–8 in.

Run	Feed		Mother liquor		Product		Heat input W
	Solids wt. per cent	*Benzene* wt. per cent	*Benzene* wt. per cent	*Rate* g/h	*Benzene* wt. per cent	*Rate* g/h	
1	26	76	67	1·6	> 99	0·6	80
2	18	69	63	4·2	99	0·9	120

large-scale commercial application to the *meta-para*-xylene system described above, this type of column crystallizer has been successfully applied to operations such as concentration of orange juice, desalting of saline water and separation of cyclohexane–methyl cyclopentane mixtures.

Zone Melting

For the removal of the last traces of impurity from a substance fractional crystallization from the melt cannot be applied with any degree of success. Apart from the fact that an almost infinite number of recrystallization steps would be necessary, there would only be a minute quantity of the pure substance left at the end of the process. High degrees of purification combined with a high yield of purified material can, however, be obtained by the technique known as zone melting or zone refining, originally developed by PFANN[15, 16] for the purification of germanium for use in transistors. Purification by zone refining can be effected when a concentration difference exists between the liquid and solid phases which are in contact during the melting or solidification of a solid solution. The method is best explained by means of a phase diagram.

Figure 6.11a shows a phase diagram for two substances A and B which form a simple solid solution; in this case the melting point of A is higher than that of B. When a homogeneous melt M of composition x is cooled, solid material will first be deposited at some temperature T. The composition of this solid, which is in equilibrium with the melt of composition x (point L on the

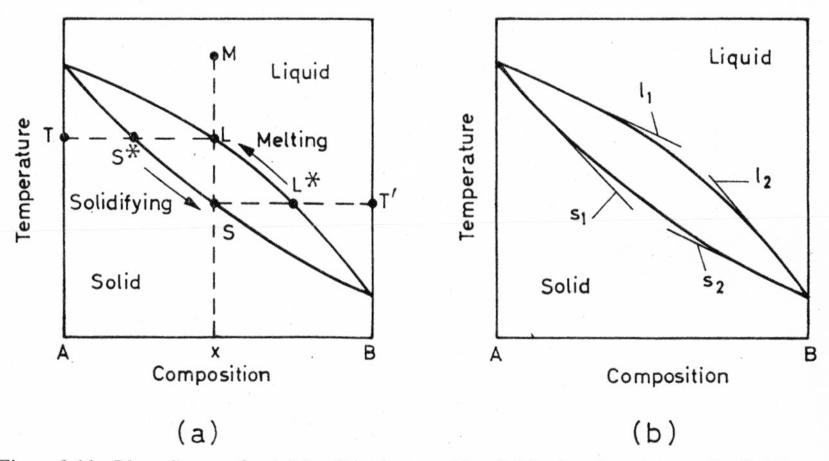

Figure 6.11. Phase diagram for simple solid solution: (a) solidification of a mixture; (b) liquidus and solidus drawn as straight lines in regions of near-pure A and B

liquidus), is given by point S^* on the solidus. As cooling proceeds, more solid is deposited, and the concentration of B in the solid increases towards S. A similar reasoning may be applied to the reversed melting procedure starting from temperature T' where a liquid of composition L^* is in equilibrium with a solid S of composition x. The solidification of a homogeneous molten mixture M yields a solid with an overall composition x, but owing to segregation the solid mixture is not homogeneous. Adjustments in the composition of the successive depositions of solid matter will not take place because of the very slow rate of diffusion in the solid state. The further apart the liquidus and solidus lines are, the greater will be the difference in concentration between the deposited solid and residual melt.

A measure of the expected efficiency of separation is given by a factor known as the segregation coefficient k. The significance of this coefficient can be seen in *Figure 6.11b* where the liquidus and solidus for a binary solid solution in the regions of low B and low A concentrations are represented by straight lines. As zone melting is only useful for the refining of substances with low impurity contents, these are the regions which are of interest. For dilute solutions the segregation coefficient k is defined by

$$k = \frac{\text{concentration of impurity in solid}}{\text{concentration of impurity in liquid}} \text{ (at equilibrium)}$$

$$= \frac{\text{slope of liquidus}}{\text{slope of solidus}} = \frac{l}{s} \qquad (3)$$

In *Figure 6.11b* it can be seen that $k_1 = l_1/s_1$ and $k_2 = l_2/s_2$, and also that $s_1 > l_1$ and $s_2 < l_2$. In general, therefore, it may be said that

$k < 1$ when the impurity lowers the melting point

and

$k > 1$ when the impurity raises the melting point.

When $k < 1$ the impurity will concentrate in the melt, when $k > 1$ in the solidifying mass. The nearer the value of k approaches unity, i.e. the closer the liquidus and solidus approach, the more difficult the segregation becomes. When $k = 1$, no zone refining is possible.

Concentration Profile

The impurity concentration at any point along a solid bar or ingot which was originally molten and then progressively cooled and solidified along its

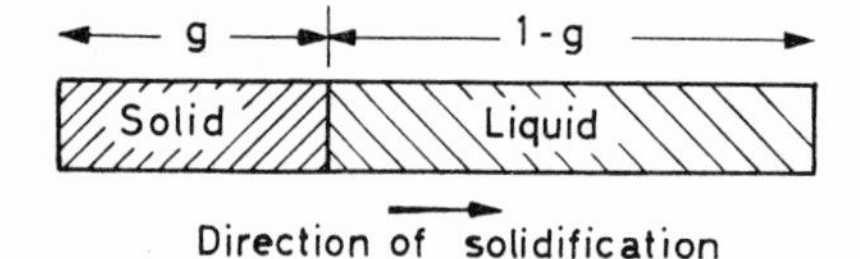

Figure 6.12. Molten bar of unit length undergoing progressive directional solidification

length (*Figure 6.12*) can be expressed in terms of the segregation coefficient. The concentration C of impurity (solute) in the solid at the solid–liquid interface is given by

$$C = kC_L \qquad (4)$$

where C_L is the concentration of impurity in the liquid phase at the interface. At any distance g along the bar of unit length, the concentration C at the solid–liquid interface is given by the equation

$$C = kC_0(1 - g)^{k-1} \qquad (5)$$

where C_0 is the initial concentration of the impurity in the homogeneous molten bar.

The distribution of impurity along a bar subjected to this process of *directional solidification* can be calculated from equation 5, and *Figure 6.13* indicates the various distributions that would be expected for different values of k. The greater the deviation of k from unity, the greater the concentration gradient along the bar.

Zone Refining

Directional solidification is not readily adaptable for purification purposes because the impurity content varies considerably along the bar at the end

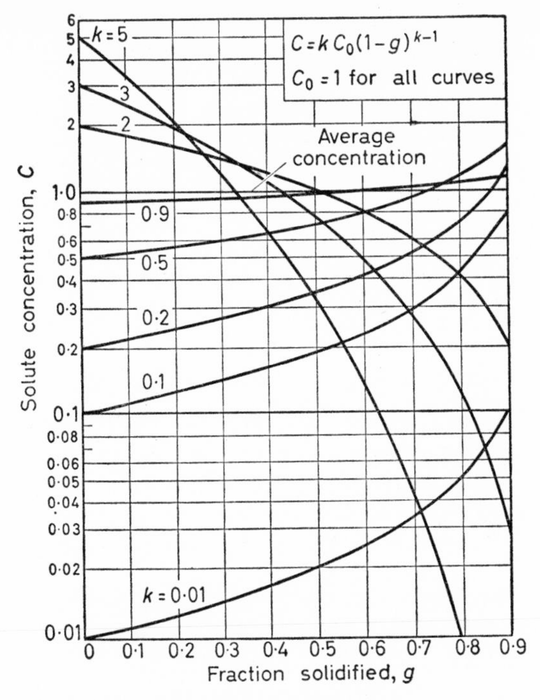

Figure 6.13. Pfann's curve for normal freezing. (*From* W. G. PFANN[13], *by courtesy of* American Institute of Mining and Metallurgical Engineers)

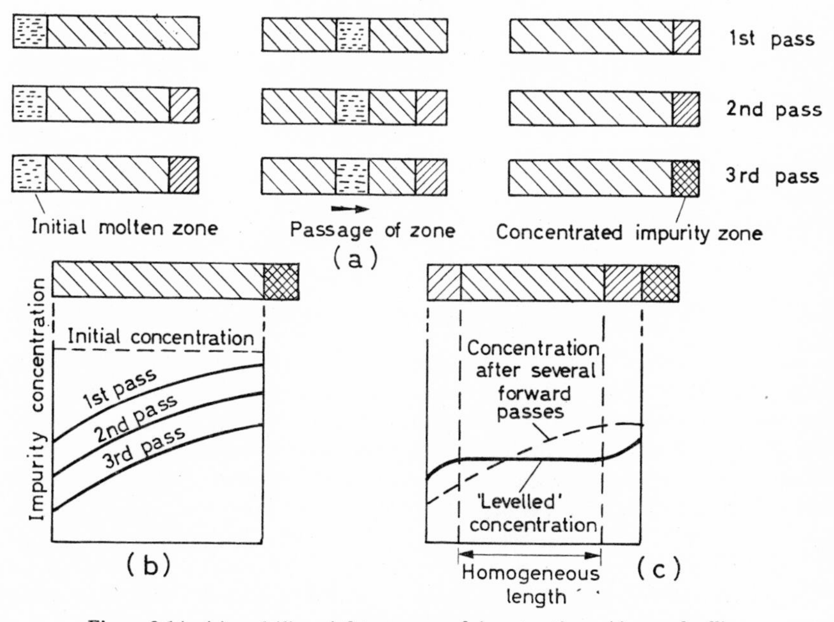

Figure 6.14. (a) and (b): A 3-pass zone-refining operation; (c) zone levelling

of the operation. The end portion of the bar, where the impurity concentration is highest, could be rejected and the process repeated after a remelting operation, but this would be extremely wasteful. However, the technique known as zone refining lends itself to repetitive purification without undue wastage. In this process a short molten zone is passed along the solid bar of material to be purified. If $k < 1$, the impurities pass into the melt and concentrate at the end of the bar. If $k > 1$, the impurities concentrate in the solid; so in this case a cooled solid zone could be passed through the melt.

The sequence of operations for molten zone refining is shown in *Figures 6.14a* and *b*. Although the impurity concentration in the bar, after the rejection of the high-impurity end portion, is much lower than initially, there is a considerable concentration gradient along the bar. In order to make the impurity concentration uniform along the bar the process known as *zone levelling* must be employed: further zoning is carried out in alternate directions until a homogeneous mid-section of the bar is obtained (*Figure 6.14c*). Both the high- and low-impurity ends of the bar are discarded.

Zone Refining Methods

A few of the basic arrangements used in zone refining[17] are shown in *Figure 6.15. Figure 6.15a* shows the material contained in a horizontal

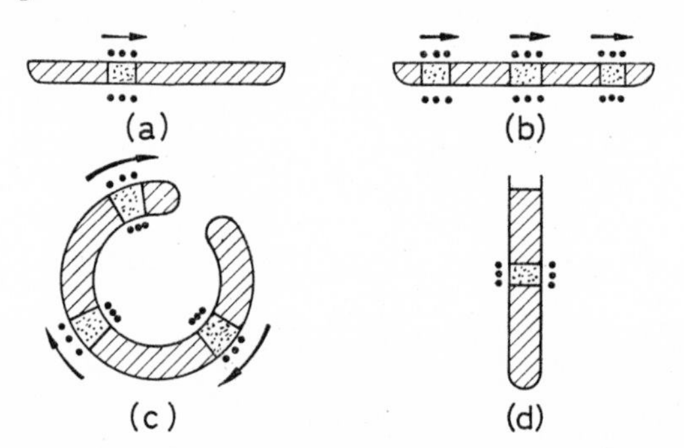

Figure 6.15. Simple methods of zone refining. (*After* N. L. PARR[17])
(*a*) *horizontal, single pass;* (*b*) *horizontal concurrent passes;* (*c*) *continuous concurrent passes in 'broken ring' crucible;* (*d*) *vertical tube, upward or downward passes*

crucible along which a heater is passed; or the crucible may be pulled through a stationary heater, and the molten zone travels through the solid. Several zones may be passed simultaneously at fixed intervals along the bar (*Figure 6.15b*) in order to reduce purification time. It is essential that the heaters are spaced at a sufficient distance to prevent any spread of the molten zones; for materials of high thermal conductivity and substances which exhibit high degrees of supercooling, alternate cooling arrangements may have to be fitted between the heaters. The ring method (*Figure 6.15c*) permits a simple multi-pass arrangement. The vertical tube method (*Figure 6.15d*) is useful when impurities which can sink or float in the melt are present. A downward zone pass can be used in the former case, an upward pass in the latter.

Different methods of heating may be used to produce molten zones; the choice of a particular method is generally governed by the physical characteristics of the material undergoing purification. Resistance heating is widely employed, and direct-flame and focused-radiation heating are quite common. Heat can be generated inside an ingot by induction heating, and this method is often used for metals and semi-conductors. Heating by electron bombardment or by electrical discharge also have their specific uses. Solar radiation, focused by lenses or reflectors, affords an automatic method for zone movement on account of the sun's motion. All these heating techniques are discussed in detail by Pfann[16].

The speed of zoning is a factor which can have a considerable effect on the efficiency of zone refining. The correct speed is that which gives a uniform zone passage and at the same time permits impurities to diffuse away from the solidifying face into the melt. If the zone speed is too fast, irregular

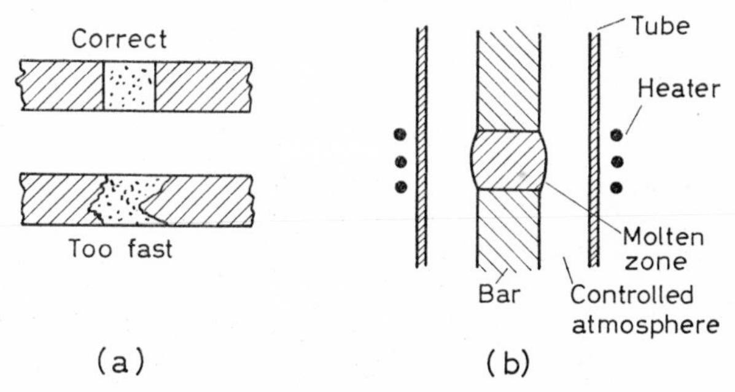

Figure 6.16. (a) Influence of zone speed; (b) floating-zone technique

crystallization will occur at the solidifying face, and impure melt will be trapped before it can diffuse into the bulk of the moving zone (*Figure 6.16a*). For the purification stages, the speed of zoning can vary from about 2 to 8 in./h, while speeds of about $\frac{1}{8}$ to $\frac{1}{2}$ in./h are common for zone levelling.

The tube or crucible which contains the material undergoing purification should not provide a source of contamination and must be capable of withstanding thermal and mechanical stresses. The purified solid must be easily removable from its container, so the melt should not wet the container walls. The choice of the container material depends on the substance to be purified. Glass and silica are commonly used for organic substances, silica for many sulphides, selenides, arsenides and antimonides, while graphite-lined silica is often used for metals.

External contamination can be prevented by dispensing with the use of a conventional container. For example, a vertical zone refining technique which uses a 'floating zone' is shown in *Figure 6.16b*. This method has been described by KECK *et al.*[18] The bar of material is fixed inside a container tube without touching the walls, and the annulus between the bar and tube may contain a controlled atmosphere. Surface tension plays a large part in preventing the collapse of the molten zone, but the control of such a zone demands a high degree of experimental skill.

Although the largest applications of zone refining are in the fields of semiconductors and metallurgy, the operation has been used very successfully in the purification of many chemical compounds, such as naphthalene, anthracene, benzoic acid, benzene, gallium and silicon halides, to name but a few. Details have been given by HERINGTON *et al.*[19, 20] of small-scale ($\sim$ 1 kg) and semi-micro scale ($\sim$ 0·5 g) units for the purification of organic compounds, and RONALD[21] has devised an automatic multi-stage semi-micro apparatus for the same purpose. Zone refining is still largely a laboratory technique, but there seems little doubt that its extension to large-scale production is only a matter of time.

De-salting of Saline Water

The possibility of using zone freezing for the purification of sea water has been considered. When salt water is partially frozen the deposited ice crystals are essentially salt-free, but owing to the ease with which brine is

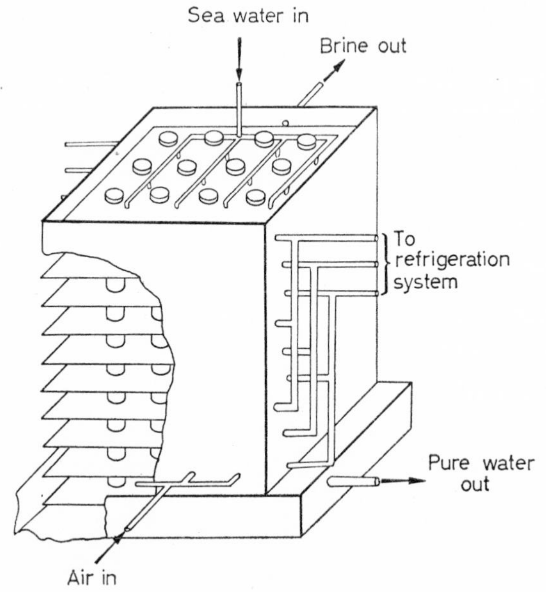

Figure 6.17. Vertical-tube de-salting process. (*From* R. C. HIMES *et al.*[22])

By courtesy of *Industrial Engineering Chemistry*

trapped in the mass of tiny crystals it is extremely difficult to separate the ice in pure form. Zone freezing, therefore, would appear to offer a solution to the problem; salt would concentrate in one phase and pure water in the other. Recent pilot-plant investigations[22] have demonstrated the economic feasibility of the process, and the large-scale production of potable water by zone freezing methods would seem to have a reasonably promising future. Two of the many processes considered for large-scale operation are shown in *Figures 6.17* and *6.18*.

In the vertical tube process (*Figure 6.17*), a number of tubes pass through closely spaced, insulated horizontal plates. The spaces between the plates are, through automatic valves, connected to both the expansion and compression parts of a refrigeration system. Thus, any space can be a 'melting', 'freezing' or 'dead' zone, as required. The bottoms of the tubes are placed in a reservoir of desalted water. Salt water is introduced near the tops of the tubes; freezing and melting zones are created and moved upwards in the tubes. A plug of ice forms in the bottom of the tube located in the purified water reservoir, and as it moves upwards air is permitted to enter the tube beneath the plug. After a certain interval of time the air entrance is closed, and the plug of ice, followed by an air space and a re-melted water zone, moves upwards. Freezing at the top of the plug and re-melting at the bottom are arranged to continue at the same rate; the re-melted water from one stage rests on the ice plug of the next. The continuous process results in an upward passage of a volume of air and a downward movement of an equal volume of water at each freezing stage. The ice plug is finally melted near the top of the tube, and salt water is introduced just above the surface of the next ice plug to wash out the brine and refill the tube.

In the spiral drum process (*Figure 6.18*) a double-walled drum is employed; the walls are so connected that the annular space takes the form of a long

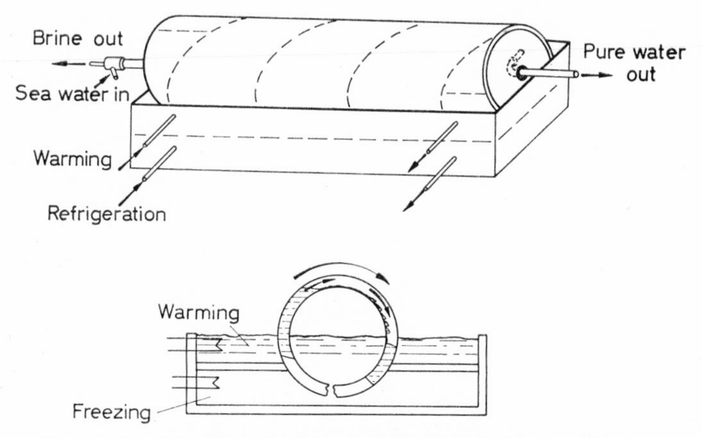

Figure 6.18. Spiral-drum de-salting process. (*From* R. C. HIMES *et al.*[22])
By courtesy of *Industrial Engineering Chemistry*

spiral passage. The drum rotates in a freezing bath; a plug of ice forms in the bottom of each coil and closes the passage. As rotation continues, water is lifted up by the plug and overflows into the next loop of the spiral where it eventually becomes a frozen zone. The ice plug leaves the freezing zone and melts, the water is lifted and transferred into the next loop, and so on. This procedure continues; the water is successively frozen and melted many times until it is discharged from the last loop in a relatively salt-free condition. When water is discharged from the loop, its place is taken by air which, because of the rotation of the drum, moves countercurrently to the water flow. Salt water is introduced into the spiral just ahead of the first ice plug, and this washes out the concentrated brine and refills the passage. The melting of the ice plugs can be effected by warm air or warm water on the outside of the drum. Preliminary cost estimates suggest that the vertical tube process would be the cheaper of the two.

Single Crystals

For many centuries the growing of large crystals has been the pastime of devoted scientists, often just for curiosity's sake. To-day, however, there is an increasing demand for perfect crystals of innumerable substances for research work on the chemical and physical properties of pure solids and, more recently, for use in the electrical industries where crystals of certain substances are required for their dielectric and piezo-electric properties. The production of large single crystals demands specialized and exacting techniques, but broadly speaking there are three general methods available, viz. growth from solutions, from melts, and from vapours. Several comprehensive reviews have recently been made of this rather special aspect of crystallization practice[23–25], and only a brief summary will be attempted here.

Slow crystallization from solution in water or organic solvents has long been a standard method for growing large pure crystals of inorganic and organic substances. Basically, a small crystal seed is immersed in a supersaturated solution of the given substance and its growth is regulated by a careful control of the temperature, concentration and degree of agitation of the system. For instance, a tiny selected crystal may be mounted on a suitable support and suspended in a vessel containing a solution of the substance, maintained at a fixed temperature. Slow rotation of the vessel will give an adequate movement of the solution around the crystal, and slow, controlled evaporation of the solvent will produce the degree of supersaturation necessary for crystal growth. Alternatively, the vessel may remain stationary and the growing crystal, or several suitably mounted crystals, may be gently rotated in the solution.

The actual operating conditions vary according to the nature of the crystallizing substance and solvent; the optimum supersaturation and solution movement past the crystal must be found by trial and error. Generally speaking, the degree of supersaturation must not be high, and in any case the solution must never be allowed to approach the labile condition. The degree of supersaturation should be kept as constant as possible to ensure a constant rate of deposition of solute on the crystal seed. HOLDEN[26] states that for salts with solubilities in the approximate range 20–50 per cent by weight, the maximum linear growth rate that can be tolerated by the fastest growing faces of a crystal is about 0·05–0·1 in. per day; faster rates tend to give imperfect crystals. He also suggests that single crystals are better grown by a cooling rather than an evaporation process, as supersaturation can then be much more closely controlled.

Another way of controlling supersaturation is to hold the crystal in a fixed position in an apparatus and make a solution of a pre-determined constant concentration flow continuously past it. The solution must be as pure as possible and all inert foreign matter, e.g. atmospheric dust, must be excluded from the system, as nucleation must be avoided at all costs.

The production of large ADP (ammonium dihydrogen phosphate) and EDT (ethylene diamine tartrate) piezo-electric crystals has become almost a small industry on its own. The growth of one ADP crystal which measured 6 × 6 × 22 in., weighed 43 lb. and took about 4 months to complete at the Bell Telephone Laboratories, has been described by BUCKLEY[23].

The record for the largest man-made single crystal, however, appears to be held by Peter Spence, Ltd., of Widnes, in whose factory a crystal of alum weighing about 240 lb. was prepared[23]. Of course, the real industrial demand is for much smaller single crystals than these. To give a rough idea of the magnitude of the production of single crystals, the 1955 output of Western Electric Co.[25] was 40,000 crystals of EDT weighing about ½ kg each.

Quartz crystal is the ideal piezo-electric material, and the demand for other, somewhat inferior substances such as EDT has arisen because of the scarcity of suitable natural quartz. Small quartz crystals have been grown in the laboratory for many years, but it is only quite recently[27] that the technique known as hydrothermal crystallization (crystallization from an aqueous caustic soda solution of SiO_2 at high temperature and pressure) has been successfully extended to pilot-plant scale. From the results reported, there would appear to be a good chance of a further extension to large-scale production in the near future.

A single crystal can be grown in a pure melt in a manner similar to that described for growth in a solution. For example, a small crystal seed could be rotated and allowed to grow in a slightly supercooled melt of the same substance. However, the most widely used methods operate on a slightly different principle. A suitable seed is dipped into the melt held at a temperature just above the melting point. When a small amount of the crystal has melted, exposing a fresh solid surface to the melt, growth is commenced and the growing crystal is pulled very slowly out of the melt at a rate which corresponds with the rate of deposition on to the contact face of the seed. If required, the whole apparatus may be operated under vacuum or in an inert atmosphere.

The various methods of pulling crystals from the melt have been described by LAWSON and NIELSEN[25]. Pull-rates depend on the temperature gradient at the crystal–melt interface and can vary from 0·1–4·0 cm/h. The steeper the gradient, the higher the growth rate and the faster the permissible growth rate. For the production of perfect crystals, the temperature of the melt and of the withdrawing crystal must be controlled very carefully. The growing crystal face in contact with the melt should be flat, and mechanical vibrations must be avoided.

The technique of zone melting is ideally suited to single crystal production, and in its simplest form the operation can be described as follows. A pure ingot of the required material is placed in a suitable crucible, one end of the ingot being in contact with the flat face of a seed crystal. A zone in the ingot is then melted and moved towards the seed crystal. When the molten zone touches the seed, the zone movement is stopped for a brief moment and then reversed. Growth then commences on the seed. Single crystals of germanium, about 1 in. diam. and 12 in. long, are readily produced by this method[16].

Crystal growth from the vapours without the intervention of the liquid phase, i.e. crystallization by the reverse of the process of sublimation, is sometimes used to produce small but very pure and almost strain-free crystals. Many different techniques can be employed, as reviewed by SHORT[24]. For example, a gas stream (e.g. nitrogen or hydrogen sulphide) may be passed over a heated container containing, say, cadmium sulphide. The sulphide vapours then pass to another part of the apparatus where they condense, in crystalline form, on a cold surface. Very pure cadmium sulphide crystals are used as sensitive elements in photocells for detecting gamma-rays.

Sublimation in a sealed tube is also used for the preparation of single crystals of many metals (e.g. zinc and cadmium) and non-melting sulphides. A quantity of the material is placed at one end of the tube, along which there is a temperature gradient, so that sublimation occurs and crystals grow at the cooler end. An electric furnace with a number of independently controlled windings is used to maintain the temperature gradient to give a rate of sublimation sufficiently slow for the growing crystals to be single and not polycrystalline.

REFERENCES

1 TIPSON, R. S., Theory and scope of methods of recrystallization, *Analyt. Chem.* 22 (1950) 628

2 TIPSON, R. S., Crystallization and recrystallization, in *Techniques of Organic Chemistry* Vol. III, Part I, A. Weissberger (Ed.), 1956. New York; Interscience

3 DOERNER, H. A. and HOSKINS, W. M., The co-precipitation of radium and barium sulphates, *J. Amer. chem. Soc.* 47 (1925) 662

4 SUNIER, A. A., A critical consideration of some schemes of fractionation, *J. phys. Chem.* 33 (1929) 577

5 GARBER, H. J. and GOODMAN, A. W., Fractional crystallization, *J. phys. Chem.* 45 (1941) 573

6 JOY, E. F. and PAYNE, J. H., Fractional precipitation or crystallization systems, *Industr. Engng Chem.* 47 (1955) 2157

7 OLIVE, T. R., Freeze concentration, *Chem. Engng* 55 (10) (1948) 118

8 GANE, R., Concentration by freezing of fruit juices, vinegars and ciders, *Food Manuf.* 23 (1948) 282

9 FORSYTH, J. S. and WOOD, J. T., Separation of organic mixtures by crystallization from the melt, *Trans. Instn chem. Engrs, Lond.* 33 (1955) 122

10 ASTON, J. G. and MASTRANGELO, S. V. R., Purification by fractional melting, *Analyt. Chem.* 22 (1950) 636

11 MASTRANGELO, S. V. R. and ASTON, J. G., Improved fractional melting apparatus, *Analyt. Chem.* 26 (1954) 764

12 COULSON, J. M. and WARNER, F. E., A problem in chemical engineering design: The manufacture of mono-nitrotoluene, *Instn chem. Engrs, London*, 1949

13 FINDLAY, R. A. and WEEDMAN, J. A., Separation and purification by crystallization, in *Advances in Petroleum Chemistry and Refining*, K. A. Kobe and J. J. McKetta (Eds.), Vol. 1,1958. New York; Interscience

14 MCKAY, D. L., DALE, G. H. and WEEDMAN, J. A., A bench-scale crystallization purification column, *Industr. Engng Chem.* 52 (1960) 197

15 PFANN, W. G., Principles of zone melting, *J. Metals, N.Y.*, 4 (1952) 747

16 PFANN, W. G., *Zone Melting*, 1958. New York; Wiley

17 PARR, N. L., Zone Refining, *Lecs, Monogrs. and Reps.* No. 3, *Roy. Inst. Chem.*, London, 1957

[18] Keck, P. H., Green, M. and Polk, M. L., Shapes of floating liquid zones between solid rods, *J. appl. Phys.* 24 (1953) 1479
[19] Herington, E. F. G., Handley, R. and Cook, A. J., Apparatus for the purification of organic compounds by zone melting, *Chem. & Ind.* No. 16 (1956) 292
[20] Handley, R. and Herington, E. F. G., Semi-micro zone melting apparatus, *Chem. & Ind.* No. 16 (1956) 304
[21] Ronald, A. P., Automatic multi-stage semi-micro zone melting apparatus, *Analyt. Chem.* 31 (1959) 965
[22] Himes, R. C., Miller, S. E., Mink, W. H. and Goering, H. L., Zone freezing in demineralizing saline waters, *Industr. Engng Chem.* 51 (1959) 1345
[23] Buckley, H. E., *Crystal Growth*, 1952. London; Chapman & Hall
[24] Short, M. A., Methods of growing crystals, *Industr. chem. Mfr* 33 (1957) 3
[25] Lawson, W. D. and Nielsen, S., *Preparation of Single Crystals*, 1958. London; Butterworths
[26] Holden, A. N., Growing single crystals from solution, *Disc. Faraday Soc.* 5 (1949) 312
[27] Laudise, R. A. and Sullivan, R. A., Pilot plant production of synthetic quartz, *Chem. Engng Progr.* 55 (5) (1959) 55

7

INDUSTRIAL CRYSTALLIZATION

CRYSTALLIZATION ranks high in the list of industrial processes devoted to the production of pure chemicals. Apart from the fact that its final product has an attractive appearance, crystallization frequently proves to be the cheapest and sometimes the easiest way in which a pure substance can be produced from an impure solution. Conventional distillation techniques cannot separate efficiently close-boiling liquids or those which form azeotropes, yet crystallization may often lead to their complete separation. There is evidence that the petroleum industry is now turning its attention to crystallization techniques to deal with difficult separations.

The methods available for crystallization are many and varied. Crystals can be grown from the liquid or the vapour phase, but in all cases the state of supersaturation has first to be achieved. The way in which supersaturation is produced depends on the characteristics of the crystallizing system; some solutes are readily deposited from their solutions merely by cooling, while others have to be evaporated to a more concentrated form. In cases of very high solubility, or for heat-labile solutions, another substance may have to be added to the system to reduce the solubility of the solute in the solvent. Again, supersaturation of the liquid or gaseous phase may be caused by the chemical reaction of two substances; one of the reaction products is then precipitated.

In this chapter an account is given of the ways in which crystallization can be performed. Little attention will be devoted to the actual types of processing equipment employed because these are discussed in greater detail in Chapter 8. Problems in the crystal-producing industries do not cease once the solid phase has been deposited out of solution; indeed, many technologists would assert that they just begin there. Brief accounts, therefore, are given of some aspects of crystal washing and the caking of crystals on storage.

Cooling and Evaporation

One of the most common ways in which the supersaturation of a liquid can be achieved is by means of a cooling process. If the solubility of the solute in the solvent decreases with a decrease in temperature, some of the solute will be deposited on cooling; a slow controlled rate of cooling in an agitated system can result in the production of crystals of regular size. The crystal yield may be slightly increased if some of the solvent evaporates during the cooling process.

If the solubility characteristics of the solute in the solvent are such that there is little change with a reduction in temperature, some of the solvent may have to be deliberately evaporated from the system in order to effect the necessary supersaturation and crystal deposition. Cooling and evaporative techniques are widely used in industrial crystallization; the majority of the solute–solvent systems of commercial importance can be processed by one or other of these methods. Descriptions of many of the cooling and evaporating crystallizers commonly encountered are given in Chapter 8.

The yield from a cooler or evaporator can be calculated, as explained in Chapter 2, from the general equation

$$Y = \frac{WR[C_1 - C_2(1 - V)]}{1 - C_2(R - 1)} \tag{1}$$

where Y = crystal yield (lb.)
W = weight of solvent present initially (lb.)
V = weight of solvent lost, either deliberately or unavoidably, by evaporation (lb. per lb. of original solvent)
R = ratio of the molecular weights of solvate (e.g. hydrate) and unsolvated (e.g. anhydrous) solute
C_1, C_2 = initial and final solution concentrations, respectively (lb. of unsolvated solute per lb. of solvent)

The yield calculated from equation 1 is the theoretical maximum, on the assumption (*a*) that C_2 refers to the equilibrium saturation at the final temperature, and (*b*) no solute is lost when the crystals are washed after being separated from the mother liquor.

Controlled Seeding

During a crystallization operation the accidental production of nuclei ('false grain') must be avoided at all costs; the solution must never be

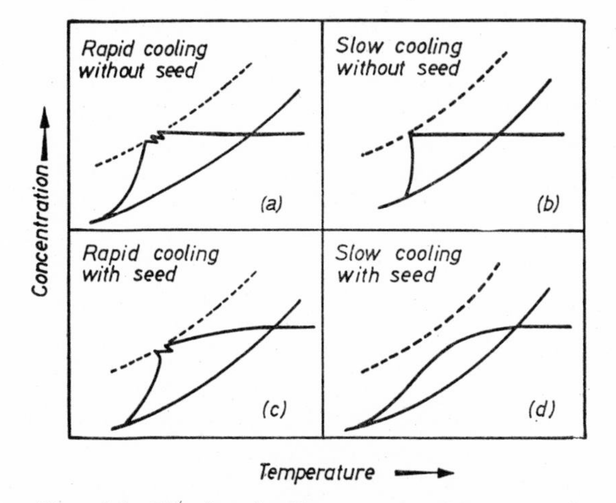

Figure 7.1. The effect of cooling rate and seeding on a crystallization operation. (*Adopted from* H. GRIFFITHS[1])

allowed to become labile. The deliberate addition of carefully selected seeds, however, is permitted so long as the deposition of crystalline matter takes place on these nuclei only. GRIFFITHS[1] investigated the crystallization of solutions by cooling, with and without the presence of seeding crystals, and his results are shown on the solubility–supersolubility diagrams in *Figure 7.1*.

The rapid cooling of an unseeded solution is represented by diagram (*a*). The metastable region is soon penetrated, and spontaneous crystallization occurs when conditions corresponding to some point on the supersolubility curve are reached; a shower of tiny crystals is suddenly deposited from the solution. The slow cooling of an unseeded solution is shown in diagram (*b*); again, crystallization cannot occur until the supersolubility curve is reached, but the rate of crystallization is slower than that in case (*a*) because the rate of heat removal is reduced. Control over the growth of crystals by this method is strictly limited and the crystals vary considerably in size. Diagram (*c*) shows the rapid cooling of a seeded solution. As soon as the solution becomes supersaturated, growth begins on the crystal seeds and the concentration of the solution decreases as more and more substance is deposited. However, because of the rapid rate of cooling, the labile condition is soon reached, and the eventual spontaneous deposition of fine crystals cannot be avoided.

The effect of a slow rate of cooling on a seeded solution is seen in *Figure 7.1d*. The temperature is controlled so that the system is kept in the metastable state throughout the operation, and the rate of growth of the small crystal seeds is governed solely by the rate of cooling. There is no sudden deposition of fine crystals because the system does not enter the labile zone. This method of operation is usually referred to as 'controlled crystallization'; crystals of a regular and pre-determined size can be grown. Many large-scale crystallization operations are carried out in this manner.

The mass of crystal seeds accidentally produced in large crystallizers may be very small indeed, yet on account of their minute size the number of seeds can be exceptionally large. If, for simplicity, spheres of diameter d and density ρ are considered, the mass of one seed is $\frac{\pi}{6}\rho d^3$. Thus 100 g of 0·1 mm ($\sim$ 150-mesh) seeds of a substance of density 2 g/cm³ will contain 100 million separate particles, and every seed is a potential crystal. *Figure 7.2* indicates the mass deposition of solute necessary for 100 g of 0·1 mm seeds to

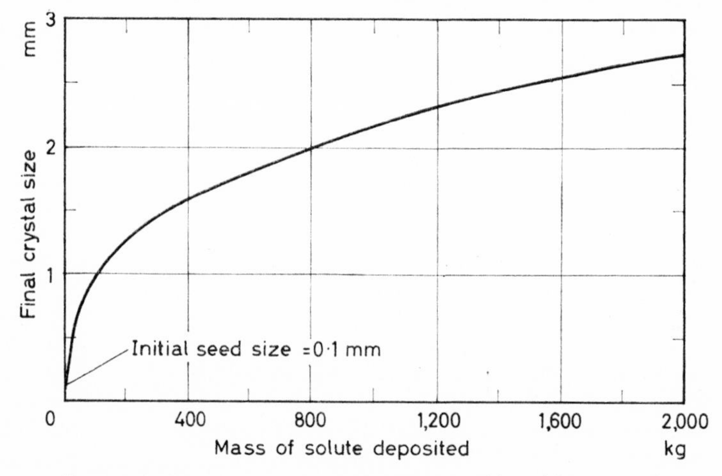

Figure 7.2. Crystal sizes and mass depositions on 100 g *of* 0·1 mm *seeds*

grow into crystals of various sizes ranging up to nearly 3 mm. It can be seen, for example, that 1 mm crystals can be grown by the deposition of only 100 kg of solute, but 800 kg has to be deposited to produce 2 mm crystals, and for $2\frac{1}{2}$ mm crystals the required mass deposition increases to more than 1,500 kg.

False grain smaller than 0·1 mm can exist in commercial crystallizers where high degrees of supersaturation may have been produced, and these seeds cannot all be allowed to grow; 1 g of 0·01 mm seeds, for example, represents about 1,000 million particles. For controlled growth, therefore, the liquor in the crystallizer must not be allowed to nucleate. Vigorous agitation, mechanical and thermal shock should be avoided, and the supersaturation should be kept to the absolute working minimum; high magma densities, up to 20 or 30 per cent suspended solids, help in this respect. If, despite these precautions, unwanted nucleation still occurs, some system of false grain removal must be operated; continuous circulation of the spent liquor through a cyclone is one such method.

Seed crystals deliberately added to a crystallizer are not always minute; their size really depends upon the type of product required. For example, for the controlled growth of 2 mm crystals in certain types of crystallizer the seeds may approach 1 mm in size. In any case, the seed crystals will rarely be uniform in size, and for these conditions McCabe[2, 3] has proposed a method for the prediction of the sieve analysis of a mass of crystals grown from a mass of crystal seeds of known size grading; the following equation was used.

$$W_p = \int_0^{W_s} \left(1 + \frac{\Delta D}{D_s}\right)^3 dW_s \tag{2}$$

W_p is the weight of the crystalline product obtained from a given weight, W_s, of seed crystals. D_s is the diameter, or some other characteristic dimension of the seeds, ΔD the increase in diameter, i.e. the growth rate. Equation 2 can be used in several ways. If the average rate of growth, ΔD, is known, the equation can be integrated graphically to give the weight of crystalline product. When, on the other hand, ΔD is not known, trial and error calculations are necessary; W_s is specified, the values of D_s of the various fractions are known, and the values of ΔD are assumed and tried until equation 2 is satisfied. The screen analysis of the product can then be computed. The trial and error calculation method is extremely tedious, but the burden can be eased by the use of a nomograph[4].

Whilst McCabe's theoretical treatment of this subject is sound, there are many difficulties in its practical application. The following assumptions are made: (*a*) the crystals are all of the same shape, (*b*) they grow invariantly, (*c*) a constant supersaturation driving force exists throughout the growth period, (*d*) no fresh nuclei are formed during the process, (*e*) there is no classifying action in the crystallizer, and (*f*) agitation is uniform, i.e. the relative velocity between the crystal surfaces and liquor remains constant. These ideal conditions can never be attained in a crystallizer; traces of impurity in the liquor, unequal agitation and temperature gradients within the crystallizing system make accurate predictions of crystal growth almost impossible. In addition to these factors, particle size measurement by sieve analysis has many inherent errors, and any error in the sieve analysis of the seeds will be greatly magnified in the final product. Nevertheless, in spite of these difficulties, McCabe's method has been widely used to predict operating conditions in industrial crystallizers.

Several workers[5–7] have presented methods for the prediction of crystal size in evaporating, vacuum and classifying crystallizers; SAEMAN[6] made a detailed analysis of the growth conditions in vacuum crystallizers of the Krystal and Swenson types, showing that the cumulative weight of crystals varies as the fourth power of their size. Thus, for example, crystals up to half the final product size constitute only $(\frac{1}{2})^4$, or one-sixteenth, of the total weight of crystals in suspension. Saeman considered the use of hydraulic classification as the sole means of controlling the growth of large crystals to be ineffective; the most positive and direct method of size control lies in the efficient segregation and elimination of the excess nuclei. Several results of trial runs on commercial crystallizers operated with and without fines-removal systems were reported, and recommendations were made for the design of classifying crystallizers.

'Vacuum' Cooling

If a hot saturated solution of a substance is fed into a vessel maintained at low pressure, some of the solvent will 'flash off' and the liquor will cool adiabatically. Supersaturation is thus achieved by both cooling and evaporation. Industrial crystallizers based on these principles are generally referred to as vacuum crystallizers; a further discussion of this method of operation is given in Chapter 8.

The yield from a vacuum crystallizer can be calculated from equation 1; before it can be applied, however, the quantity V, the amount of evaporation, must be known. It is possible to predict the evaporation capacity of a vacuum crystallizer, given certain information, as follows.

The amount of solvent evaporated depends upon the heat made available during the operation of the crystallizer. This is the sum of the sensible heat drop of the solution, which cools from the feed temperature to the equilibrium temperature in the vessel, and the heat of crystallization liberated. The heat balance, therefore, will be

$$\begin{pmatrix}\text{Solvent}\\ \text{evaporated}\end{pmatrix} \times \begin{pmatrix}\text{Latent}\\ \text{heat}\end{pmatrix} = \text{Sensible heat drop} + \text{heat of crystallization}$$

Let l_v = latent heat of evaporation of solvent (B.t.u./lb.)
q = heat of crystallization of product (B.t.u./lb.)
t_1 = initial temperature of solution (°F)
t_2 = final temperature of solution (°F)
c = heat capacity of solution—assumed constant (B.t.u./lb. °F)

In addition, let the symbols used in equation 1 be used, viz. C_1 and C_2 = initial and final solution concentrations (lb./lb. of solvent), W = initial weight of solvent (lb.), V = loss of solvent by evaporation (lb./lb. of original solvent),

R = ratio of the molecular weights of the hydrate and anhydrous salt, and
Y = crystal yield (lb.). Then

$$VWl_v = c(t_1 - t_2)W(1 + C_1) + qY$$

Substituting for the value of Y from equation 1 and simplifying

$$V = \frac{qR(C_1 - C_2) + c(t_1 - t_2)(1 + C_1)[1 - C_2(R - 1)]}{l_v[1 - C_2(R - 1)] - qRC_2} \tag{3}$$

Once the value of the evaporation capacity V is determined from equation 3, the yield from the crystallizer can be calculated from equation 1.

Example—Determine the yield of sodium acetate crystals ($CH_3COONa \cdot 3H_2O$) obtainable from a vacuum crystallizer operating with an internal pressure of 10 mm Hg, when it is supplied with 5,000 lb./h of a 40 per cent aqueous solution of sodium acetate at 180° F. The boiling point elevation of the solution may be taken as 20° F.

Data

Heat of crystallization, q	= 62 B.t.u./lb. of $CH_3COONa \cdot 3H_2O$
Heat capacity of the solution, c	= 0·83 B.t.u./lb. °F
Latent heat of water at 10 mm Hg, l_v	= 1,065 B.t.u./lb.
Boiling point of water at 10 mm Hg	= 52·4° F

Solubilities of sodium acetate at 32, 50, 68, 86 and 104° F are 36·3, 40·8, 46·5, 54·5 and 65·5 lb. anhydrous salt per 100 lb. water, respectively.

Solution

Equilibrium temperature of liquor in the crystallizer	= 52·4 + 20 = 72·4° F
Initial concentration, C_1	= 0·4/0·6 = 0·667 lb./lb. of solvent
Final concentration at 72·4° F, C_2	= 0·494 lb./lb. of solvent
Initial weight of water, W	= 0·61 × 5,000 = 3,000 lb.
Ratio of molecular weights, R	= 136/82 = 1·66

Vaporization, V, calculated from equation 3

$$V = \frac{62 \times 1{\cdot}66(0{\cdot}667 - 0{\cdot}494) + 0{\cdot}83 \times 107{\cdot}6 \times 1{\cdot}667[1 - 0{\cdot}494(0{\cdot}66)]}{1{,}065[1 - 0{\cdot}494(0{\cdot}66)] - 62 \times 1{\cdot}66 \times 0{\cdot}494}$$

= 0·165 lb./lb. of water present originally.

This value of V, substituted in equation 1, gives the yield of crystals, Y

$$Y = \frac{3{,}000 \times 1{\cdot}66[0{\cdot}667 - 0{\cdot}494(1 - 0{\cdot}165)]}{1 - 0{\cdot}494(0{\cdot}66)}$$

= 1,868 lb. of $CH_3COONa \cdot 3H_2O$

Salting-out

Another way in which the supersaturation of a solution can be effected is by the addition to the system of some substance which reduces the solubility of the solute in the solvent. The added substance, which may be a liquid, solid or gas, is often referred to as a diluent or precipitant. Liquid diluents are most frequently used. Such a process is known as salting-out, precipitation or crystallization by dilution. The properties required of the diluent are that it should be miscible with the solvent of the original solution, at least over the ranges of concentration encountered, that the solute be relatively insoluble in it, and also that the final solvent–diluent mixture should be capable of easy separation, e.g. by distillation.

Although salting-out is widely employed industrially, relatively few published data are available regarding its use in crystallization operations. The process is commonly encountered, for instance, in the crystallization of organic substances from water-miscible organic solvents by the controlled addition of water to the solution; the term 'watering-out' is used in this connection. Some of the advantages of salting-out or dilution crystallization are as follows. Very concentrated initial solutions can be prepared, often with great ease, by dissolving the impure crystalline mass in a suitable solvent. A high solute recovery can be made by cooling the solution as well as salting it out. If the solute is very soluble in the initial solvent, high dissolution temperatures are not necessary, and the temperature of the batch during the crystallization operation can be kept low; this is advantageous when heat-labile substances are being processed. Purification is sometimes greatly simplified when the mother liquor retains undesirable impurities owing to their greater solubility in the solvent–diluent mixture. Probably the biggest disadvantage of dilution crystallization is the need for a recovery unit to handle fairly large quantities of the mother liquor in order to separate solvent and diluent, one or both of which may be valuable.

There is also evidence that salting-out crystallization induced by an added organic solvent may be useful for the preparation of pure inorganic substances, despite the fact that the organic diluents are usually much more expensive than the required inorganic products. An example of this use of salting-out crystallization was demonstrated by THOMPSON and MOLSTAD[8] who determined solubility and density isotherms for potassium and ammonium nitrates in aqueous isopropanol solutions over the temperature range 25–70° C. Both salts are very soluble in water, and their solubilities increase greatly with increasing temperature. The work was carried out with a view to investigating the possibility of increasing the purity, as well as the crystal yield, of these substances. It was shown, for example, that 15 lb. of isopropanol added to 100 lb. of a saturated aqueous solution of KNO_3 at 40° C would result in the precipitation of 44 per cent of the dissolved salt. The salt recovery would be increased to 68 per cent if 50 lb. of isopropanol were added.

GEE *et al.*[9] presented a detailed account of a pilot plant (2 ton/day) investigation into the commercial possibilities of the production of iron-free alum by the use of ethanol as a precipitant for aqueous solutions of crude alum. Under optimum conditions, alcohol losses from the system of less than $\frac{1}{2}$ per cent were reported. A detailed cost evaluation showed that the large-scale production of pure alum (20 ton/day) by this method is economically feasible.

The crystal yield, Y, from a salting-out operation can be calculated, if no solvate is formed, from the simple equation

$$Y = W(C_1 - C_2') \tag{4}$$

where W = weight of initial solvent, C_1 and C_2' = initial and final concentrations of the solute (lb./lb. of original solvent). The presence of the diluent affects the value of C_2'; this will be very different from that of the equilibrium solubility of the solute in the original solvent at the final temperature. Solubility data for mixed solvents are rarely found in the literature, but in any case it is always wise to determine the solubility characteristics of the solute with the actual liquors to be encountered in the crystallization process; small traces of impurity can affect solubility to a marked degree. If evaporation and solute formation also occur in a salting-out crystallization, the crystal yield can be calculated by means of a suitably modified form of equation 1.

As previously mentioned, the salting-out substance need not be a liquid. A suitable solid can be added to the solution, provided that it meets the above requirement, i.e. it must be highly soluble in the original solvent and must not react with the required solute. To avoid crystal deposition occurring on the added solid, the added substance could be introduced in the form of a concentrated solution in the main solvent. The use of solid substances as salting-out aids is not very common, but the addition of solid salts can be used to precipitate other salts from solution as a result of the formation of a stable salt pair. This behaviour is encountered when two solutes, AX and BY, usually without a common ion, react in solution and undergo a double decomposition

$$AX + BY \rightleftharpoons AY + BX$$

The four salts AX, BY, AY and BX constitute a reciprocal salt pair. One of these pairs AX, BY or AY, BX is a stable salt pair (compatible salts), the other an unstable salt pair (incompatible salts); the stable pair comprises the two salts that can co-exist in solution in stable equilibrium. Reciprocal salt pairs are discussed in Chapter 4 where the graphical representation of these systems is also described. PURDON and SLATER[10] give several examples of the commercial importance of salt-pair formation in the recovery of pure crystalline solids from complex solutions. GARRETT[11] described the large-scale production of potassium sulphate from glasserite ($3K_2SO_4 \cdot Na_2SO_4$) and potassium chloride; glasserite is added to a concentrated solution of KCl, conversion to a stable salt-pair occurs, and the Na_2SO_4 and KCl remain in solution while the K_2SO_4 is deposited.

Crystallization by Reaction

The precipitation of a solid product as the result of the chemical reaction between gases and/or liquids is a standard method for the preparation of many industrial chemicals. Precipitation occurs because the gaseous or liquid phase becomes supersaturated with respect to the solid component.

A crude precipitation operation, therefore, can be transformed into a crystallization process by controlling carefully the degree of supersaturation. Reaction-crystallization is practised widely, especially in industries where valuable waste gases are produced. For instance, ammonia can be recovered from coke oven gases by converting it into ammonium sulphate by reaction with sulphuric acid.

Only one example will be given here of the reaction method of crystallization. *Figure 7.3* shows a carbonation tower[11] used for the production of sodium bicarbonate by the interaction between brine and flue gases containing about 10–20 per cent of carbon dioxide. The 50 ft. high tower is kept full of brine, and the flue gas, which enters at the bottom of the tower, flows upwards countercurrently to the brine flow. Carbonated brine is pumped

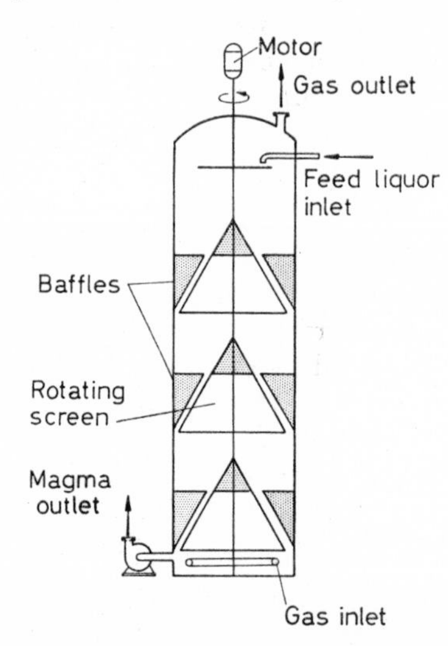

Figure 7.3. Reaction crystallizer—a carbonation tower for the production of sodium bicarbonate from brine and flue gas. (*After* D. E. GARRETT[11])

continuously out of the bottom of the tower. To effect efficient absorption of CO_2, three large rotating screens continually re-disperse the gas stream in the form of tiny bubbles in the liquor. The operating temperature is controlled at about 100° F, which has been found from experience to give both good absorption and crystal growth.

Extractive Crystallization

As described in Chapters 4 and 6, fractional crystallization of a binary system which forms a eutectic mixture can only produce one of the components in pure form. Many hydrocarbon isomers or close-boiling mixtures, such as those encountered in the petroleum industry, do form eutectics, and this has presented a limiting factor in the application of conventional crystallization techniques for these separations. However, if the solid–liquid phase relationships can be altered by introducing a third component into a binary system, it is frequently possible by a series of crystallization steps to separate the two desired components in pure form. The added third component is usually a liquid, called the solvent, and the process is known by the name 'extractive crystallization'.

CHIVATE and SHAH[12] discussed the use of extractive crystallization for the separation of mixtures of *m*- and *p*-cresol, a system in which two eutectics are formed. Acetic acid was used as the extraction solvent. Details of the relevant phase equilibria encountered in the various combinations of systems were given together with an account of the laboratory investigations. Whilst it was shown that acetic acid is not a particularly good solvent for the separation process in question, it was clearly indicated that extractive crystallization, provided a suitable solvent is chosen, has a large number of potential applications.

A procedure that could be adopted for an industrial extractive crystallization process has been described by FINDLAY and WEEDMAN[13]. A hypothetical

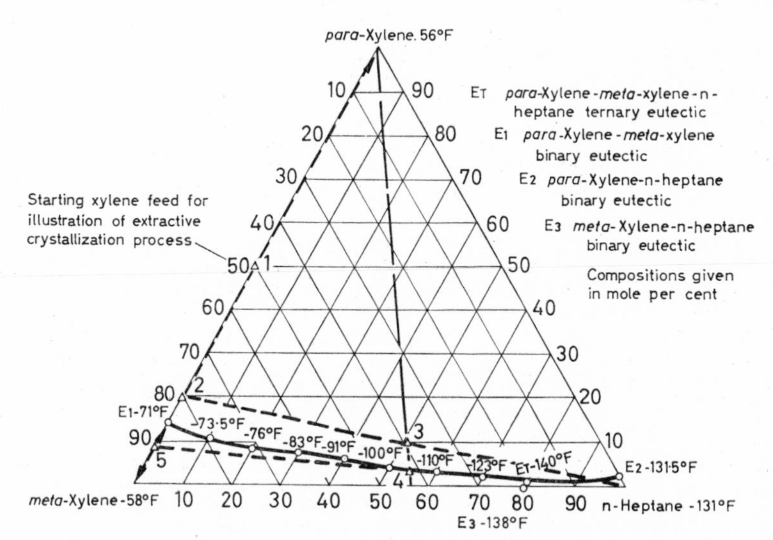

Figure 7.4. Solid–liquid phase diagram for p-xylene–m-xylene–n-heptane system, illustrating extractive crystallization. (*From* R. A. FINDLAY *and* J. A. WEEDMAN[13])
By courtesy of *Interscience*

separation is considered, and the process can be followed from the phase diagram in *Figure 7.4* and the flow diagram in *Figure 7.5*. Suppose, for example, that it is required to separate completely 100 lb./h of an equal mixture of *m*- and *p*-xylene. These two isomers form a binary eutectic. For this particular process n-heptane is chosen as the solvent, and a 7 : 1 solvent ratio based on the *m*-xylene product is used. The sequence of operations is given by points 1 to 5 in *Figure 7.4*. The 50 per cent *meta-para* mixture (1) is first cooled to − 55° F to give pure *para* crystals and a liquid (2) containing 80 per cent *meta*. The crystals are removed and the required quantity of solvent (n-heptane) is added to the liquor, thus giving a ternary mixture (3); this is cooled to about − 105° F (4) when more *para* is deposited. Cooling beyond about − 105° F would result in the co-deposition of both *m*- and *p*-xylene. The crystals and liquor are separated, and the n-heptane is removed from the mother liquor, e.g. by distillation, leaving a liquor (5) containing 92 per cent of *m*-xylene, which is cooled to yield pure *m*-xylene crystals and a liquor. After separation the mother liquor is returned to liquor (2), making the process continuous.

The flow sheet in *Figure 7.5* indicates the cycle of operations in this theoretical extractive crystallization process. For simplicity, the solid–liquid separators are assumed to be 100 per cent efficient. The flow rates have been calculated from material balances based on the information obtained from the phase diagram in *Figure 7.4*. No industrial applications of this process have yet been reported, but as Findlay and Weedman point out it could be used for the recovery of valuable components of mixtures that could not be

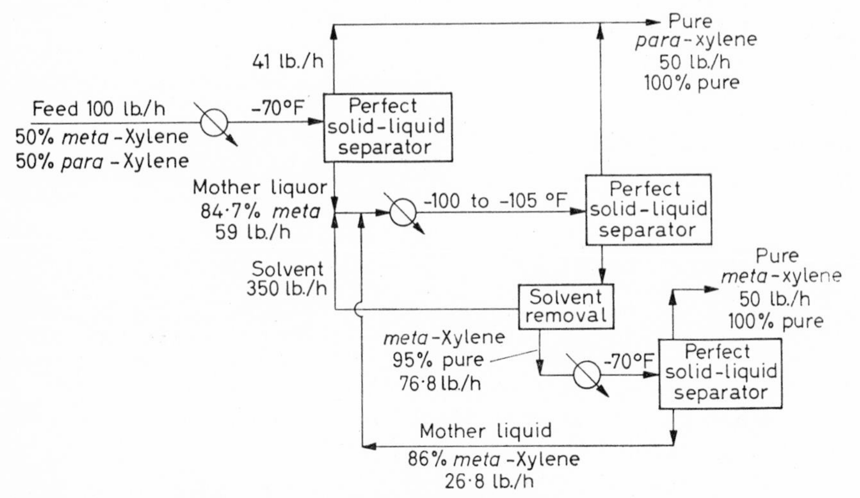

Figure 7.5. Theoretical extractive crystallization process. (*From* R. A. FINDLAY *and* J. A. WEEDMAN[13]) By courtesy of *Interscience*

separated by simple crystallization. The process would, of course, be more expensive than a conventional crystallization because of the large amounts of cooling necessary and the several recovery and recycle stages required.

Adductive Crystallization

The separation of mixtures of substances which form eutectics, and hence cannot be separated by conventional crystallization, can often be effected by methods based on the well known phenomenon of compound formation (see Chapter 4). A typical sequence of operations is as follows. A certain substance X is added to a given binary mixture of components A and B. A solid complex, say $A.X$, is precipitated, leaving component B in solution. The solid and liquid phases are separated, and the complex is split into pure A and X by the application of heat or by dissolution in some suitable solvent.

The best known example of compound formation in solvent–solute systems is the formation of hydrates, but other solvates, e.g. with methanol, ethanol and acetic acid, are known. In these cases the ratio of the molecules of the two components in the solid solvate can usually be expressed in terms

Mile End Library

of small integers, e.g. $CuSO_4 \cdot 5H_2O$ or $(C_6H_5)_2NH \cdot (C_6H_5)_2CO$ [diphenylamine·benzophenone]. There are, however, several other types of molecular complex that can be formed which are not necessarily expressed in terms of simple ratios. These complexes are best considered not as chemical compounds but as strongly bound physical mixtures. Clathrate compounds are of this type; molecules of one substance are trapped in the open structure of molecules of another. Hydroquinone forms clathrate compounds with SO_2 and methanol, for example. Urea and thiourea also have the property of forming complexes, known as adducts, with certain types of hydrocarbons. In these cases, molecules of the hydrocarbons fit into 'holes' or 'channels' in the crystals of urea or thiourea; the shape and size of the molecules determine whether they will be adducted or not.

It is open to question whether adduct formation can really be considered as a true crystallization process, but the methods of operation employed are often indistinguishable from crystallization methods. Several different names have been given to separation techniques based on the formation of adducts, but the name adductive crystallization[13] is probably the best and will be used here. Several possible commercial applications of adductive crystallization have been reported in recent years[13–16].

The system *m*-cresol–*p*-cresol forms two eutectics over the complete range of composition, and the separation of the pure components cannot be made by conventional crystallization. SAVITT and OTHMER[14] described the separation of mixtures of these two components by the use of benzidine to form a solid addition compound with *p*-cresol. Actually both *m*-cresol and *p*-cresol form addition compounds with benzidine, but the *meta* compound melts at a lower temperature than the *para*. If the process is carried out at an elevated temperature the formation of the *meta* complex can be avoided. The method consists of adding benzidine to the *meta-para* mixture at 110° C. The *p*-cresol·benzidine crystallizes out, leaving the *m*-cresol in solution. The crystals are filtered off, washed with benzene to remove *m*-cresol, and the washed cake is distilled under reduced pressure (100 mm Hg) to yield a 98 per cent pure *p*-cresol.

In a somewhat similar manner *p*-xylene can be separated from a mixture of *m*- and *p*-xylene; this binary system forms a eutectic. Carbon tetrachloride produces an equimolecular solid compound with *p*-xylene, but not with *o*- or *m*-xylene. EGAN and LUTHY[15] reported on a plant for the production of pure *p*-xylene by crystallizing *meta-para* xylene mixtures in the presence of carbon tetrachloride. Up to 90 per cent of the *para* isomer was recovered by distillation after splitting the separated solid complex. The *meta* isomer was recovered by fractionally crystallizing the CCl_4-free mother liquor. Perfect separation of *p*-xylene is not possible because the ternary system CCl_4/*m*-xylene/$CCl_4 \cdot$*p*-xylene forms a eutectic, but fortunately the concentration of the complex $CCl_4 \cdot$*p*-xylene in this eutectic is very low.

Separation processes based on the formation of adducts with urea and thiourea have been described by several authors[13–16]. Urea forms addition complexes with straight-chain, or nearly straight-chain, organic compounds such as paraffin and unsaturated hydrocarbons ($>$ 6 carbon atoms), acids, esters and ketones. Thiourea forms rather less stable complexes with branched-chain hydrocarbons and naphthenes, e.g. cyclohexane. For example[13], if a saturated aqueous solution of urea in methanol is added to an agitated mixture of cetane and isooctane, a solid complex of cetane·urea is formed almost immediately, deposited from the solution in the form of fine needle crystals. The isooctane is left in solution. After filtration and washing the complex is heated or dissolved in water, and pure cetane is recovered by distillation. If thiourea had been used instead of urea, isooctane could have been recovered, leaving the cetane in solution.

Bailey *et al.*[16] have reported the separation of n-paraffins and olefines from petroleum fractions ranging from gasoline to heavy lubricating oil, by

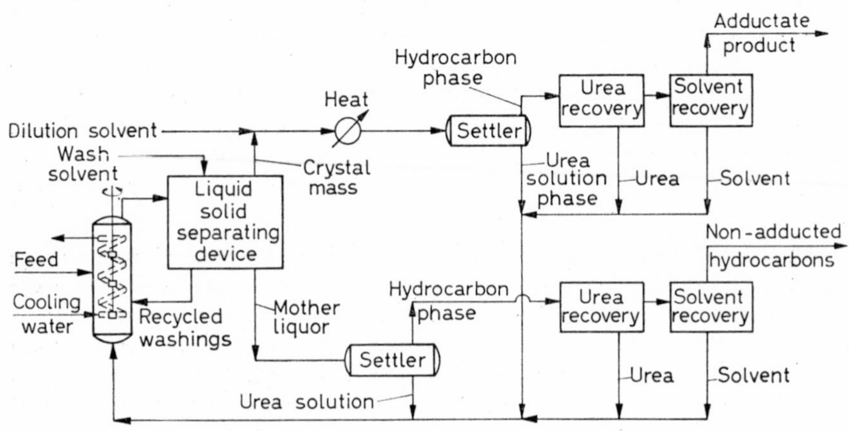

Figure 7.6. Generalized flow diagram for urea-adduct process. (*From* R. A. Findlay *and* J. A. Weedman[13])

By courtesy of *Interscience*

selective reaction with urea. They called the process 'extractive' crystallization, but this nomenclature has been criticized because the term extraction more aptly describes the solvent extraction method of crystallization described above. A simplified flow diagram of a possible urea-adductive crystallization process[13] is shown in *Figure 7.6*. The hydrocarbon feedstock is fed continuously to a stirred reactor where it is contacted with a solution of urea in a suitable solvent. The reacted slurry, containing tiny crystals of urea adducts with n-paraffins and/or straight-chain olefines, passes to a solid–liquid separator. The rest of the separation and recovery stages in this hypothetical process can be traced from the flow diagram.

So far, no large-scale application of urea-adductive crystallization has been reported. The process would tend to be rather costly on account of the relatively expensive urea and solvents, e.g. methanol or methyl isobutylketone, and the need for a large number of separation and recovery units in the complete plant.

Spray Crystallization

The term 'spray' crystallization is really a misnomer; strictly speaking, crystals are not grown by this method, solid is simply deposited from a very concentrated solution by a technique similar to that used in spray drying. Size and shape of the solid particles depend to a large extent on those of the spray droplets. The spray method is often employed when difficulties are encountered in the conventional crystallization techniques, or if a product with better storage and handling properties can be produced.

HOLLAND[17] described a spray crystallizer for the production of anhydrous sodium sulphate. Below 32·4° C sodium sulphate crystallizes from solution in the form of the decahydrate, above this temperature the anhydrous salt is formed. However, anhydrous sodium sulphate has an inverted temperature–solubility characteristic (see p. 23), and trouble is encountered with scale formation on the heat transfer surfaces of conventional evaporating crystallizers when using operating temperatures in excess of 32·4° C. In the plant described by Holland, a concentrated solution of sodium sulphate was sprayed, or splashed, in the form of tiny droplets into a chamber through which hot gases flowed. The gases entered at about 1,600–1,800° F. The continuously operated unit produced a powdered anhydrous product.

A type of spray crystallization is often employed in the manufacture of ammonium nitrate. The name 'prilling' is used to describe this process, and the solid spherical granules are called prills. Ammonium nitrate, which presents a potential explosion hazard, is produced commercially in particulate form by three main methods, viz. conventional crystallization, graining and prilling[18]. The first process carried out, for example, in vacuum classifying crystallizers requires rather expensive equipment, but as the operating temperatures never exceed 150–160° F the process is quite safe. Slightly rounded crystals of about 16 to 35-mesh are generally produced. In the graining process, solutions of the salt are evaporated in open pans to a moisture content of about 2 per cent, and this material, which is almost a molten salt, is passed to graining kettles where it is solidified into rounded granules, 12 to 35-mesh, under the action of heavy ploughs. The high operating temperatures in this process, 305–310° F, render the process liable to be hazardous. Both equipment and operating costs are high.

The prilling process is cheapest in capital and operating costs, but intermediate between conventional crystallization and graining in potential hazards. The method of operation is as follows[18]. A solution of ammonium nitrate is concentrated to about 5 per cent by weight of water and sprayed at about 280° F into the top of large towers, 100 ft. high and 20 ft. in diameter. Cool air, at about 80° F, enters the base of the tower and flows upwards, countercurrently to the falling droplets. The droplets, suddenly chilled when they meet the air stream, solidify. The prills are removed from the bottom of the tower, at about 170° F, but as they still contain about 4–5 per cent moisture they are dried at a temperature not exceeding 180° F to ensure that no phase transition occurs (see p. 14); they may then be dusted with diatomaceous earth or some other coating agent. The prills made by this process are reported to have twice the crushing strength of those formed by the spray drying of fused, water-free ammonium nitrate.

A rather unique method of prilling has been developed recently[19] for the processing of calcium nitrate in a form suitable for use as a fertilizer. Calcium nitrate, which is hygroscopic, is produced as a by-product in the manufacture of nitrophosphate fertilizer. The process consists of crystallizing droplets of calcium nitrate in the form of prills in a mineral oil to which seed crystals have been added. A spray of droplets of the concentrated solution, formed by allowing jets of the liquid at 140° C to fall on to a rotating cup, fall into an oil bath kept at 50–80° C. The prills are removed, centrifuged to remove surplus oil, and packed into bags. Because of the thin film of oil which remains on the particles, the material is less hygroscopic than it would normally be, there is no dust problem on handling, and the tendency to cake on storage is minimized.

Sublimation

So far, in the discussion of industrial crystallization processes, the deposition of a solid phase from a supersaturated or supercooled liquid phase only has been considered. The crystallization of a solid substance can, however, be induced from a supersaturated vapour in the process generally known as 'sublimation'. Strictly speaking, the term sublimation refers to the phase change

$$\text{solid} \rightarrow \text{vapour}$$

without the intervention of the liquid phase. In its industrial application, however, the term usually refers to the condensation process as well, i.e.

$$\text{solid} \rightarrow \text{vapour} \rightarrow \text{solid}$$

In practice, for heat transfer reasons, it is often desirable to vaporize the substance from the liquid state, so the complete series of phase changes in an industrial sublimation process can be

$$\text{solid} \rightarrow \text{liquid} \rightarrow \text{vapour} \rightarrow \text{solid}$$

It is on the condensation side of the process that the appearance of the liquid phase is prohibited. The supersaturated vapour must condense directly to the crystalline solid state. Two recent reviews have been made[20, 21] of the industrial applications of sublimation techniques.

The mechanism of a sublimation process can be described with reference to the pressure–temperature phase diagram in *Figure 7.7*. The significance of the P–T diagram applied to one-component systems has already been discussed in Chapters 3 and 4. The phase diagram is divided into three regions, solid, liquid and vapour, by the sublimation, vaporization and fusion curves. These three curves intersect at the triple point T. The position of the triple point in the diagram is of the utmost importance; if it occurs at a pressure above atmospheric, the solid cannot melt under normal atmospheric conditions, and true sublimation, i.e. solid → vapour, is easy to achieve. The triple point for carbon dioxide, for example, is − 57° C and 5 atm., so liquid CO_2 cannot result from the heating of solid CO_2 at atmospheric pressure: the solid simply vaporizes. However, if the triple point occurs at a pressure less than atmospheric, certain precautions are necessary if the phase changes solid → liquid → vapour → *liquid* → solid are not to take place. If the solid → liquid stage is permitted before vaporization the operation is often called 'pseudo-sublimation'.

In *Figure 7.7* both true and pseudo-sublimation cycles are depicted. For the case of a substance with a triple point at a pressure greater than atmospheric, true sublimation occurs. The complete cycle is given by path *ABCDE*. The original solid *A* is heated to some temperature represented by point *B*. The increase in the vapour pressure of the substance is traced along the sublimation curve from *A* to *B*. The condensation side of the process is represented by the broken line *BCDE*. As the vapour passes out of the vaporizer into the condenser it may cool slightly, and it may become diluted as it mixes with some inert gas such as air. Point *C*, therefore, representing a temperature and partial pressure slightly lower than point *B*, can be taken as the condition at the inlet to the condenser. After entering the condenser

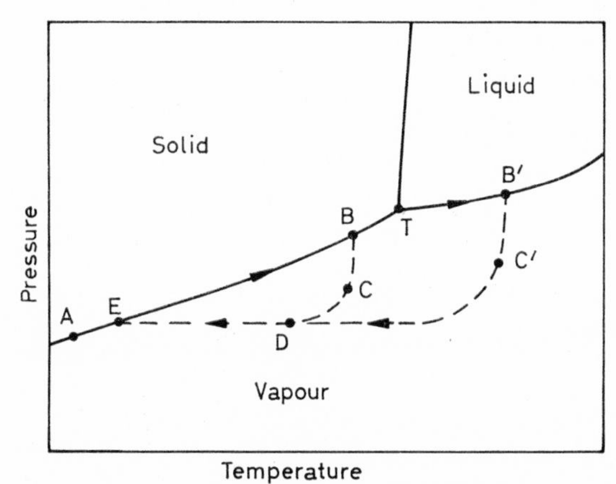

Figure 7.7. True and pseudo-sublimation cycles

the vapour mixes with more inert gas, and the partial pressure of the substance and its temperature will drop to some point *D*. Thereafter the vapour cools essentially at constant pressure to the conditions represented by point *E*, the temperature of the condenser.

When the triple point of the substance occurs at a pressure less than atmospheric, the heating of the solid may easily result in its temperature and vapour pressure exceeding the triple point conditions. The solid will then melt in the vaporizer; path *A* to *B'* in *Figure 7.7* represents such a process. However, great care must be taken in the condensation stage; the partial pressure of the substance in the vapour stream entering the condenser must be reduced below the triple point pressure to prevent initial condensation to a liquid. The required partial pressure reduction can be brought about by diluting the vapours with an inert gas, but the frictional pressure drop in the vapour lines is generally sufficient in itself. Point *C'* in *Figure 7.7* represents the conditions at the point of entry into the condenser, and the condensation path is represented by *C'DE*.

Sublimation techniques can be classified conveniently into three types: simple, vacuum and entrainer. In simple sublimation the solid material is heated and vaporized, and the vapours diffuse towards the condenser, the driving force for diffusion being the partial pressure difference between the vaporizing and the condensing surfaces. The vapour path between vaporizer and condenser should be as short as possible to reduce the resistance to flow. Simple sublimation has been practised for centuries; ammonium chloride, iodine and 'flowers' of sulphur have all been sublimed in this manner, often in the crudest of equipment.

Vacuum sublimation is a natural follow-on from simple sublimation. The transfer of vapour from the vaporizer to the condenser is enhanced by reducing the pressure in the condenser, thus increasing the partial pressure driving force. Iodine, pyrogallol and many metals have been purified by this type of process. The exit gases from the condenser usually pass through a cyclone or scrubber to protect the vacuum-raising equipment and to minimize the loss of product.

An inert gas may be blown into the vaporization chamber of a sublimer to increase the rate of flow of vapours to the condensing equipment and thus increase the yield. Such a process is known as 'entrainer' or 'carrier' sublimation. Air is the most commonly used entrainer, but superheated steam can be employed for substances which are relatively insoluble in water, e.g. anthracene. When steam is used as the entrainer, the vapours may be cooled and condensed by direct contact with a spray of cold water. In this manner an efficient recovery of the sublimate is made, but the product is obtained in the wet state.

The industrial purification of salicylic acid provides a good example of the use of entrainer sublimation. Air may be used as the carrier gas, but as

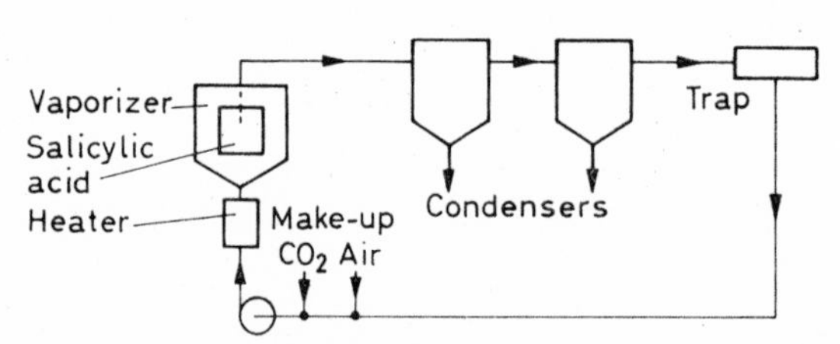

Figure 7.8. An entrainer sublimation process used for salicylic acid purification

salicylic acid can be decarboxylated in hot air, a mixture of air and CO_2 is often preferred. The process shown in *Figure 7.8* is carried out batchwise. A 5–10 per cent mixture of CO_2 in air is recycled through the plant, passing over heater coils before passing over the containers, e.g. bins or trays, holding the impure salicylic acid in the vaporizer. The vapours then pass to a series of air-cooled chambers where the sublimed salicylic acid is deposited. A trap removes any entrained sublimate before the gas stream is returned to the heater. Make-up CO_2 and air are introduced to the system as required, and the process continues until the containers are emptied of all volatile matter.

A typical example of a continuous sublimation plant is shown in *Figure 7.9*. The impure material is pulverized in a mill, and hot air or any other suitable gas mixture blows the fine particles, which readily volatilize, into a series of separators, e.g. cyclones, where non-volatile solid impurities are removed; a filter may also be fitted in the vapour lines to remove final traces of impurity.

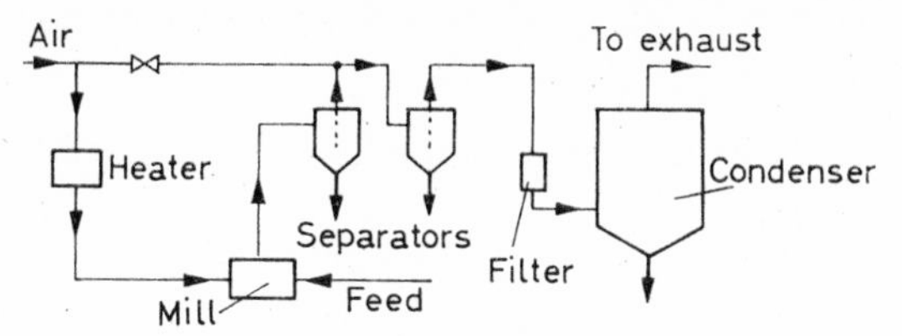

Figure 7.9. A typical continuous sublimation unit

The vapours then pass to a series of condensers. The exhaust gases can be recycled or passed to atmosphere through a cyclone or wet scrubber.

The yield from an entrainer sublimation process can be estimated as follows. If the inert gas mass flow rate is denoted by G, the mass rate of sublimation, S, will be given by

$$\frac{G}{S} = \frac{\rho_G p_G}{\rho_S p_S} \tag{5}$$

where p_G and p_S are the partial pressures of the inert gas and vaporized substance, respectively, in the vapour stream, ρ_G and ρ_S are their vapour densities at the temperature of vapour stream. The total pressure, P, of the system will be the sum of the partial pressures of the components

$$P = p_G + p_S$$

so equation 5 can be written

$$S = G\left(\frac{\rho_S}{\rho_G}\right)\left(\frac{p_S}{P - p_S}\right) \tag{6}$$

or, in terms of the molecular weights of the inert gas, M_G, and the material being sublimed, M_S,

$$S = G\left(\frac{M_S}{M_G}\right)\left(\frac{p_S}{P - p_S}\right) \tag{7}$$

The use of equation 7 for predicting the yield from an entrainer sublimation process is illustrated in the following example.

Example—It is proposed to purify salicylic acid (m.p. 159° C) by entrainer sublimation with air at 150° C. The vapours pass to a series of condensers, the internal temperature and pressure of the last being 40° C and atm. (760 mm Hg). The air flow rate is to be 5,000 lb./h, the expected pressure drop between vaporizer and last condenser, 6 in. water gauge. The vapour pressures of salicylic acid at 150 and 40° C are 10·8 and 0·017 mm Hg, respectively. Calculate the maximum possible rate of sublimation and the quantity of salicylic acid remaining uncondensed in the exit gases.

Solution—The maximum rate of sublimation is calculated by assuming that the air stream leaving the vaporizer is saturated with salicylic acid vapour.

Molecular weights: salicylic acid = 138, air = 29.

(*a*) *Vaporization stage* (150° C)

The total pressure, P, in the vaporizer is equal to 760 mm Hg plus the pressure drop (6 in. water gauge = 11 mm Hg). Therefore, $P = 771$ mm Hg; $p_S = 10{\cdot}8$ mm Hg, and $G = 5{,}000$ lb./h. Substituting these values in equation 7

$$S = 5{,}000\left(\frac{138}{29}\right)\left(\frac{10{\cdot}8}{771 - 10{\cdot}8}\right)$$
$$= 338 \text{ lb./h}$$

(*b*) *Condensation stage* (40° C)

At this temperature $p_S = 0{\cdot}017$ mm Hg, total pressure $P = 760$ mm Hg. Again, using equation 7

$$S = 5{,}000\left(\frac{138}{29}\right)\left(\frac{0{\cdot}017}{760 - 0{\cdot}017}\right)$$
$$= 0{\cdot}53 \text{ lb./h}$$

Therefore, the rate of sublimation is 338 lb./h, while the loss from the condenser exit gases is only 0·53 lb./h.

It must be noted, however, that this maximum yield will only be obtained if the air is saturated with salicylic acid vapour at 150° C, and saturation will only be approached if the air and salicylic acid are contacted for a sufficient period of time at the required temperature. A fluidized-bed vaporizer may approach these optimum conditions, but if air is simply blown over bins or trays containing the solid, saturation will not be reached and the actual rate of sublimation will be less than calculated. In some cases the degree of saturation achieved may be as low as 10 per cent of the possible value. Little comment is necessary, therefore, on the importance of designing an efficient vaporizer.

Only the minimum loss of product in the exit gases can be calculated. Any other losses due to entrainment, which might be considerable, will depend on the design of the condenser and cannot be calculated theoretically. An efficient scrubber can, of course, minimize or eliminate these losses.

Sublimer condensers usually take the form of large air-cooled chambers which generally give very low heat transfer rates, probably not greater than 1 or 2 B.t.u./h·ft.^{2} °F. This is only to be expected because sublimate deposits on the condenser walls and acts as a lagging. In addition, vapour velocities within the chambers are generally very low. Quenching of the vapours with cold air in the chamber may increase the rate of heat removal, but excessive nucleation is liable to occur. If this happens the material is deposited as a fine snow. The condenser walls may be kept clear of solid by using internal scrapers or brushes, or even swinging weights. All vapour lines in sublimation units should be of a wide bore, adequately insulated and, if necessary, provided with a heating source to minimize blockage due to the build-up of sublimate.

One of the main hazards of air-entrainment sublimation is the risk of fire; many substances that are considered to be quite safe in their normal state can produce explosive mixtures with air. All the electrical equipment, lights, etc., should be flame-proof, and every precaution should be taken to avoid the accidental production of a spark. Personnel should wear rubber-soled shoes, and all handling and maintenance tools should be made of non-sparking materials. In some cases the vapour lines and condenser chambers can become charged with static electricity, so it is essential that these items of equipment should be earthed efficiently.

In sublimation plants it is usually necessary to have a battery of condensers linked in series. The internal temperatures of the condensers will decrease down the line, and it is generally found that the crystal form of the product alters from condenser to condenser. In the first chamber, where the temperature is fairly high, the rate of crystal growth may exceed the rate of condensation, and large crystals, frequently acicular, will grow. As the vapours proceed through the condensers the crystal form will reduce in size, and a fine fluffy product may be deposited in the last. In the absence of a market demand for different crystal formations, a blending operation will be necessary. If the first condenser in the battery is too cold, the entering vapours become heavily supersaturated and spontaneous nucleation will occur, with the result that a light flocculent mass (snow) is deposited. The operating conditions necessary for the control of crystal growth in sublimers can only be found from experience.

Washing and Lixiviation

The actual crystals produced in a crystallizer are themselves essentially pure, yet after they have been separated from their mother liquor and dried, the resultant mass may be relatively impure. The reason for this is that the removal of mother liquor is frequently inadequate. Crystals retain a small quantity of mother liquor on their surfaces by adsorption, and a larger amount within the voids of the mass due to capillary attraction. If the crystals are irregular the amount of mother liquor retention within the crevices may be considerable; crystal clusters and agglomerates are notorious in this respect.

All crystallization processes that utilize a liquid working medium must thus be followed by an efficient liquid–solid separation. Centrifugal filtration can reduce the mother liquor content to about 1 per cent by weight in certain cases, but regular crystalline masses are usually discharged from a centrifuge containing 5–10 per cent. For irregular fine crystals, however, even centrifuging may yield a product containing up to 50 per cent by weight of mother liquor. It is extremely important, therefore, not only to use the most efficient type of filter, but to produce regular crystals in the crystallizer.

After filtration, the product is usually given a wash to reduce still further the amount of impure mother liquor retained. On a centrifuge, for example, the filter cake is best washed with the basket rotating at a speed slower than used during the filtration stage. Hoses and watering cans are often employed for this purpose, but these methods are most inefficient. Uniform washing is best achieved by the use of a series of nozzles which direct an even spray over the depth of the cake. A suitable wash liquor tank can be installed above the machine so that, on the operation of a simple on–off valve, each basket load receives a fixed quantity of wash.

Cakes of soluble crystalline substances should not be too thick if a washing operation is required; otherwise the wash liquor becomes saturated long before it passes through the mass, and the mother liquor impurities are not removed effectively; prolonged washing would, of course, reduce the final yield. The washing operation is followed by maximum-speed spinning to remove the wash liquor. A two-way outlet can be fitted to the centrifuge casing so that the strong mother liquor and weak wash liquor are directed to separate collection points.

If the crystals are very soluble in the mother solvent, another liquid in which the substance is relatively insoluble may be used for washing purposes. The wash liquid should be miscible with the mother solvent. For example, water could be used for washing substances crystallized from methanol, and methanol could be used for washing substances crystallized from benzene. Unfortunately, this 'two-solvent' method usually means that a solvent recovery unit is required. Alternatively, a wash liquor consisting of a cold saturated solution of the pure substance in the pure mother solvent could be used if the crystals were appreciably soluble. The contaminated wash liquor could then be recycled or re-used in some other way.

In extreme cases when very pure crystals are required, or when simple washing is inadequate, two mother liquor removal stages may be necessary. The wet crystals could be removed from the filter, re-dispersed in pure cold solvent, and then filtered off again. There may, of course, be an appreciable loss of yield after such a washing or lixiviation process, but this would be less than that after a complete recrystallization.

When crystallization has been carried out by the reaction method (p. 165) the crystalline material may be relatively insoluble in the working solvent. The mother liquor, however, may contain large quantities of soluble material, and simple filtration and washing may be quite inadequate for its complete removal, especially if the crystalline particles are very fine. For example, barium sulphate can be precipitated as a crystalline product by mixing hot solutions of barium chloride and sodium sulphate, but the $BaSO_4$ crystals are usually very small indeed, and difficulties are encountered in both the filtration and washing operations. In this case the sodium chloride in solution could be removed by a washing and decantation technique carried out either batchwise or continuously.

The wash liquor requirements for decantation washing can be deduced rapidly by a graphical method proposed by KIRBY[22] who has also described a submerged filter for increasing the efficiency of the process[23]. The theory of decantation washing outlined by Kirby is similar to the theory of leaching proposed by DONALD[24]. The assumptions are that all the soluble impurity is in solution and that no concentration gradients exist in the agitated vessel.

Let Y_0, Y_1 and Y_n be the impurity contents of the material (lb. of impurity per lb. of product) initially, after stage 1 and after stage n, respectively. Let

F be the fraction of liquid removed at each decantation. Then mass balances over the various stages give

stage 1: $$Y_1 = Y_0 - FY_0 = Y_0(1 - F)$$

stage 2: $$Y_2 = Y_0(1 - F)^2$$

stage n: $$Y_n = Y_0(1 - F)^n \qquad (8)$$

Equation 8 may be rewritten in the form

$$\ln (Y_n/Y_0) = n . \ln (1 - F) \qquad (9)$$

This equation is plotted in *Figure 7.10*.

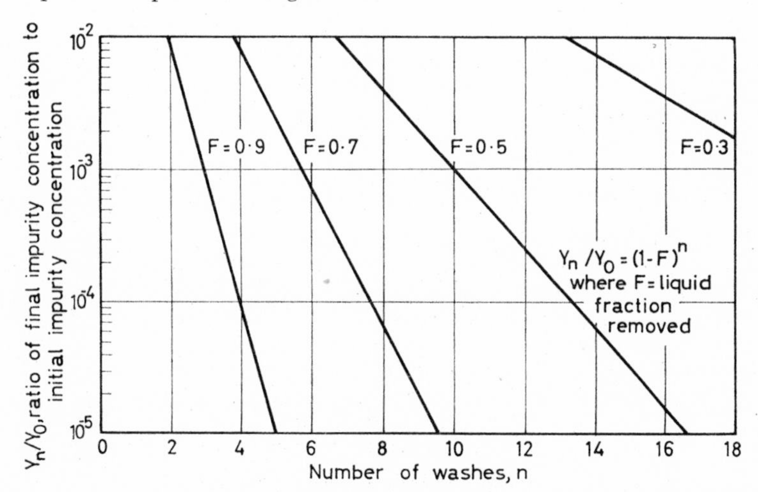

Figure 7.10. Graphical method for estimating the number of washes required in batch lixiviation. (*After* T. KIRBY[22])

For washing on a continuous basis, where fresh wash liquid enters the vessel continuously and liquor is continually being withdrawn through a filter screen, a mass balance over the unit gives

$$VdY = -YdW$$

or

$$\frac{W}{V} = -\ln \frac{Y_n}{Y_0} \qquad (10)$$

where Y_0 and Y_n are the initial and final impurity concentrations, V and W the volumes of the liquor in the vessel and wash water, respectively. *Figure 7.11* is drawn on the basis of $V = 1{,}000$ gal. From equations 9 and 10 we get

$$n . \ln (1 - F) = -\frac{W}{V} \qquad (11)$$

or, rearranging and dividing by F

$$\frac{W}{nFV} = \frac{-\ln (1 - F)}{F} \qquad (12)$$

In batch washing, the quantity nFV represents the amount of wash liquor used. The following example[22] illustrates the use of the above equations and of *Figures 7.10* and *7.11*.

Example—265 lb. of zinc carbonate is produced by the reaction

$$\begin{array}{ccc} Zn(NO_3)_2 + Na_2CO_3 & \rightarrow & ZnCO_3 + 2NaNO_3 \\ 400 + 224 & & 265 + 359 \end{array}$$

(*a*) Determine the number of washes required to reduce the $NaNO_3$ concentration to 0·01 per cent in the final product. The fraction of liquid

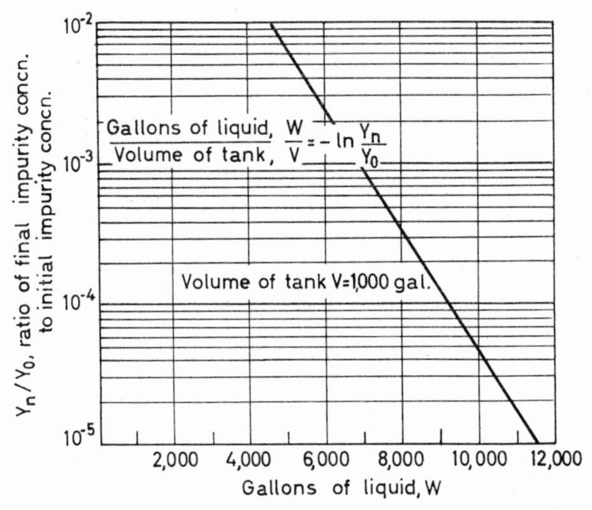

Figure 7.11. Graphical method for estimating the volume of wash liquor required in a continuous lixiviation. (*After* T. KIRBY[22])

removed in each decantation is determined by observing the settling of the slurry in a graduated cylinder. In this case, $F = 0{\cdot}7$.

(*b*) Calculate the volume of wash water required if 100 gal. of the reaction mixture is to be washed continuously to produce the same product as in (*a*).

(*c*) Compare the wash water consumptions of the batch and continuous processes.

Solution—(*a*)

$$\begin{aligned} Y_0 &= 359/265 = 1{\cdot}35 \\ Y_n &= 0{\cdot}0001 \\ Y_n/Y_0 &= 0{\cdot}000074 \\ F &= 0{\cdot}7 \end{aligned}$$

From *Figure 7.10* it can be seen that for these values of F and Y_n/Y_0, $n = 7{\cdot}9$. Therefore, eight washes would be required.

(*b*) For a 1,000 gal. charge and a value of $Y_n/Y_0 = 0{\cdot}000074$, the corresponding value of W can be read off *Figure 7.11* as 9,500 gal. Therefore, for a 100 gal. charge, $W = 950$ gal. Thus 950 gal. of liquor has to be drawn through the filter before the $NaNO_3$ concentration falls to the required value.

(*c*) The ratio of water consumption for the two methods can be determined from equation 12 written in the form

$$\frac{W_{\mathrm{cont}}}{W_{\mathrm{batch}}} = \frac{\ln(1-F)}{F} = \frac{\ln(1-0\cdot7)}{0\cdot7} = 1\cdot72$$

In this case, therefore, the continuous method uses 72 per cent more water than the batch method.

Caking of Crystals

One of the most troublesome properties of crystalline materials is their tendency to bind together, or cake, on storage. Most crystalline products are required in a free-flowing form; they should, for example, flow readily out of containers, e.g. sugar and table salt, or be capable of being distributed evenly over surfaces, e.g. fertilizers. Handling, packaging, tabletting and many other operations are all made easier if the crystalline mass remains in a particulate state. Caking not only destroys the free-flowing nature of the product but also necessitates some crushing operation, either manual or mechanical, before it can be used.

The causes of caking may vary for different materials. Crystal size, shape, moisture content, the pressure under which the product is stored, temperature and humidity variations during storage and storage time can all contribute to the compacting of a granular crystalline product into a solid lump. In general caking is caused by the crystal surfaces becoming damp; the solution which is formed evaporates later and unites adjacent crystals with a cement of recrystallized solid. Crystal surfaces can become damp in a number of ways—the product may contain traces of solvent left behind due to inefficient drying, or the moisture may come from external sources.

Take, for example, the case of a water-soluble substance. If the partial pressure of water vapour in the atmosphere is greater than the vapour pressure which would be exerted by a saturated aqueous solution of the pure substance at that temperature, water will be absorbed by the crystals. If, later, the atmospheric moisture content is reduced to give a partial pressure below the vapour pressure of the saturated solution, the crystals will dry out and bind together. Small fluctuations in atmospheric temperature and humidity, sufficient to bring about these changes, can occur several times in one day.

A simple test for determining the atmospheric humidities at which a particular mass of crystals will absorb moisture consists of placing samples of the crystals in desiccators containing atmospheres of different known moisture content. Solutions of various strengths of sulphuric acid may be placed in the base of the desiccator; the equivalent relative humidities of the resulting atmospheres can be calculated from vapour pressure data. Atmospheres of constant relative humidity can also be obtained by using saturated solutions of various salts. *Table 7.1* gives a list of the percentage relative humidities of the atmospheres above various saturated salt solutions at 60° F.

Table 7.1. Percentage Relative Humidities of the Atmospheres above Saturated Solutions of Various Pure Salts at 60° F

Substance	*Formula*	*Stable phase at* 60° F	*Percentage relative humidity*
Lead nitrate	$Pb(NO_3)_2$	anhyd.	98
Disodium phosphate	Na_2HPO_4	$12H_2O$	95
Sodium sulphate	Na_2SO_4	$10H_2O$	93
Sodium bromate	$NaBrO_3$	anhyd.	92
Dipotassium phosphate	K_2HPO_4	anhyd.	92
Potassium nitrate	KNO_3	anhyd.	92
Sodium carbonate	Na_2CO_3	$10H_2O$	90
Zinc sulphate	$ZnSO_4$	$7H_2O$	90
Potassium chromate	K_2CrO_4	anhyd.	88
Barium chloride	$BaCl_2$	$2H_2O$	88
Potassium bisulphate	$KHSO_4$	anhyd.	86
Potassium bromide	KBr	anhyd.	84
Ammonium sulphate	$(NH_4)_2SO_4$	anhyd.	81
Ammonium chloride	NH_4Cl	anhyd.	79
Sodium chloride	$NaCl$	anhyd.	78
Sodium nitrate	$NaNO_3$	anhyd.	77
Sodium acetate	$C_2H_3O_2Na$	$3H_2O$	76
Sodium chlorate	$NaClO_3$	anhyd.	75
Ammonium nitrate	NH_4NO_3	anhyd.	67
Sodium nitrite	$NaNO_2$	anhyd.	66
Magnesium acetate	$Mg(C_2H_3O_2)_2$	$4H_2O$	65
Sodium bromide	$NaBr$	$2H_2O$	58
Sodium dichromate	$Na_2Cr_2O_7$	$2H_2O$	52
Potassium nitrite	KNO_2	anhyd.	45
Potassium carbonate	K_2CO_3	$2H_2O$	43
Calcium chloride	$CaCl_2$	$6H_2O$	32
Lithium chloride	$LiCl$	H_2O	15

The percentages listed in *Table 7.1* can also be considered to be the critical humidities of the salts at 60° F; if the atmospheric humidity is greater than the critical the salt becomes hygroscopic. The term 'deliquescence' is usually reserved for the case where the substance absorbs atmospheric moisture and continues to do so until it becomes completely dissolved in the absorbed water. This will only occur if the vapour pressure of the salt solution always remains lower than the partial pressure of the water vapour in the atmosphere. Calcium chloride, with a critical humidity of 32 per cent at 60° F, is a well known example of a deliquescent salt. The term 'efflorescence' refers to the loss of water of crystallization from a salt hydrate; this occurs when the vapour pressure exerted by the hydrate exceeds the partial pressure of water vapour in the atmosphere. Sodium sulphate decahydrate, with a critical humidity of 93 per cent, is an example of an efflorescent salt.

Commercial crystalline salts frequently exhibit hygroscopy at atmospheric humidities lower than those given for the pure salts in *Table 7.1*. Usually impurities present in the product cause the trouble. For example, traces of calcium chloride in sodium chloride will render the crystals damp at very low atmospheric humidities. Removal of the hygroscopic impurity would be the answer, but this is not always economical. Coating of the crystals with a fine inert dust will often prevent the mass becoming damp; table salt, for instance, can be coated with magnesium carbonate.

A number of precautionary measures can be taken to minimize the possibility of caking. One obvious method would be to pack the perfectly dry crystals in a dry atmosphere, store them in an air-tight container and prevent any pressure being applied during storage. These desirable conditions, however, cannot always be obtained. Caking can also be minimized by reducing the number of contacts between the crystals, and this can be done by endeavouring to produce granular crystals of a uniform size. The crystals should be as large as possible; the smaller they are the larger will be the exposed surface area per unit mass. The actual size, however, is only

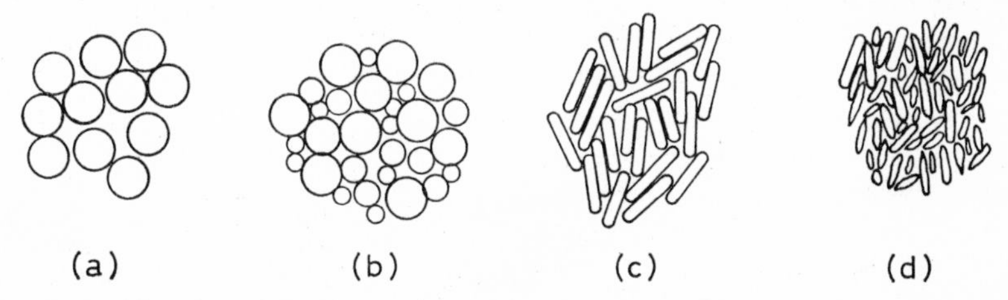

Figure 7.12. Effect of particle shape on the caking of crystals: (a) large uniform granular crystals (good); (b) non-uniform granular crystals (poor); (c) large uniform elongated crystals (poor); (d) non-uniform elongated crystals (very bad)

of secondary importance in this respect. Shape and uniformity of the crystals, on the other hand, are extremely important factors affecting caking, as indicated in *Figure 7.12.*

The minimum caking tendency is shown by uniform granular crystals (*Figure 7.12a*). Even if caking did occur, the mass could easily be broken up because of the open structure and the relatively few points of contact per unit volume. Non-uniform granular crystals (*Figure 7.12b*) are more prone to caking; the voids in between the large granules are filled with the smaller particles, resulting in a larger number of points of contact per unit volume, and the caked mass will not be broken up as easily as in the former case. Although the crystals depicted in *Figure 7.12c* are uniform, they are also elongated and may tend to cake badly. Here there are areas, as well as points, of contact; the needles can pack together and set hard on caking. The condition shown in *Figure 7.12d* is even worse. The crystals are both elongated and non-uniform; they can pack together very tightly into a mass with negligible voidage. When this sort of mass cakes, it is often quite impossible to transform it back to its original particulate state. Plate-like crystals also have bad caking characteristics.

Shape and uniformity of the particles also affect the behaviour of the product under storage. If the crystals are packed in bags and stacked on top of one another, the pressure in the bags near the bottom of the pile tends to force the crystals into closer contact; with non-uniform crystals this compaction may be quite severe and, in extreme cases, many of the crystals may be crushed. If the solubility of the salt in water increases with an increase in pressure, traces of solution may be formed under the high local pressure at the points of contact. The solution will then tend to flow into the voids, where the pressure is lower, and crystallize. Storage of crystalline materials under pressure should always be avoided if possible.

Controlled crystallization, coupled with some form of classifying action in the crystallizer, helps to produce crystals of uniform size. The production of granular crystals, however, may demand the careful control of other conditions of crystallization to modify the crystal habit (see Chapter 5); the rate of cooling, the degree of supersaturation and the pH of the crystallizing solution can exert considerable influence. The deliberate addition of traces of impurity, in the form of active ions or surface-active agents, may also help to produce the right type of crystal. BUCKLEY[25] reported a large number of cases of induced habit modification in a wide variety of substances. Unfortunately, there is not a great deal of published information on the control of crystal habit in industrial crystallizers, but a recent review on this subject, by GARRETT[26], includes many interesting examples and gives a clear indication of the possibilities of habit control on a large scale.

The use of coating agents has already been mentioned above in connection with table salt, where magnesium carbonate is frequently used. Icing sugar can be coated with about 0·5 per cent of tricalcium phosphate or cornflour to prevent caking. Many other anti-caking and flow-conditioning agents are used for industrial crystalline materials; chalk, zinc oxide, calcium sulphate, kaolin, diatomaceous earth, magnesium aluminium silicate, aluminium powder, synthetic resins and paraffin wax are a few examples of materials that have been used. Finely divided substances used as dusting agents must have a good 'covering power' so that very small quantities will produce the desired effect.

Several comprehensive accounts have been given of the methods employed for conditioning crystals[18, 27, 28] and of testing procedures[28–31] used to determine the flow properties and caking tendency of crystalline materials. Much of the published work is concerned with ammonium nitrate, manufactured on a very large scale for use in explosives and as a fertilizer, which presents a rather special case on account of its potential explosion hazard[32]; organic coating agents are not recommended in this case as these have been shown to increase the tendency for the salt to detonate.

WHETSTONE[27, 33, 34] proposed a novel method for preventing caking in water-soluble salts whereby the crystals are treated with a very dilute solution of a habit-modifying dyestuff, e.g. Acid Magenta. Should the crystals cake at some later stage, the cement of crystalline material, now consisting of a modified habit, will be very weak and the crystal mass will easily break down on handling. This method has proved effective with salts such as ammonium nitrate, ammonium sulphate and potassium nitrate.

REFERENCES

[1] GRIFFITHS, H., Mechanical crystallization, *J. Soc. chem. Ind., Lond.* 44 (1925) 7T

[2] McCABE, W. L., Crystal growth in aqueous solutions, *Industr. Engng Chem.* 21 (1929) 30

[3] McCABE, W. L., Crystallization, in *Chemical Engineers' Handbook*, J. H. Perry (Ed.), 3rd ed., 1950. New York; McGraw-Hill

[4] HOOKS, I. J. and KERZE, F., Nomograph for crystal size prediction, *Chem. metall. Engng* 53 (7) (1946) 140

[5] MILLER, P. and SAEMAN, W. C., Continuous vacuum crystallization of ammonium nitrate, *Chem. Engng Progr.* 43 (1947) 667

[6] SAEMAN, W. C., Crystal size distribution in mixed suspensions, *Amer. Instn chem. Engrs J.* 2 (1956) 107

[7] ROBINSON, J. N. and ROBERTS, J. E., Crystal growth in a cascade of agitators, *Can. J. chem. Engng* 35 (Oct., 1957) 105

[8] THOMPSON, A. R. and MOLSTAD, M. C., Solubility and density isotherms for potassium and ammonium nitrates in isopropanol solutions, *Industr. Engng Chem.* 37 (1945) 1244

[9] GEE, E. A., CUNNINGHAM, W. K. and HEINDL, R. A., Production of iron-free alum, *Industr. Engng Chem.* 39 (1947) 1178

[10] PURDON, F. F. and SLATER, V. W., *Aqueous Solution and the Phase Diagram*, 1946. London; Arnold

[11] GARRETT, D. E., Industrial crystallization at Trona, *Chem. Engng Progr.* 54 (12) (1958) 65

[12] CHIVATE, M. R. and SHAH, S. M., Separation of *m*-cresol and *p*-cresol by extractive crystallization, *Chem. Engng Sci.* 5 (1956) 232

[13] FINDLAY, R. A. and WEEDMAN, J. A., Separation and purification by crystallization, in *Advances in Petroleum Chemistry and Refining*, Vol. 1, K. A. Kobe and J. J. McKetta (Eds.), 1958. New York; Interscience

[14] SAVITT, S. A. and OTHMER, D. F., Separation of *m*- and *p*-cresols from their mixtures, *Industr. Engng Chem.* 44 (1952) 2428

[15] EGAN, C. J. and LUTHY, R. V., Separation of xylenes, *Industr. Engng Chem.* 47 (1955) 250

[16] BAILEY, W. A., BANNERDOT, R. A., FETTERLY, L. C. and SMITH, A. G., Urea extractive crystallization of straight-chain hydrocarbons, *Industr. Engng Chem.* 43 (1951) 2125

[17] HOLLAND, A. A., A new type of evaporator, *Chem. Engng* 58 (1) (1951) 106

[18] SHEARON, W. H. and DUNWOODY, W. B., Ammonium nitrate, *Industr. Engng Chem.* 45 (1953) 496

[19] VAN DEN BERG, P. J. and HALLIE, G., New developments in granulation techniques *Proc. Fertilizer Soc., Lond.*, 1960

[20] MULLIN, J. W., Sublimation in theory and practice, *Industr. Chem. Mfr* 31 (1955) 540

[21] KEMP, S. D., Sublimation and vacuum freeze drying, in *Chemical Engineering Practice*, H. W. Cremer and T. Davies (Eds.), Vol. 6, 1959. London; Butterworths

[22] KIRBY, T., Wash liquor requirements, *Chem. Engng* 66 (8) (1959) 169

[23] KIRBY, T. and FEORINO, J., Lamp-shade filter to increase the efficiency of batch decantation washing, *Chem. Engng Progr.* 55 (11) (1959) 174

[24] DONALD, M. B., Percolation leaching in theory and practice, *Trans. Instn chem. Engrs, Lond.* 15 (1937) 77

[25] BUCKLEY, H. E., *Crystal Growth*, 1952. London; Chapman & Hall

[26] GARRETT, D. E., Industrial crystallization—the influence of chemical environment, *Brit. chem. Engng* 4 (1959) 673

[27] WHETSTONE, J., Anti-caking treatment for ammonium nitrate, *Industr. Chem. Mfr* 25 (1949) 401

[28] IRANI, R. R., CALLIS, C. F. and LIU, T., Flow conditioning and anti-caking agents, *Industr. Engng Chem.* 51 (1959) 1285

[29] HARDESTY, J. O. and ROSS, W. H., Factors affecting the granulation of fertilizers, *Industr. Engng Chem.* 30 (1938) 668

[30] ADAMS, J. R. and ROSS, W. H., Relative caking tendencies of fertilizers, *Industr. Engng Chem.* 33 (1941) 121
[31] MILLER, P. and SAEMAN, W. C., Properties of monocrystalline ammonium nitrate, *Industr. Engng Chem.* 40 (1948) 154
[32] WHETSTONE, J. and HOLMES, A. W., Explosion and fire hazards with ammonium nitrate, *Industr. Chem. Mfr* 23 (1947) 717
[33] WHETSTONE, J., The effect of crystal habit modification on the setting of inorganic oxy-salts, *Disc. Faraday Soc.* 5 (1949) 261
[34] BUTCHART, A. and WHETSTONE, J., The effect of dyes on the crystal habits of some oxy-salts, *Disc. Faraday Soc.* 5 (1949) 254

8

CRYSTALLIZATION EQUIPMENT

THERE are many ways in which the various types of industrial crystallizers can be classified. Self-explanatory headings such as 'batch' and 'continuous', or 'agitated' and 'non-agitated' equipment may be used, but these descriptions are too generalized for most purposes. The units may be classified according to the degree of control that can be exercised over the final product size; names such as 'controlled' and 'uncontrolled', or 'classifying' and 'non-classifying' crystallizers are commonly encountered.

Another method of classification is based on the way in which the necessary supersaturation of the solution or melt is achieved, e.g. by cooling, evaporation or reaction between two phases. If cooling is effected by flash evaporation at low pressure, the term 'vacuum' crystallizer may be applied. The manner in which the growing crystals are brought into contact with the supersaturated liquor has also been suggested for classification purposes[1]; according to this system the three headings 'non-agitated', 'circulating liquor' and 'circulating magma' crystallizers are used.

In the following section the various items of equipment will be divided into three main groups: cooling, evaporating and vacuum crystallizers, but reference will be made to the other classification systems. Much of this information has already been given in another publication[2].

COOLING CRYSTALLIZERS

Open Tank Crystallizers

In its simplest form an open tank crystallizer consists of a smooth-walled vessel with no mechanical moving parts, which permits a large surface area of solution to come into contact with the atmosphere. A hot concentrated solution of the solute is poured into the tank where it cools by natural convection and evaporation over a given period of time, often as long as a few days. A gentle circulation of the liquor usually occurs owing to temperature and concentration gradients within the vessel. Cooling may be aided by the passage of cold air over the surface or by bubbling cold air through the liquid.

Certain melts and aqueous solutions can be processed in this manner, and the batch may be given an occasional stir with a hand paddle to prevent the deposition of a hard crystalline block on the bottom of the crystallizer. Sometimes thin rods are hung in the solution; crystals grow on the rods and are thus prevented from falling to the bottom of the crystallizer. No seeding is required because nucleation usually occurs quite readily due to the presence of atmospheric dust and crystalline fragments left from previous batches.

The crystal magma can be transferred by hand, or mechanical grab, to a filtration unit. Alternatively, a large proportion of the mother liquor may be drained off at the bottom of the tank, thus minimizing the quantity of liquid to be handled by the filter. In batteries of large effluent tanks used for by-product recovery, sufficient settling time may be allowed to permit continuous entry of feedstock and continuous outflow of crystal-free mother liquor; when a given tank is full of crystalline product, it is by-passed and emptied.

No control over the crystal size is possible in open tank crystallizers but experience will indicate the required size of the vessel and the cooling time necessary to produce the desired type of crystalline mass. Because of the slow rate of cooling, a high percentage of large interlocking crystals is usually obtained, but variation in size will be considerable, ranging from fine dust to large lumps. The irregularity of the crystals results in a high retention of mother liquor within the mass after filtration. Consequently, the dried crystals will almost invariably be impure.

For small production of commercial grade crystals, and for by-product recovery purposes, the open tank method of crystallization may be the only economical one. The capital outlay is usually small, and maintenance and operating costs are generally negligible. On the other hand, handling labour costs can be rather high, in terms of cost per pound of product, compared with other crystallization methods. Reagent-grade chemicals are usually manufactured in open-pan crystallizers[3].

Agitated Tanks and Vessels

Undesirable temperature gradients in an open tank crystallizer can be minimized if a mechanical stirrer is employed; smaller and more uniform crystals will be deposited and the cooling time cycle may be reduced. A somewhat purer product will be obtained as a result of the smaller retention of mother liquor by the crystalline mass, and more efficient washing of the crystals is possible during the filtration stage of the process.

The agitated vessel may be equipped with a water jacket, or coils, to aid the cooling process. Jackets are preferred to coils because these tend to become coated with a hard crystalline deposit which reduces very considerably the rate of heat transfer. For very high cooling duties, however, jackets are usually inadequate, and coils have to be used to provide the necessary heat transfer area. One useful rule to remember is that temperature differences greater than 10° C between the cooling surface and the liquor are undesirable because high local degrees of supersaturation lead to excessive nucleation and crystal deposition.

If a cooling jacket is employed, the inner surfaces of the crystallizer should be as smooth and as flat as possible. This minimizes crystal build-up on the cold surfaces or, when deposition does occur, facilitates its removal. Crystals should never be chipped away from the wall of a crystallizer because tiny scratches on the surface become undesirable 'seed centres'. Dissolution is the only safe method of crystal scale removal. Polished stainless steel and glass-lined mild steel are good materials of construction for the inner surfaces of crystallizers.

Although operating and maintenance costs for an agitated cooler are higher than for a simple tank crystallizer, they are small in comparison with the financial advantages gained by the quicker throughput and better product. Labour costs for handling the crystals, however, may still be rather high. The main disadvantage of both tank and agitated crystallizers is that the equipment is generally bulky and occupies much valuable floor space.

No general design for tank crystallizers, stirred or otherwise, can be proposed; the vessels will, according to the needs of the particular process, vary from small shallow pans to very large cylindrical tanks. When non-aqueous solvents are involved, the equipment will take the form of an enclosed agitated vessel fitted with coils or a cooling jacket and a vent leading to a water-cooled condenser.

A typical operating sequence for a batch agitated crystallizer is as follows. The hot solution is cooled as rapidly as possible until supersaturation is achieved. The cooling rate is then reduced to prevent the solution becoming labile, and a pre-determined quantity of suitable seed crystals is added. Seeding may not be necessary in all cases, but it is generally safer to seed than to rely on spontaneous nucleation which may be uncontrolled and excessive. Once crystallization commences the temperature of the charge will tend to rise due to the liberation of the heat of crystallization. The cooling rate is therefore adjusted to allow crystallization to occur at a fixed, or slowly decreasing, temperature. Slow cooling is continued until a large proportion of the solute has been deposited on the seeds; thereafter the rate of cooling can be increased until the required discharge temperature has been achieved. Trial and error on the actual plant will decide the finer details of the procedure. For the crystallization of substances which oxidize in contact with air, an inert or reducing atmosphere may be introduced into the closed vessel. A sulphur dioxide atmosphere, for example, is used for hydroquinone crystallization in agitated vessels to prevent the darkening of the product.

A description of modern agitated tank crystallizers used for the large-scale production of borax from natural brines at Trona, California, has been given by several authors[4, 5]. The units used in this particular case are of the Pachuca-type growth crystallizers, as shown in *Figure 8.1.* The large tall

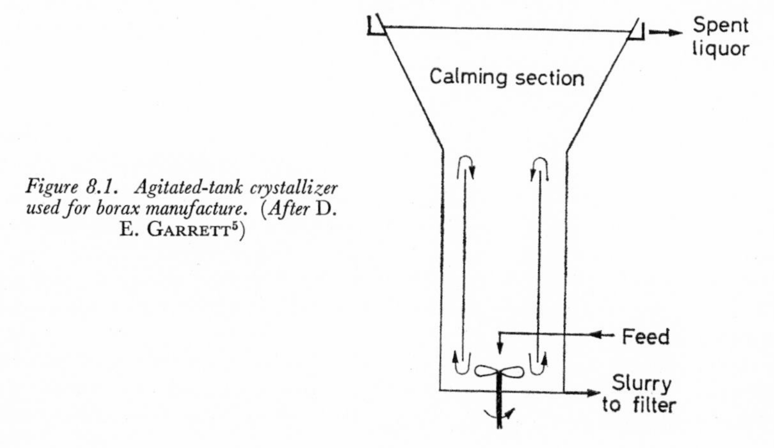

Figure 8.1. Agitated-tank crystallizer used for borax manufacture. (*After* D. E. Garrett[5])

cylindrical vessels have an upper classifying section which permits an outflow of liquor free from crystals of an appreciable size and maintains a predetermined sludge density in the growth zone. The under-driven agitator maintains a circulation of magma through the shroud as indicated. Feed solution enters continuously at a point on the suction side of the agitator. Fairly large uniform crystals can be grown by this method.

The supersaturation, and hence the crystallization, of certain solutions can be effected by the addition of a suitable third component, e.g. another liquid, to the system. This method of crystallization, known by names such as salting-out, watering-out, dilution crystallization and precipitation, has already been discussed in Chapter 7. Careful control of batch temperature, rate of diluent addition and, sometimes, deliberate seeding permit a considerable degree of influence to be exercised over size and regularity of the crystalline product. As in the case of the other cooling crystallizers the types of equipment used vary widely. Some form of closed, jacketed and agitated vessel is normally employed, and the vessels may have up to 1,000 gal. or more capacity.

Trough Crystallizers

Crystallizers consisting of a long shallow trough with an internal agitator and a cooling system are frequently encountered in industries where concentrated and viscous solutions are to be crystallized. The use of the trough

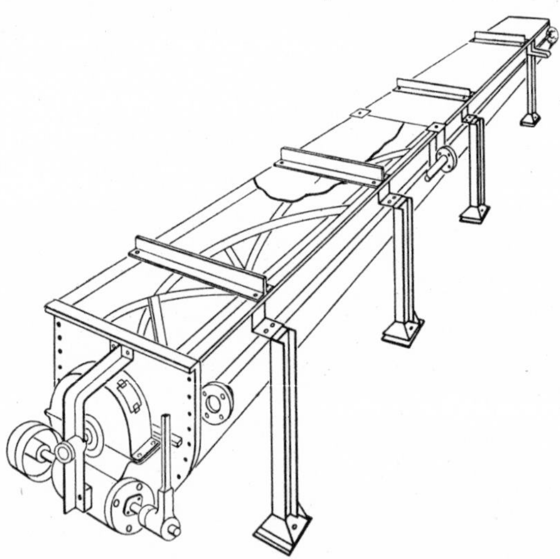

Figure 8.2 Swenson–Walker crystallizer. (*After* G. E. SEAVOY and H. R. CALDWELL[6])

crystallizer is, of course, not restricted to viscous solutions only. A well known piece of equipment belonging to this class is the Swenson–Walker crystallizer[6] illustrated in *Figure 8.2*. One unit of this crystallizer consists of a semi-cylindrical trough about 2 ft. wide and 10–15 ft. long, generally fitted with a water-cooled jacket. The trough may be open or closed, depending upon whether additional atmospheric cooling is required or not. A helical stirrer rotates at a slow speed (5–10 rev/min) inside the trough to aid the growth of the crystals by lifting them and then allowing them to fall back through the solution. The system is kept in gentle agitation and the crystals are conveyed along the trough. This crystallizer, therefore, may be considered as belonging to the 'circulating magma' type.

A number of units can be joined together to provide as long a crystallization path as desired. Alternatively, to save space, the units can be mounted above one another, but excessive nucleation may occur at the point where the discharge from one unit falls into the lower one, resulting in the production of many fine crystals. Cold water, or refrigerated brine in the later stages, can be passed through the jackets countercurrently to the flow of crystals. At the discharge end the crystal magma is usually delivered direct to a filtration unit. Moderately sized and fairly uniform crystals can be obtained from this type of crystallizer. A variety of materials of construction can be used depending on the substances being handled.

A 40 ft. run is the maximum length that can safely be driven from one stirrer shaft: an effective heat transfer area per foot run of crystallizer of the above dimensions is about 3 $ft.^2$, and overall heat transfer coefficients of 10–25 (B.t.u./h · $ft.^2$ °F), based on a logarithmic mean temperature difference between solution and cooling water, may be expected[7]. Amongst other things, the viscosity of the batch exerts a considerable influence on the rate of heat transfer; the analysis presented by KOOL[8] indicates the complex nature of the problem.

Very low heat transfer coefficients will be obtained if the inner cooling surface is not kept free from deposited crystals, and the helical stirrer can be arranged in close contact with the vessel wall to act as a scraper. Scraping action produces crystal attrition, however, and a high percentage of fines will result in the finished product. For many purposes this is most undesirable, and the helix may have to be located at a distance of $\frac{1}{2}$–1 in. away from the wall.

Crystallizers similar to the Swenson–Walker type with semi-cylindrical (U-type) or nearly cylindrical (O-type) cross-section have been used for many years in the sugar industry for the crystallization of concentrated molasses. Units fitted with water-cooled jacket are still employed, but these show poor heat transfer characteristics; scraper-stirrers are not permitted on account of the damage they do to the crystals. Accordingly, many other types of cooling arrangement have been tried[9]. For example, separate banks of water-cooled tubes have been inserted in between slowly rotating paddles in the trough. Alternatively, the water-cooled tubes can function as stirrers (*Figure 8.3*), but they must be of sturdy construction to prevent water leakage. Both stationary and rotating cooling discs have been employed quite successfully in trough crystallizers.

Another continuous trough crystallizer of the circulating magma type is the Wulff–Bock unit, sometimes called a crystallizing cradle or rocking crystallizer (*Figure 8.4*). This consists of a long shallow trough which can be rocked on supporting rollers. The solution to be crystallized is fed in at one end and the crystals are discharged at the other end, continuously. Transverse baffles may be fitted inside the trough to prevent longitudinal surging of the liquor, so that the charge flows in zigzag fashion along the unit. The slope of the trough, towards the discharge end, is varied according to the required residence time of the liquor in the crystallizer.

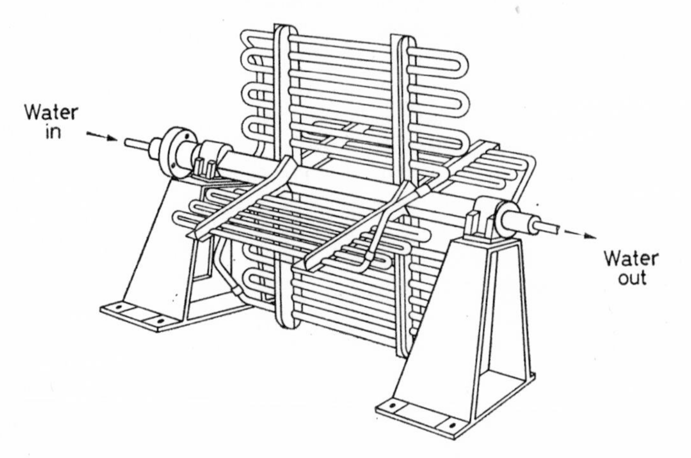

Figure 8.3. Rotating-tube cooling system for a trough crystallizer. (*After* G. C. DE BRUYN[9])

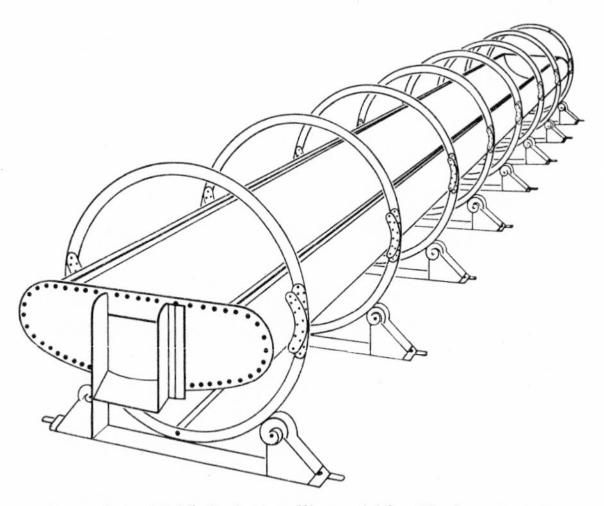

Figure 8.4. Wulff–Bock crystallizer. (*After* H. GRIFFITHS[10])

One of the great advantages of the Wulff–Bock crystallizer is the absence of moving parts within the crystallization zone. Corrosion problems, therefore, are greatly minimized; the trough can easily be rubber-lined or similarly protected if corrosive liquors have to be handled. The method is not suitable for organic solvent systems or for the crystallization of oxidizable substances because the crystallizer is open to the atmosphere. A number of units may be joined together; assemblies up to 100 ft. in total length have been installed. The power requirement for driving a typical 50 ft. by 3 ft. unit is about $1\frac{1}{2}$ h.p.

No external cooling is employed; heat is lost by natural connection to the atmosphere. High degrees of supersaturation, therefore, are not encountered at any point within the unit, and crystallization occurs slowly. The gentle agitation prevents crystal attrition as well as formation of a skin of crystals at the surface of the liquor. Large, uniform crystals can be grown in Wulff–Bock crystallizers, and production rates of 3 ton/day of $\frac{1}{2}$ in. crystals have been reported[10]. Potassium chloride, potassium permanganate, sodium thiosulphate and sodium sulphate have been produced commercially in this manner.

Double-pipe Crystallizer

A typical double-pipe crystallizer, or scraped-surface chiller, is the Votator apparatus shown in *Figure 8.5*. The unit is essentially a double-pipe heat

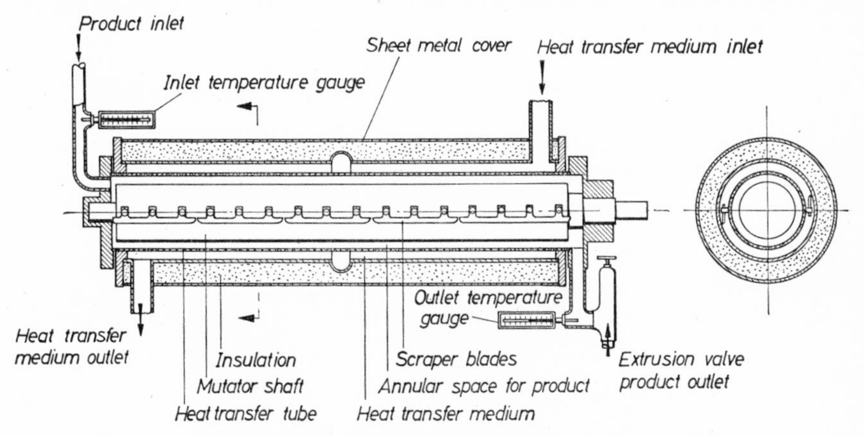

Figure 8.5. Double-pipe crystallizer (Votator apparatus)
By courtesy of *A. Johnson & Co. (London) Ltd.*

exchanger fitted with an internal scraping device to keep the heat transfer surfaces clean. The annulus between the two concentric tubes contains the cooling fluid which moves countercurrently to the crystallizing solution flowing in the central pipe. Located in the central pipe is a shaft upon which scraper blades, generally spring-loaded, are fixed. The solution is pumped through the unit and, due to the high degree of turbulence, fairly high heat transfer coefficients can be recorded.

Double-pipe crystallizers are employed for the crystallization of paraffin wax and the processing of viscous materials such as lard and margarine, but very small crystals are produced since a considerable degree of attrition and nucleation is unavoidable. In hydrocarbon separation processes, e.g. the recovery of *p*-xylene from solvent xylene mixtures (p. 144), ethylene or propane-cooled Votator units may be used in series with open tank crystallizers; the Votator effects heat removal and nucleation, the tank permits the slow growth of the crystals. One advantage of the double-pipe crystallizer is that the amount of liquor hold-up is very low.

The units, which can be fabricated in many materials, vary in diameter from about 3 to 24 in. and in length from about 1 to 10 ft. Rotor speeds in the range 300 to 2,000 rev/min may be used. Overall heat transfer coefficients, for crystallization operations, in the range 150–650 B.t.u./h · ft.^{2} °F may be obtained, but their prediction presents a difficult problem. Rotor speeds, blade clearances, liquor properties, etc. all play an important role. Heat transfer studies on Votators have been reported by several authors[11, 12].

Krystal Cooling Crystallizer

Towards the end of the First World War, investigations were carried out in Norway by Isaachsen and Jeremiassen into the problems associated with the continuous production of large uniform crystals, in particular of sodium chloride. These investigations subsequently lead to the development, by Jeremiassen, of a method for obtaining a stable suspension of crystals within the growth zone of a crystallizer. The practical application of this method has been incorporated in a continuous classifying crystallizer which is known by the names Oslo, Jeremiassen or Krystal apparatus[13, 14]. The latter name is now almost exclusively used, and the manufacturing rights of Krystal crystallizers in the United Kingdom, British Commonwealth and many other countries are held by the Power–Gas Corporation, Ltd. The Struthers–Wells organization manufacture these crystallizers in the U.S. Bamforth[15–17] and Svanoe[18] have given several concise accounts of the design and uses of these versatile and compact units.

There are three basic forms of the Krystal apparatus, viz. the cooling crystallizer (*Figure 8.6*), the evaporating crystallizer and the vacuum crystallizer. The last two forms are described under their appropriate headings on

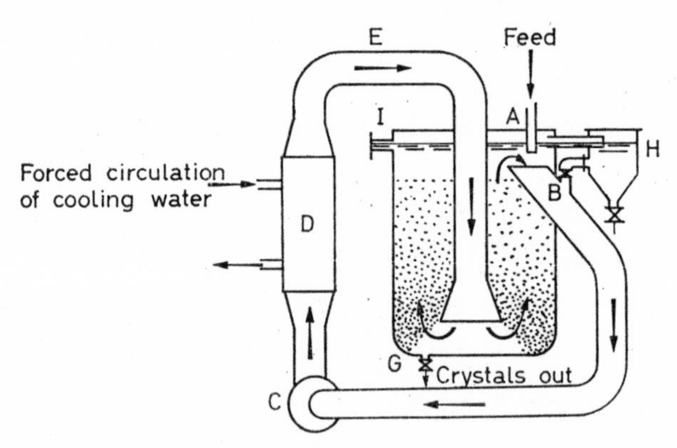

Figure 8.6. Krystal cooling crystallizer

pp. 201 and 207. More than twenty distinct modifications of these three basic types have been introduced for specific processes[15], but all crystallizers based on the original Jeremiassen process have one feature in common—a concentrated solution, which is continuously cycled through the crystallizer, is supersaturated in one part of the apparatus, and the supersaturated solution is conveyed to another part where it is gently released into a mass of growing crystals. These units, therefore, belong to the 'circulating liquor' type of crystallizer.

The operation of the Krystal cooling crystallizer (*Figure 8.6*) may be described as follows. A small quantity of warm concentrated feed solution enters the crystallizer vessel at point *A*, located directly above the inlet to the circulation pipe *B*. Saturated solution from the upper regions of the vessel, together with the small amount of feed liquor, is circulated by pump *C* through the tubes of heat exchanger *D*, which is cooled rapidly by a forced circulation of water or brine. On cooling, the solution becomes supersaturated, but not sufficiently for spontaneous nucleation to occur, i.e. metastable, and great care is taken to prevent it entering the labile condition. The supersaturated solution flows down pipe *E* and emerges from the outlet *F*, located near the bottom of the crystallizer vessel, directly into a mass of crystals growing in the vessel. The rate of liquor circulation is such that the crystals are maintained in a fluidized state in the vessel, and classification occurs by a hindered settling process. Crystals which have grown to the required size fall to the bottom of the vessel and are discharged from outlet *G*, continuously or at regular intervals. Any excess fine crystals floating near the surface of the solution in the crystallizer vessel are removed in a small cyclone separator *H*, and the clear liquor is introduced back into the system through the circulation pipe. A mother liquor overflow pipe is located at point *I*.

Like all other cooling crystallizers, this unit can only be used to advantage when the solute shows an appreciable reduction in solubility with decrease in temperature. Examples of some of the salts which can be crystallized in this manner are sodium acetate, sodium thiosulphate, saltpetre, silver nitrate, copper sulphate, magnesium sulphate and nickel sulphate.

Other Cooling Classifiers

McCabe[7] described a continuous cooling crystallizer which also acts as a classifier. This unit, called the Howard crystallizer (*Figure 8.7*), comprises three conical sections, each provided with its own cooling system. Hot concentrated feed solution enters near the bottom of the crystallizer and flows upwards, countercurrently to the flow of cooling water. The central section, the main crystallization zone, is provided with an inner cooling cone of the 'cold finger' type. The crystals are kept suspended in the upward flowing stream until they grow to a size which permits them to fall into the bottom discharge chamber. The large upper conical section contains cooling coils and an overflow weir; the spent solution flows through this section at a low velocity, thus minimizing any carry-over of fine crystals. A steady state is achieved in the Howard crystallizer; nucleation is controlled in the middle zone, and most of the crystal growth occurs in this section.

A pulsed-column cooling crystallizer, based on an invention by A. E. Zdansky, has recently been reported[19]. Hot saturated solution enters the top of a tall column which is divided into sections by means of a series of conical (apex downward) perforated plates. Also within the column are a number of vertical water-cooled tubes of the 'cold finger' type which give a flow of water countercurrent to the downward flowing solution. The feedstock is pulsed down the column at the rate of 1 pulse/sec and the bottom discharge outlet is opened briefly at 6–10 sec intervals for ejecting the cooled magma. The pulse action tends to re-disperse the crystal nuclei so that they

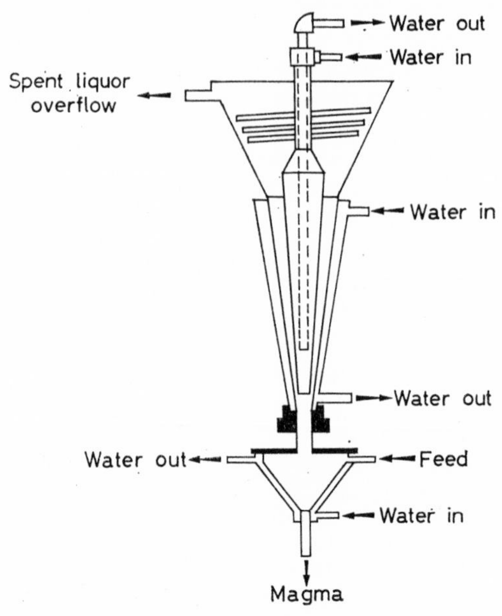

Figure 8.7. The Howard cooling and classifying crystallizer (*After* W. L. McCabe[7])

continue to grow, and the unit acts as a classifier. Production rates of 12–15 ton/day of 0·5 mm crystals and cooling rates of 200,000 kcal/h are claimed.

EVAPORATING CRYSTALLIZERS

When the solubility of a solute in a solvent is not appreciably decreased by a reduction in temperature, supersaturation of the solution can be achieved by removal of some of the solvent. Evaporation techniques for the crystallization of salts have been used for centuries, and the simplest method, the utilization of solar heat, is still a commercial proposition in many parts of the world. Novomeysky[20] has given a detailed account of the development of a large solar evaporation plant for the production of salts from Dead Sea waters, and Peiasach[21] described the recovery of pure sodium chloride and sodium sulphate from South African natural brines by a process involving concentration, refrigeration and solar evaporation.

Fishery salt, required in the form of hard saucer-shaped crystals, $\frac{1}{4}$–$\frac{1}{2}$ in. diameter, is produced in large quantities by the evaporation of brine in long, shallow open pans heated by direct fire, hot gases or steam coils. Evaporation may be aided by a gentle draught of air over the surface of the pans. Crystals which appear on the surface of the brine continue to grow until they become too heavy to be supported on the surface, whereupon they fall to the bottom of the pan and are removed by rakes. Labour costs per pound of product are high and the efficiency of heat utilization is low, yet no other method has so far been discovered which will produce this peculiar type of hard crystal necessary for use in the fisheries industry.

HESTER and DIAMOND[22] described the production of grainer salt, similar in shape to fishery salt but smaller in size, in open troughs 120 ft. long, 16 ft. wide and 2 ft. deep. The brine is kept at about 210° F by recirculation at 1,000–2,000 gal/min through external steam-heated heat exchangers. Each pan produces about 30 ton/day of the hopper-shaped crystals. This coarse salt, which has a high surface area per unit mass, must be handled with care to prevent breakage. After screening, several grades are produced, the coarsest being 6 to 28-mesh. Grainer salt, which has a high rate of solubility in water, is used in the manufacture of cheese and butter.

For most other purposes common salt is produced from brine in enclosed calandria evaporators—appropriately called salting evaporators. No control over the crystal size is possible and fine crystals, rarely larger than $\frac{1}{16}$ in. in diameter, are formed. The use of reduced pressure in an evaporator to aid the removal of solvent, to minimize the heat consumption or to decrease the operating temperature of the solution, is common practice. Calandria or coil evaporators, often in multi-effect series, are used in the sugar and common salt industries. Sometimes, as in sugar refining, the concentrated solution leaving the last effect of an evaporator system is charged into a separate evaporator where the liquor is 'struck', by skilled control of the vacuum, to produce a mass of small regular crystals.

Salt and sugar crystallization may be regarded as rather special cases, not because their governing principles are in any way different from the general principles of evaporative crystallization, but because of the considerable amount of technical 'know-how' retained in the various large commercial organizations. Several publications[22–25] deal in some detail with the practices adopted in these industries. A brief account will be given here of some of the equipment commonly used as evaporating crystallizers, and their methods of operation.

Most evaporation units are steam-heated, although resistance heating and heating by passing hot gases through liquors can be used. Steam coils, once widely used in sugar pans, are gradually becoming less and less favoured, but they are by no means obsolescent yet. Two typical evaporator bodies used in evaporative crystallization[26] are shown in *Figure 8.8*. Both are of the short-tube vertical type, heated by steam which condenses on the outside of the tubes. *Figure 8.8a* shows a steam chest, or calandria, with a large central downcomer which allows the magma to circulate through the tubes; it is arranged so that during operation the tops of the tubes are just covered with liquor. To increase the rate of heat transfer, especially when dealing with viscous liquors, a forced circulation of liquor may be effected by installing an impeller in the downcomer. In the basket-type steam chest (*Figure 8.8b*) the annulus between the evaporator body and the steam chest acts as the downcomer. Again, steam condenses on the outside of the tubes. Basket-type evaporators cannot be fabricated in very large sizes.

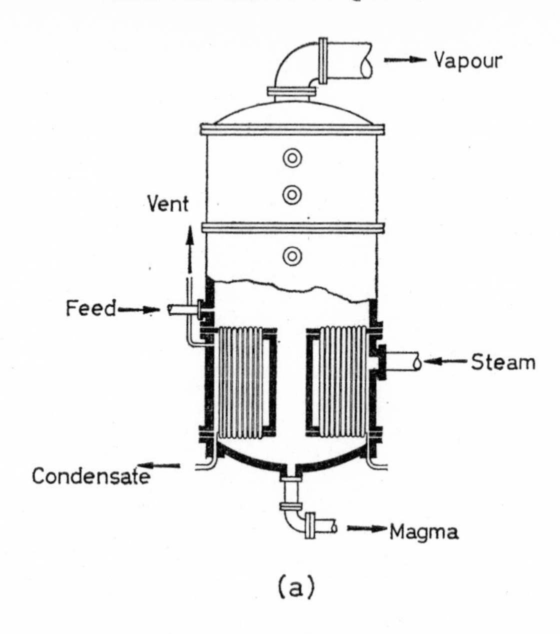

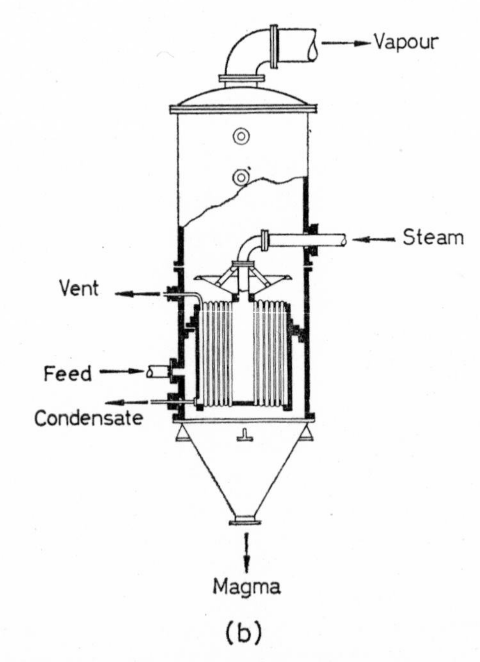

Figure 8.8. Two typical evaporator bodies: (a) calandria with large central downcomer; (b) basket-type steam chest with annular downcomer. (After G. MATZ[26])

Multiple-effect Evaporation

Low pressure steam, i.e. < 40 p.s.i.g., is normally used in evaporators, and frequently by-product steam (~ 10 p.s.i.g.) from some other process is employed. Nevertheless, one pound of steam cannot evaporate more than one pound of water from a liquor, and for very high evaporation duties the use of process steam as the sole heat source can be very costly. However, if the vapour from one evaporator is passed into the steam chest of a second evaporator, a great saving can be achieved. This is the principle of the

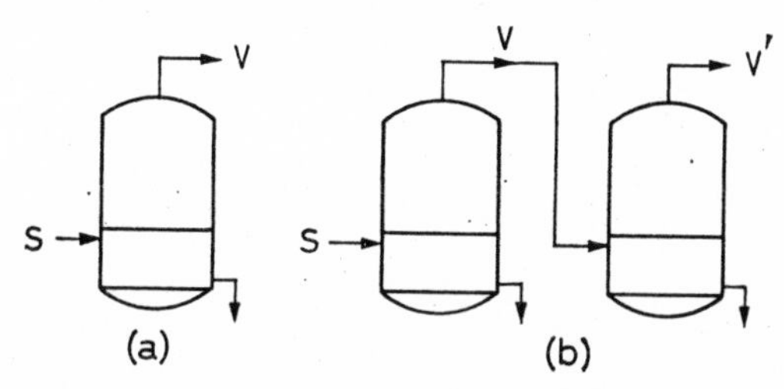

Figure 8.9. Principle of multiple-effect evaporation: (a) single-effect; (b) double-effect

method of operation known as multiple-effect evaporation (*Figure 8.9*). As many as six effects may be used in practice.

It is beyond the scope of this book to deal in any detail with the subject of multiple-effect evaporation; all standard chemical engineering textbooks contain adequate accounts of this method of operation, but two important points may be made here. First, multiple-effect evaporation increases the *efficiency* of steam utilization (lb. of water evaporated per lb. of steam used) but reduces the *capacity* of the system (lb. of water evaporated). The well known equation for heat transfer may be written

$$Q = UA\Delta T$$

where Q is the rate of heat transfer, U the overall heat transfer coefficient, A the area of the heat exchanger, and ΔT the driving force, the temperature difference across the heat transfer septum. The area A is usually fixed, so the variables to consider are Q, U and ΔT. Q will be reduced by heat losses from the equipment, U by sluggish liquor movement and scaling in the tubes of the calandria, ΔT by the increase in the boiling point of the liquor, as it gets more and more concentrated. These and many other factors prevent the achievement of the ideal condition

1 effect: 1 lb. steam $\rightarrow$ 1 lb. vapour
2 effects: 1 lb. steam $\rightarrow$ 2 lb. vapour
3 effects: 1 lb. steam $\rightarrow$ 3 lb. vapour, etc.

Nevertheless, a close approach to this ideality can often be produced. Some of the evaporators in a multiple-effect system are frequently operated under reduced pressure to reduce the boiling point of the liquor and thereby increase the available ΔT.

Before leaving this brief account of multiple-effect evaporation, mention may be made of the various methods of feeding that can be employed. *Figure 8.10* shows, in diagrammatic form, the possible feed arrangements. In all cases, fresh process steam enters effect number one. S denotes live steam, F the feed solution, V vapour passing to the condenser system, and L the thick liquor, or crystalline magma, passing to a cooling system or direct to a centrifuge. Pumps are indicated on these diagrams to indicate the number required for each of the systems. In all cases the vapours flow in the direction of effects $1 \rightarrow 2 \rightarrow 3 \rightarrow$, etc.

In the forward-feed arrangement (*Figure 8.10a*) liquor as well as vapour pass from effect $1 \rightarrow 2 \rightarrow 3 \rightarrow$, etc. As the last effect is usually operated under reduced pressure, a pump is required to remove the thick liquor; a feed pump is also required. The transfer of liquor between the intermediate effects is automatic as the pressure decreases in each successive effect. Suitable control valves are installed in the liquor lines. The disadvantages of a forward-feed arrangement are (*a*) the feed may enter cold and consequently require a considerable amount of live steam to heat it to its boiling point, (*b*) the thick liquor, which flows sluggishly, is produced in the last effect where the available ΔT is lowest, (*c*) the liquor pipe-lines can easily become steam-blocked as the liquor flashes into the evaporator body.

In the backward-feed arrangement (*Figure 8.10b*) the thick liquor is produced in effect 1 where the ΔT is highest; the liquor is more mobile on account of the higher operating temperature. Any feed pre-heating is done in the last effect, where low-quality steam (vapour) is being utilized. More pumps will be required in backward feeding than in forward feeding; the liquor passes into each effect in the direction of increasing pressure. No feed pump is necessary, as the last effect is under reduced pressure. Liquor does not flash as it enters an evaporator body, so small-bore liquor lines can be used. Backward feeding is best for cold feed liquor.

In the parallel-feed arrangement (*Figure 8.10c*) one feed pump is required, and pre-determined flows F_1, F_2 and F_3 are passed into the corresponding effects. A thick liquor product is taken from each effect; for this purpose pumps are generally necessary. Parallel feeding is often encountered in crystallization practice, e.g. in the salt industry, and it is useful if a concentrated feedstock is being processed.

In *Figure 8.10d*, a 5-effect system is chosen to illustrate the mixed-feed arrangement. Many different sequences are possible; the one demonstrated is as follows. Feed enters an intermediate effect (number 3 in this case) wherefrom the liquor flows in forward-feed arrangement to the last effect, is then pumped to effect number 2 and from there flows in backward-feed arrangement to the first effect. Some of the advantages of this method of feeding are (*a*) fewer pumps are required compared with backward feeding, (*b*) the final evaporation is effected at the highest operating temperature, and (*c*) frothing and scaling problems are claimed to be minimized. Caustic soda evaporation is often carried out on a mixed-feed basis.

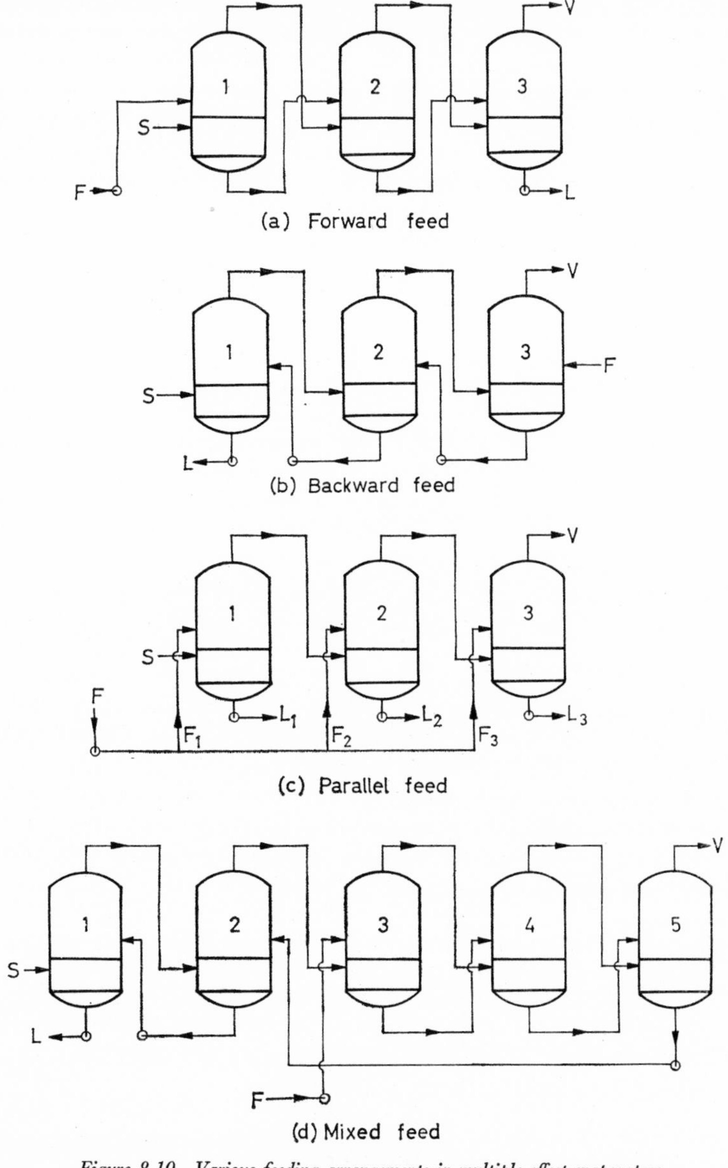

Figure 8.10. Various feeding arrangements in multiple-effect evaporators

Krystal Evaporating Crystallizer

The principles of the Krystal process, already referred to in connection with cooling crystallizers, can also be applied to evaporative crystallization. One form of the Krystal evaporating crystallizer is shown in *Figure 8.11.* Feed liquor enters at point *A* where it meets saturated liquor leaving the crystallizer vessel *B*. The mixed streams flow downwards under the influence of circulating pump *C* through the tubes of a steam-heated exchanger *D*.

The hot liquor is pumped up to the vaporizing chamber *E* where flash evaporation occurs, and the vapour leaves through outlet *F*. The solution, which has become supersaturated in chamber *E*, flows down pipe *G* to the bottom of the crystallizer vessel where it is released into a suspended mass of growing crystals. The action at this point is identical with that described for the Krystal cooling crystallizer. Crystals are removed from the system through outlet *H*, and excess fine crystals floating near the surface of the liquor are

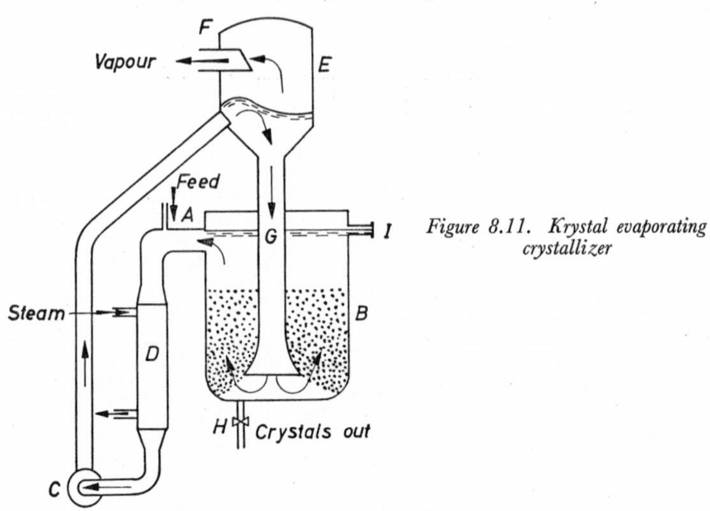

Figure 8.11. Krystal evaporating crystallizer

removed at point *I*. The evaporation rate can be increased, if desired, by operating the unit under vacuum.

Constructional and operating details of the Sindri, India, plant incorporating Krystal evaporating crystallizers in triple-effect series for the production of 1,200 ton/day of ammonium sulphate crystals (0·2 to 2 mm diameter) have been given by BAMFORTH[15]. The same author[17] also described a triple-effect unit for the production of 300 ton/day of 99·9 per cent sodium chloride crystals (0·75 to 1 mm diam.).

Krystal evaporating crystallizers have also been used for the manufacture of sodium dichromate, sodium sesquicarbonate, ammonium sulphate, ammonium nitrate and oxalic acid. In a slightly modified form[16] these units can also deal successfully with scale-forming salts such as calcium sulphate, anhydrous sodium sulphate and anhydrous sodium sulphite.

Wetted-wall Evaporative Crystallizer

A somewhat unusual application of the wetted-wall column, frequently used in gas–liquid mass transfer operations, has been reported by CHANDLER[27]. A hot concentrated solution is fed into a horizontal pipe, and cold air is blown in concurrently at a velocity of about 100 ft./sec. The liquid stream spreads over the internal surface of the pipe and cools, mainly by evaporation (see *Figure 8.12*). The crystal slurry and air leave from the same end of the pipe. Only small crystals can be produced by this method, and because of the evaporative loss of solvent only aqueous solutions can be handled.

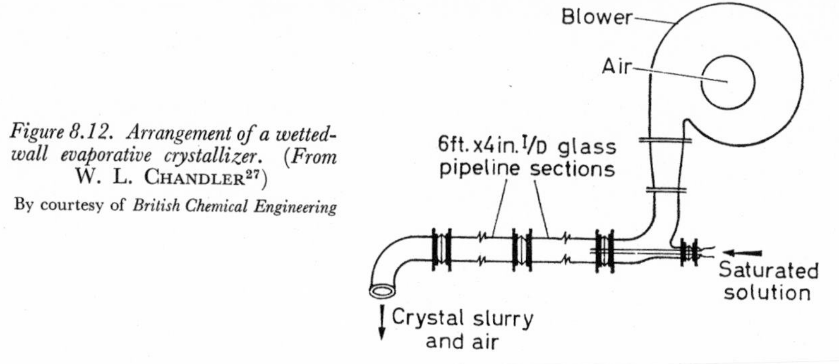

Figure 8.12. Arrangement of a wetted-wall evaporative crystallizer. (*From* W. L. CHANDLER[27])

By courtesy of *British Chemical Engineering*

Nevertheless, the equipment required is extremely simple and cheap, apart from the blower.

Though the pilot plant work described by Chandler was carried out on a relatively small unit and confined to the crystallization of sodium chloride, the wetted-wall crystallizer could be scaled up to larger sizes and used for other systems. The potential throughput of this small unit can be quite high, depending on the temperature–solubility characteristics of the solute–solvent system, as indicated in *Table 8.1.*

Table 8.1. Approximate Theoretical Capacity of a 4-in.-diam. *Wetted-wall Evaporative Crystallizer, based on a* 3,000 lb./h *Solution Flow Rate and a Solution Temperature Drop from* 212–152° F

(*After* J. L. CHANDLER[27])

Solute	*Crystal yield,* lb./h *of anhydrous salt*			*Deposited crystalline phase*
	Due to cooling	*Due to evaporation*	*Total*	
NaCl	66	60	126	anhyd.
$CuSO_4$	912	115	1,027	$5H_2O$
$CaCl_2$	570	240	810	$2H_2O$
$CuCl_2$	480	161	641	$2H_2O$
$Al_2(SO_4)_3$	750	137	887	$18H_2O$
$BaCl_2$	313	90	403	$2H_2O$
$Ba(NO_3)_2$	352	52	404	anhyd.
K_2SO_4	150	37	187	anhyd.
KNO_3	3,480	375	3,855	anhyd.
$CH_3 \cdot COOK$	1,560	625	2,185	$\frac{1}{2}H_2O$
$K_2SO_4 \cdot Al_2(SO_4)_3$	3,600	227	3,827	$24H_2O$
$MgSO_4$	165	65	230	$6H_2O$
$MgCl_2$	300	111	411	$6H_2O$
$MnCl_2$	159	175	334	$2H_2O$
$(NH_4)_2SO_4$	398	157	555	anhyd.
NH_4Cl	580	117	697	anhyd.
Na_3PO_4	1,350	180	1,530	$12H_2O$
Na_2HPO_4	486	155	641	$2H_2O$
NaH_2PO_4	1,830	372	2,202	anhyd.
$CH_3 \cdot COONa$	811	258	1,069	anhyd.

VACUUM CRYSTALLIZERS

The term 'vacuum' crystallization is capable of being interpreted in many ways; any crystallizer that is operated under reduced pressure could be called a vacuum crystallizer. Some of the evaporators described above could be classified in this manner, but these units are better, and more correctly, described as reduced-pressure evaporating crystallizers. The true vacuum crystallizer operates on a slightly different principle: supersaturation is achieved by simultaneous evaporation and adiabatic cooling of the feed solution. These units, therefore, act as both evaporators and coolers.

To demonstrate the operating principles of these units, consider a hot saturated solution introduced into a lagged vessel which is maintained under vacuum. If the feed temperature is higher than that at which the solution would boil under the low pressure existing in the vessel, the feed solution cools adiabatically to this temperature. The sensible heat liberated by the solution, together with any heat of crystallization liberated owing to the deposition of crystals at the lower temperature, causes the evaporation of a small amount of the solvent, which in turn results in the deposition of more crystals due to the increased concentration.

In a continuously operated vacuum crystallizer, the feed solution should reach the surface of the liquor in the vessel quickly, otherwise evaporation and cooling will not take place because, due to the hydrostatic head of solution, the boiling point elevation becomes appreciable at the low pressures (5–15 mm Hg) used in these vessels, and the feed solution will tend to migrate down towards the bottom outlet. Care must be taken, therefore, either to introduce the feed near the surface of the liquor in the vessel or to provide some form of agitation.

Swenson Batch Vacuum Crystallizer

Figure 8.13 shows a typical batch vacuum crystallizer[6], which consists of a lagged vertical cylindrical vessel with a top vapour outlet leading to a condenser and vacuum unit, and a conical bottom section fitted with a

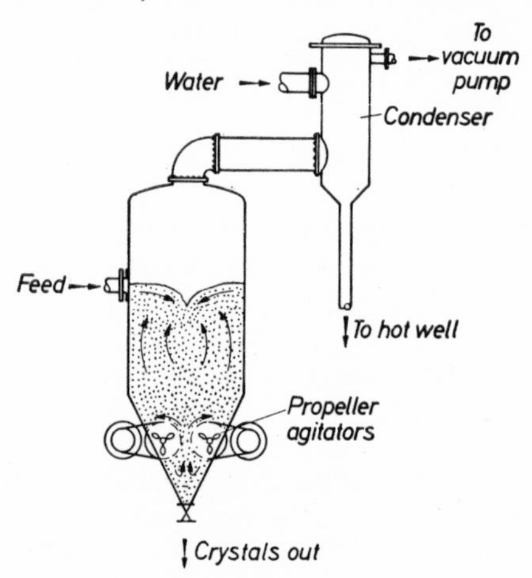

Figure 8.13. Swenson batch crystallizer[6]

discharge outlet. The vessel is charged to a pre-determined level with a hot concentrated solution, and the propeller agitators, vacuum and condensing equipment are put into operation. As the pressure inside the vessel is reduced, the solution begins to boil and cool until the limit of the condensing equipment is reached. In order to increase the capacity of the condenser, and thereby increase the crystal yield, the vapour leaving the vessel can be compressed before condensation by the use of a steam-jet booster (see *Figure 8.14*). The vacuum equipment usually consists of a two-stage steam ejector.

The swirling motion produced by the agitators helps to maintain the batch at a fairly uniform temperature and to keep the crystals suspended in the liquor. Crystalline deposits around the upper portions of the inner walls of the vessel cause little inconvenience; as the unit is operated batch-wise, the next charge will redissolve the deposit. When the batch reaches the required temperature, i.e. the desired degree of crystallization, it is discharged to a filtration unit. Small crystals, rarely much larger than about 60-mesh, are obtained from this type of crystallizer.

Swenson Continuous Vacuum Crystallizer

The batch unit described above can be adapted for continuous operation[6], as shown in *Figure 8.14*. Hot concentrated feed solution is introduced

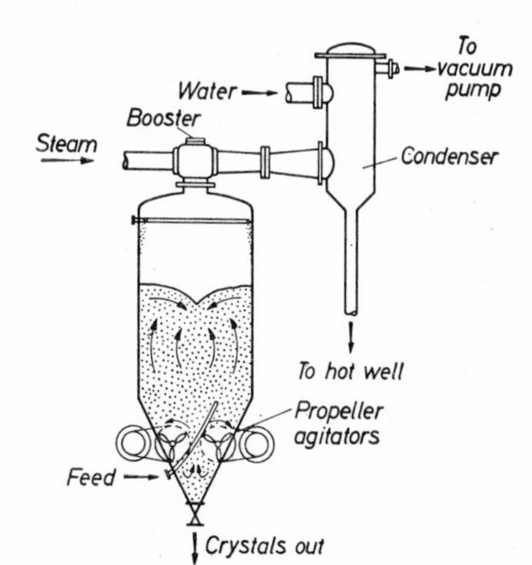

Figure 8.14. Swenson continuous vacuum crystallizer. (*After* G. E. SEAVOY and H. B. CALDWELL[6])

continuously through an insulated nozzle at a velocity such that it is delivered to the surface of the liquor in the vessel. The agitators perform the same duty as in the batch unit. The product is discharged continuously through the bottom outlet. A steam-jet booster compresses the vapours leaving the vessel before they enter the condenser.

As in the batch unit, a crystalline deposit builds up on the upper walls of the vessel; one way of overcoming this troublesome feature is to allow a small quantity of water to flow film-wise down the walls. A water rate of 1–2 gal./min is found adequate for vessels 10 ft. in diameter, and as this is less than the normal vaporization rate no serious dilution of the charge occurs.

Another type of continuous crystallizer, of the circulating magma type, in which agitation inside the vessel is caused by recycling the charge, is shown in *Figure 8.15.* An axial-flow pump takes the crystal magma from the conical base portion and re-introduces it tangentially into the vessel, just below the liquor level, creating a swirling action. Hot concentrated feed solution enters continuously at a point in the circulating pipe on the suction side of the pump, and crystal magma is discharged continuously from an outlet below the conical base. The discharge pipe is directed upwards to prevent blockage if for some reason the product take-off is stopped for a short while.

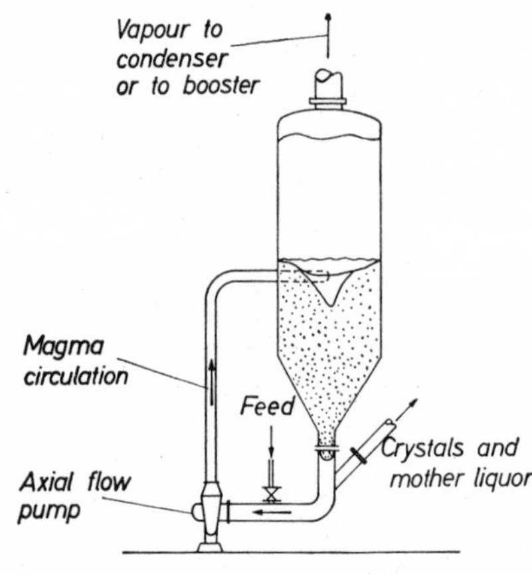

Figure 8.15. Swenson continuous vacuum crystallizer with agitation by pump circulation. (From W. L. McCabe *and* J. C. Smith[28], *by courtesy of* McGraw-Hill)

Vapours leave the vessel through a top outlet, as described for the previous units. Slightly larger-sized crystals, 20 to 30-mesh, are reported for this modified unit[28].

Draft-tube Baffle Crystallizer

The three vacuum crystallizers described above cannot be used for the production of large uniform crystals. Even though certain salts have been grown to 20 or 30-mesh size in a forced-circulation unit such as shown in *Figure 8.15*, the achievement of this size is rarely an easy matter. A more typical product size is about 60-mesh, with an appreciable proportion of material finer than 100-mesh. The main difficulties are as follows. First, propeller agitation or pump circulation can be rather violent, and mechanical shock is a well known method of inducing nucleation, so excessive quantities of unwanted crystal nuclei can be produced within the circulating magma. Secondly, the sudden flashing of vapour from the circulating liquor as soon as it enters the crystallizer causes the liquor to become supersaturated; if it becomes labile, spontaneous nucleation will occur. On the other hand, even if the labile zone is not penetrated, the degree of supersaturation may be too high for the required growth rate on the crystal seeds located near the surface of the liquor in the vessel. Vigorous agitation, of course, would bring more small crystals into the highly supersaturated region, but the dangers of violent circulation have already been pointed out.

In view of these difficulties, a recent modification has been made by the Swenson Evaporator Co. to enable vacuum crystallizers to produce large crystals. The resulting equipment is called the draft-tube baffle (DTB) crystallizer (*Figure 8.16*). A comprehensive account of this and the older types of Swenson crystallizers has been given by NEWMAN and BENNETT[1] who discuss their application to the crystallization of potassium chloride. A change-over from a unit such as that shown in *Figure 8.14* to a DTB type increased the percentage of material retained on a 28-mesh screen from 0·1 to 94.

The principles of the DTB crystallizer, described with reference to *Figure 8.16*, are as follows. A propeller agitator is located centrally in the lower section of the crystallizer body; a shroud, or draft-tube, surrounds the propeller and shaft up to a few inches below the surface of the liquor. Feed liquor is directed into the shroud. The gentle, but positive, movement of slurry to the surface reduces flashing and produces only about 1° F supercooling. The boiling action is also spread uniformly over the whole exposed surface. Because of the non-violent flashing, no salt build-up is produced on the upper walls of the vessel. High magma densities can be permitted and longer residence times are possible, thus allowing the growth of large uniform crystals.

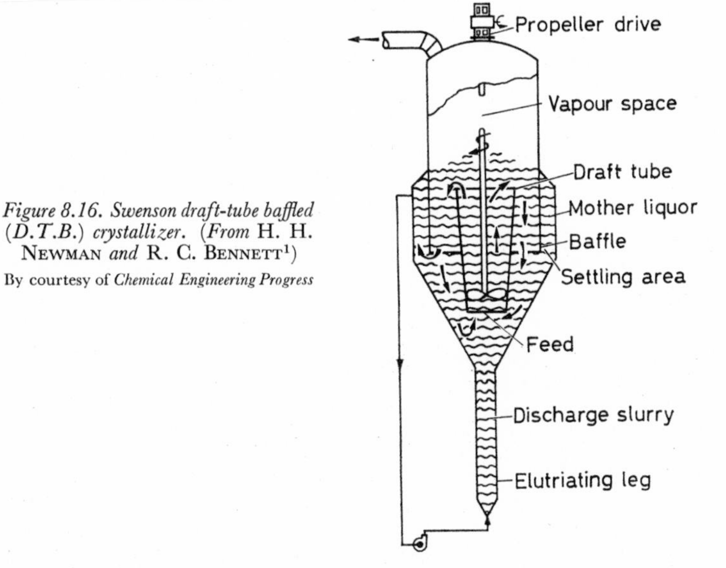

Figure 8.16. Swenson draft-tube baffled (D.T.B.) crystallizer. (From H. H. NEWMAN *and* R. C. BENNETT[1])
By courtesy of *Chemical Engineering Progress*

Krystal Vacuum Crystallizer

The Krystal vacuum continuous crystallizer (*Figure 8.17*) is in some respects similar to the evaporating crystallizer shown in *Figure 8.11*, the difference being that in the present case no external heat is supplied to the circulating liquor. Hot concentrated feed solution enters at point *A* in the crystallizing vessel *B*. Under the influence of pump *C*, the feed plus the circulating liquor from *B* are delivered via *D* to the vapour chamber *E* where adiabatic flash evaporation occurs. The vapours leave chamber E through outlet F on their way to the condenser–vacuum system. The cooled supersaturated liquor flows down pipe G and emerges into the mass of growing crystals in B. The operation thereafter is the same as described for the previous Krystal units. Fine crystals floating near the surface of the vessel are removed at point I. When the heat of crystallization of the product is not particularly high, the

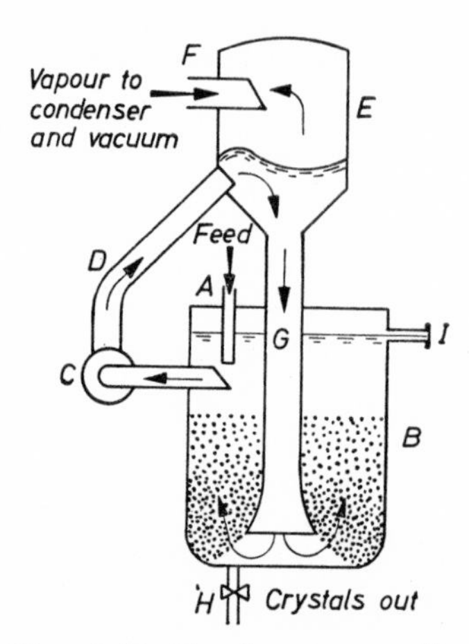

Figure 8.17. Krystal vacuum crystallizer

solvent evaporation rate may be increased by supplying some external heat to the liquor as it passes up the circulation pipe D.

Miller and Saeman[29] gave extensive data of the operation of a pilot plant, consisting of a 2 ft. diameter continuous vacuum Krystal crystallizer, for the production of 50 lb./h of ammonium nitrate. The crystals produced by this unit were reported to have good storage properties and resistance to shattering[30]. The information gained from this pilot-plant study was later used in the design of a plant with a production capacity of 525 ton/day[31]. Other uses of the Krystal vacuum crystallizer include the production of ammonium sulphate from coke-oven gas and the large-scale production of adipic acid.

Saeman[32] investigated the control of crystal size in Krystal and Swenson vacuum crystallizers, in particular under conditions of excessive nucleation in the crystallizer. The prime factor governing the final product size was found to be the segregation time of nuclei in the fines-removal system; a mathematical analysis of the problem was presented.

Condensing Systems

The vapours leaving a vacuum crystallizer or evaporator have to be condensed, either in a conventional tubular heat exchanger or by direct contact with a spray of water in a jet or contact condenser. The latter method is probably the most widely used. The low-temperature vapours produced in the flash chamber are usually compressed by means of a steam booster to facilitate their condensation in the cooling system. The temperature in the flash chamber depends on that of the available cooling water; the partial pressure of water vapour in the flash chamber is equal to the vapour pressure of water at the temperature of the mixture of cooling water and condensate leaving the jet condenser.

In chemical plants concentrated aqueous solutions of salts are frequently available, and as salt solutions have lower vapour pressures than water at the same temperature, these brines may provide a better source of cooling media than water; less compression, and hence less steam, would be required in the booster, for instance. WITTENBERG[33] described the problems encountered in the operation of vacuum coolers and crystallizers in the Dead Sea area where cooling water supplies are scarce, and often as warm as 30° C. He pointed out that if brine with a vapour pressure of 9 mm Hg at 20° C were used in a jet condenser it would be more effective than pure water at 10° C.

Figure 8.18 shows a typical layout of a plant used for the production of carnallite ($KCl \cdot MgCl_2 \cdot 6H_2O$) from Dead Sea brine, where saline mother

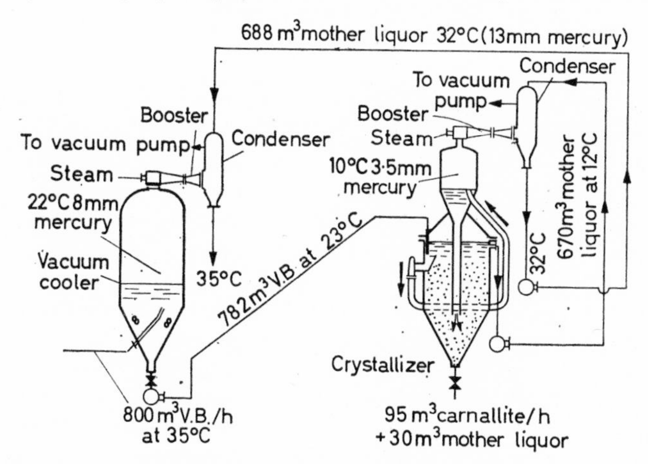

Figure 8.18. Use of brine as a condensing medium in the recovery of carnallite. (*From* D. WITTENBERG[33])
By courtesy of *Industrial Chemist & Chemical Manufacturer*

liquor is used as the condensing medium. In this particular plant the mother liquor from the crystallizer, free from carnallite crystals, is used in the condensers of the crystallizer and the vacuum cooler in turn. Other sources of brine can, of course, be used. As Wittenberg points out, a brine taken from the Dead Sea at a depth of 50 ft. has a fairly constant temperature of about 20° C, specific gravity of 1·22 and a vapour pressure of only 9 mm Hg. This brine would be quite suitable for cooling purposes.

CHOICE OF CRYSTALLIZER

Solutions which require crystallizing generally fall into one of two categories: those which will deposit appreciable quantities of crystals on cooling, and those which will not. The temperature–solubility relationship between the required product and the solvent is therefore of prime importance in the choice of a crystallizer. For solutions in the second category, an evaporating crystallizer would normally be used, although the salting-out method could be employed in specific cases. For solutions which deposit appreciable quantities of crystals on cooling, the choice of equipment will lie between a simple cooling crystallizer and a vacuum crystallizer.

Unfortunately, few performance and cost data are available for the various units commonly used in practice. SEAVOY and CALDWELL[6] discussed in some detail the relative merits of a cooling (Swenson–Walker type) and a vacuum crystallizer (Swenson type), and McCABE[7] presented American production and installation costs for several units. A more recent cost survey, again of American origin, has been made by GARRETT and ROSENBAUM[34] who compare several standard types of Krystal, growth-type Pachuca, forced-circulation and mechanical crystallizers. Broadly speaking, for rough estimation purposes the six-tenths rule

$$\text{Capital cost} \propto (\text{capacity})^{0{\cdot}6}$$

can be applied for capacities in the range 10 to 1,000 ton/day.

BAMFORTH[16] has also dealt with the economics of crystallization, using for comparison purposes a typical batch tank crystallizer operated on an evaporation–cooling cycle and a small continuous Krystal evaporating crystallizer operated under reduced pressure. Even with a production as small as 1 ton of crystals per day, the continuous unit was found to be more economical in operating costs, but its greatest advantage was the fact that whereas about 40 per cent of the liquor from the tank crystallizer required reworking, the quantity for the continuous unit was only about 7 per cent. Labour costs per pound of product for both units were similar but decrease considerably for the continuous unit with increasing production capacity. For example, a continuous crystallizer producing say 1 ton/day would require substantially the same labour force as a unit producing say 100 ton/day.

There are, of course, many other advantages associated with crystallizers that also act as classifiers: product size is more regular, efficient filtration and washing are facilitated, thus tending to give a purer product, drying costs are reduced because of the lower amount of retained solvent, and screening of the product is frequently unnecessary.

Although some of the simpler cooling crystallizers, especially the open tanks, are relatively inexpensive, the initial cost of a mechanical unit can be fairly high. The maintenance costs of a mechanical crystallizer can also be quite appreciable. On the other hand, no costly vacuum-producing or condensing equipment is required. Very dense crystal slurries can be handled in cooling units not requiring liquor circulation. Unfortunately, the inner cooling surfaces often become coated with a hard crust of crystals, and the outer water-sides of these surfaces readily become fouled, with the result that cooling efficiency is reduced considerably. Similar problems can also be encountered in the operation of evaporating crystallizers; coils and calandrias often become coated with scale. As the true vacuum crystallizers have no heat-exchange surfaces they do not suffer from these scaling problems, but they cannot be used when the liquor has a high boiling point elevation. Vacuum and evaporating crystallizers require a considerable amount of head room, but the floor space needed is usually very much less than that required by evaporating pans and cooling tanks to produce the same amount of finished product. In an example quoted by BAMFORTH[16] a continuous Krystal vacuum crystallizer occupied one-fifth of the floor area required by a batch tank crystallizer of similar capacity.

Once all the above points have been considered, a certain class of crystallizer can be decided upon. However, the choice of a specific unit belonging to this particular class will depend upon many factors, such as initial, operating and maintenance costs, space availability, production rate, type and size of crystals required, physical characteristics of feed liquor and crystal slurry, need for corrosion resistance and so on. The supply of feed liquor from the production plant to the crystallizer will generally be the deciding factor in the choice between a batch and continuous unit. Feed rates in excess of about 5,000 gal./h are probably best handled on a continuous basis.

Scale-up Problems

The design of a large-scale crystallizer from data obtained on a small-scale unit is never a simple matter. Even after the equipment is finally constructed, much trial and error is necessary before the exact operating conditions are determined for the production of a given type of crystal. One must not give the impression that crystallizer design is largely a matter of guess-work; nothing would be farther from the truth. But it must be appreciated that the scaling-up of crystallization equipment is probably more difficult than that for any of the other unit operations of chemical engineering.

One of the first attempts to generalize the design procedure for continuous crystallizers was made by GRIFFITHS[35]. He suggested a separation-intensity (S.I.) factor as a criterion of performance; this factor is defined as the mass output of crystalline material per unit time, per unit volume occupied by the loosely packed crystal content in the crystallizer. The S.I. factor is a function of the supersaturation of the liquor; if the value of S.I. is increased beyond a certain limit there will be a copious precipitation of tiny nuclei due to the labile state being reached.

Despite the fact that the S.I. factor is influenced by the solubility characteristics of the solute, the presence of impurities, temperature, viscosity, the degree of agitation and many other variables, Griffiths found that values of the factor fell within a very narrow range for a wide variety of crystalline substances produced under various conditions in different industrial equipment. For sodium thiosulphate and copper sulphate, S.I. values of 300 ± 50 kg/h $\cdot$ m^3 and 170 ± 40 kg/h $\cdot$ m^3, respectively, were obtained for 1mm equivalent grains at 30° C; the expression '1 mm equivalent grain' refers to crystals, not necessarily uniform, which have approximately the same surface area per unit mass as regular 1 mm grains.

The above S.I. factors were obtained experimentally, employing the minimum degree of agitation required to prevent agglomeration of the crystals during the crystallization process. However, when these values were compared with those calculated from the results of large-scale crystallizations, remarkably good agreement was found. For instance, 5 mm equivalent grain sodium thiosulphate and copper sulphate gave S.I. factors of 60 and 40 kg/h · m^3, respectively. In addition to copper sulphate and sodium thiosulphate, plant data were obtained on sodium sulphite, sodium phosphate, sodium sulphate, sodium ferrocyanide, magnesium sulphate, alum and many other products.

S.I. factors for all the substances investigated (calculated back to 1 mm equivalent grains) fell within the limits 100–500 kg/h · m^3. The lower values were obtained for crystallizers working at 10–20° C, the larger values at higher working temperatures. Further evidence of the applicability of the S.I. factor can be deduced from some recent results[36] reported for a semi-technical scale continuous vacuum crystallizer of the Oslo type; the operating temperature was 73° C, and ~ 2 mm crystals of sodium chloride were produced. A recalculation of the data shows that values of S.I. ~ 400 kg/h · m^3 were obtained.

For crystals smaller than 1 mm the S.I. factors calculated by Griffiths were found to be unreliable, probably owing to the fact that comparable conditions of agitation could not be achieved in the different crystallizers. It may be stressed that empirical rules of this kind can only be applied with extreme caution, but because of the fairly narrow range of S.I. values encountered in large-scale equipment, a preliminary design may be based on this factor. It would be desirable, of course, to ascertain by small-scale trials that no complications are likely to be encountered.

If a small crystallizer produces the required type of product, then the proposed large crystallizer must simulate a large number of different conditions obtaining in the small unit. The four most important conditions are

(*a*) identical flow characteristics of liquid and solid particles
(*b*) identical degrees of supersaturation in all equivalent regions of the crystallizer
(*c*) identical initial seed sizes and magma densities
(*d*) identical contact times between growing crystals and supersaturated liquor.

The scaling-up of agitation equipment has long been recognized as a difficult problem, and the two dimensionless numbers most frequently encountered in the analysis of stirrers and agitators are the Reynolds number, Re, and the Froude number, Fr. The former gives a value of the ratio of the inertia and viscous forces, the latter the ratio of centrifugal acceleration and acceleration due to gravity. For a stirrer blade of diameter d, rotating at n rev per unit time in a liquid of density ρ and absolute viscosity η, these two numbers may be written as

$$\mathrm{Re} = \frac{\rho n d^2}{\eta} \tag{1}$$

$$\mathrm{Fr} = \frac{n^2 d}{g} \tag{2}$$

where g = acceleration due to gravity.

For scale-up purposes, values of both Re and Fr should be kept constant, but this is exceptionally difficult, if not impossible in most cases. For example, if the stirrer diameter is increased by a factor of 4, the stirrer speed must be decreased by a factor of 2 if the Froude number is to be kept constant ($n \propto 1/\sqrt{d}$), but by a factor of 16 if the Reynolds number is to remain the same ($n \propto 1/d^2$).

If the conditions in two agitated systems are to be similar in all respects, conditions of geometrical, kinematic and dynamic similarity must be ensured. By the application of the technique of dimensional analysis the dimensionless equation can be derived[37]

$$\frac{P}{\rho n^3 d^5} = \phi(\text{Re}, \text{Fr}) \tag{3}$$

P is the power input to the stirrer (e.g. ft. · lb./sec). The group $P/\rho n^3 d^5$ is known as the Power number, P_N. If it is assumed that $P \propto \rho n^3 d^5$ then it can be seen, for example, that if the speed of a given paddle is doubled, the power input would have to be increased 8-fold. Again, if the paddle diameter is doubled, its speed remaining constant, the power input would have to be increased by a factor of 32.

The significance of these dimensionless groups can now be discussed in relation to the problems of crystallizer scale-up. NEWMAN and BENNETT[1]

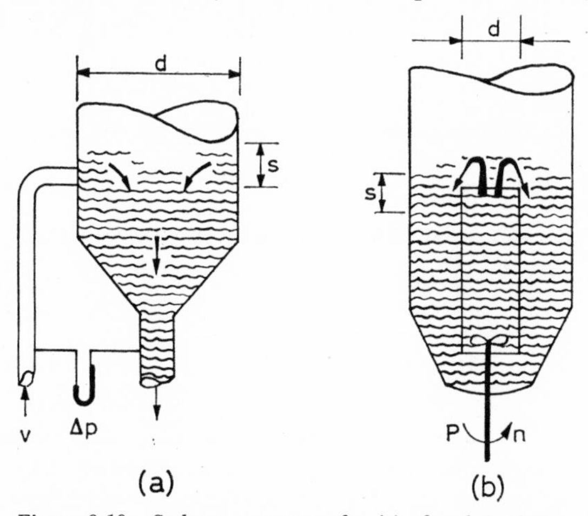

Figure 8.19. Scale-up parameter for (a) forced-circulation, (b) draft-tube (D.T.B.) crystallizer. (After H. H. NEWMAN *and* R. C. BENNETT[1])

showed that it is much easier to scale up successfully a crystallizer of draft-tube DTB type (e.g. as in *Figure 8.16*) than one of forced-circulation type (e.g. as in *Figure 8.14*). *Figure 8.19* shows the variables that have to be considered in each case. For the DTB type, the Reynolds, Froude, and Power numbers are given by equations 1 to 3. For the forced-circulation type, the corresponding numbers are given by

$$\text{Re} = \frac{\rho v d}{\eta} \qquad \text{Fr} = \frac{v^2}{dg} \qquad P_N = \frac{g \Delta p}{\rho v^2} \tag{4}$$

where d = diameter of the crystallizer body, v = inlet velocity of the liquor, Δp = pressure drop. Another dimensionless ratio d/s was also defined, where s denotes the submergence required to suppress vapour formation in the superheated liquid entering the vessel (forced circulation) or rising to the boiling surface (DTB).

When both viscous and gravitational forces apply, e.g. in the forced-circulation crystallizer where vortexing and toroidal circulation occur, both the Reynolds and Froude numbers are important factors. Because Fr $\propto v^2$, Re $\propto v$, and $P_N \propto 1/v^2$, it is very difficult to reproduce the same hydraulic régime in different vessels. The submergence ratio is also an important scale-up factor. In the DTB type of crystallizer, however, vortexing does not occur and Froude number and submergence ratio are not important variables. Equivalent hydraulic régimes can be produced in large equipment with the same Power number at very different Reynolds numbers. The agitator, of course, must operate at a speed well below that at which attrition and nucleation are induced. The claim that a DTB crystallizer is more easily scaled up than a forced-circulation type is substantiated by experimental evidence[1].

REFERENCES

[1] NEWMAN, H. H. and BENNETT, R. C., Circulating magma crystallizers, *Chem. Engng Progr.* 55 (3) (1959) 65

[2] MULLIN, J. W., Crystallization, in *Chemical Engineering Practice*, H. W. Cremer and T. Davies (Eds.), Vol. 6, 1959. London; Butterworths

[3] TAEBEL, W. A. and ANDERSON, W. F., Reagent-grade chemicals, in *Modern Chemical Processes*, Vol. 2, 1952. New York; Reinhold

[4] BIXLER, G. H. and SAWYER, D. L., Boron chemicals from Searles Lake brines, *Industr. Engng Chem.* 49 (1957) 322

[5] GARRETT, D. E., Industrial crystallization at Trona, *Chem. Engng Progr.* 54 (12) (1958) 65

[6] SEAVOY, G. E. and CALDWELL, H. B., Vacuum and mechanical crystallizers, *Industr. Engng Chem.* 32 (1940) 627

[7] McCABE, W. L., Crystallization, in *Chemical Engineers' Handbook*, J. H. Perry (Ed.) 1950. New York; McGraw-Hill

[8] KOOL, J., Heat transfer in scraped vessels and pipes handling viscous materials, *Trans. Instn chem. Engrs, Lond.* 36 (1958) 253

[9] DE BRUYN, G. C., Crystallization of massecuites by cooling, in *Principles of Sugar Technology*, P. Honig (Ed.), Vol. 2, 1959. Amsterdam; Elsevier

[10] GRIFFITHS, H., Mechanical crystallization, *J. Soc. chem. Ind., Lond.* 44 (1925) 7T

[11] HOULTON, H. G., Heat transfer in the Votator, *Industr. Engng Chem.* 36 (1944) 522

[12] SKELLAND, A. H. P., Correlation of scraped-film heat transfer in the Votator, *Chem. Engng Sci.* 7 (1958) 166

[13] *Brit. Pat.* 171,370 (1920); 194,676 (1922); 195,597 (1922); 219,301 (1923); 240,164 (1924); 240,170 (1924); 260,133 (1926); 290,369 (1927); 392,829 (1932); 418,349 (1933); 457,301 (1936); 616,351 (1939)

[14] JEREMIASSEN, F. and SVANOE, H., Supersaturation control attains close crystal sizing, *Chem. metall. Engng* 39 (1932) 594

[15] BAMFORTH, A. W., Controlled crystallization, *Industr. Chem. Mfr* 25 (1949) 81

[16] BAMFORTH, A. W., The economics of crystallization, *Times Rev. Ind.*, November 1949

[17] BAMFORTH, A. W., Modern plant for salt production, *Times Rev. Ind.*, May 1953

[18] SVANOE, H., Krystal classifying crystallizer, *Industr. Engng Chem.* 32 (1940) 636

[19] Continuous flow crystallizer, *Engineer* 206 (1958) 984

[20] Novomeysky, M. A., The Dead Sea—a storehouse of chemicals, *Trans. Instn chem. Engrs, Lond.* 14 (1936) 60
[21] Peiasach, M., *S. Afr. industr. Chem.* 9 (1955) 210
[22] Diamond, H. W., and Hester, A. S., Salt manufacture, *Industr. Engng Chem.* 47 (1955) 672
[23] Honig, P. (Ed.), *Principles of Sugar Technology*, Vol. 2, 1959. Amsterdam; Elsevier
[24] Lyle, O., *The Efficient Use of Steam*, 1947. London; H.M. Stationery Office
[25] Watkins, S. B., MacMurray, H. D. and Forker, K. G., Evaporation practice, in *Chemical Engineering Practice*, Vol. 6, H. W. Cremer and T. Davies (Eds.), 1959. London; Butterworths
[26] Matz, G., *Die Kristallisation in der Verfahrenstechnik*, 1954. Berlin; Springer
[27] Chandler, J. L., The wetted-wall column as an evaporative crystallizer, *Brit. chem. Engng* 4 (1959) 83
[28] McCabe, W. L. and Smith, J. C., *Unit Operations of Chemical Engineering*, 1956. New York; McGraw-Hill
[29] Miller, P. and Saeman, W. C., Continuous vacuum crystallization of ammonium nitrate, *Chem. Engng Progr.* 43 (1947) 667
[30] Miller, P. and Saeman, W. C., Properties of monocrystalline ammonium nitrate fertilizer, *Industr. Engng Chem.* 40 (1948) 154
[31] Saeman, W. C., McCamy, I. W. and Houston, E. C., Production of ammonium nitrate by continuous vacuum crystallization, *Industr. Engng Chem.* 44 (1952) 1912
[32] Saeman, W. C., Crystal size distribution in mixed suspensions, *Amer. Instn chem. Engrs J.* 2 (1956) 107
[33] Wittenberg, D., The use of brine in vacuum refrigeration and evaporation, *Industr. Chem. Mfr* 28 (1952) 535
[34] Garrett, D. E. and Rosenbaum, G. P., Crystallization, *Chem. Engng* 65 (Aug. 11, 1958) 127
[35] Griffiths, H., Crystallization, *Trans. Instn chem. Engrs, Lond.* 25 (1947) xiv
[36] Rumford, F. and Bain, J., The controlled crystallization of sodium chloride, *Trans. Instn chem. Engrs, Lond.* 38 (1960) 10
[37] Rushton, J. H., Costich, E. W. and Everitt, H. J., Power characteristics of mixing impellers, *Chem. Engng Progr.* 46 (1950) 395, 467

9

SIZE GRADING OF CRYSTALS

THE MOST widely employed physical test applied to a crystalline product is the one by means of which an estimate may be made of the particle size distribution. Product specifications invariably incorporate a clause which defines, often quite stringently, the degree of fineness or coarseness of the material. For most industrial purposes the demand is for a small range of particle size; regularity results in the crystalline product having good storage and transportation properties, a free-flowing nature and, above all, a pleasant appearance. Manufacturers are well aware of the enhanced selling potential of a nice looking product.

Size analyses are indispensable in the routine control of crystallization plant. Economical performances of such operations as filtration, washing and drying largely depend on particle size. Here again, uniformity of size plays an important role. The production of excessive quantities of crystals smaller or greater than the desired size entails the installation of a screening operation, and unless there is an outlet for these 'fines' or 'roughs' a wasteful recrystallization process may have to be considered. If there is a demand for many different size grades of the same crystalline product, complex screening arrangements have to be made.

Sampling

The physical and chemical characteristics of a bulk quantity of crystalline material are determined by means of tests on small samples. These test samples must be truly representative of the bulk quantity, otherwise any results obtained will be grossly misleading or completely useless. Inefficient sampling followed by detailed analyses in the laboratory constitutes a waste of everyone's time and effort.

Sampling, which is a highly specialized skill, should be carried out by conscientious, well trained personnel who are fully aware of the tests that are to be made on the sample, without having any direct interest in the outcome of the analyses. Far too often the task is delegated to the most junior laboratory assistant who regards it as a chore, or to a busy process operator whose main aim is to send a 'good' sample—from his point of view—to the analyst.

Although a sample should represent, as closely as possible, all the characteristics of the bulk quantity, it is impossible to achieve the ideal condition of a sample being identical in all respects with the parent lot. In general, however, it may be stated that the larger the sample taken, the greater will be the probability that the error in the value of the measured property is small. The complex theories of sampling, based on the laws of probability, have been discussed by HASSIALIS and BEHRE[1] who also present a comprehensive account of a large number of possible sampling methods.

The actual technique employed for sampling will depend on many factors, such as the nature of the bulk quantity of material, its location, the properties to be tested, the accuracy required in the test, and so on. Difficulties may be encountered in the sampling of solids in containers after transportation, owing to the partial segregation of fine and coarse particles; the fines tend to migrate towards the bottom of the container, and thorough re-mixing may be the only answer. Similar problems caused by segregation may be met in the sampling of solids flowing down chutes or through outlets.

Automatic sampling by mechanical means is to be preferred to sampling by hand, and is also the method best suited to continuous processes. Hand sampling, widely employed for batchwise produced materials, is a time-consuming operation and prone to error, but its use cannot always be avoided. One common method of hand sampling involves the use of a sample gun or 'thief'; this simply consists of a piece of pipe with a sharp bottom edge, which is plunged vertically downwards into the full depth of the material. It is then withdrawn and the sample removed. This operation can be performed at fixed or random intervals in the bulk quantity. Sampling at intervals by means of scoops or shovels, known as grab sampling, is also widely used, but serious errors can be encountered when dealing with non-homogeneous materials.

The above methods and many others are designed to produce a bulk sample representative of the bulk quantity. Bulk samples may range from a pound or so for small batches to a hundredweight or more for large tonnage lots. The next step is the preparation of a series of test samples which are each representative of the bulk sample. This operation may be carried out by hand or with the aid of a sample divider.

The best known hand method is that of coning and quartering. The sequence of operations, carried out on a clean, smooth surface, or on black glossy paper for small quantities in the laboratory, is shown in *Figure 9.1*.

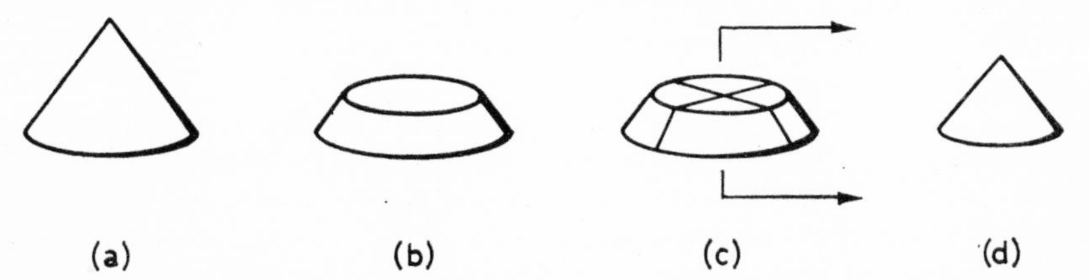

Figure 9.1. Method of coning and quartering: (a) bulk sample in a conical heap; (b) flattened heap; (c) flattened heap quartered; (d) two opposite quarters mixed together and piled into a conical heap

The bulk sample is thoroughly mixed and piled into a conical heap. The pile is then flattened and the truncated cone divided into four equal quarters (*Figure 9.1c*). This may be done, for example, with a sharp-edged wooden or sheet metal cross pressed into the heap. One pair of opposite quarters are rejected, the other pair are thoroughly mixed together and piled into a conical heap, the procedure being repeated until the required laboratory sample is obtained.

If several tests are to be carried out on the reduced laboratory sample, a number of test samples may be prepared by scooping the crystalline material into a series of sample tins in the following manner[2]. If, for example, four test samples are required, four tins are placed in a row. A quantity of the laboratory sample is scooped up and, as near as can be judged, one-quarter of the scoopful is discharged into each tin. Fresh scoopfuls are taken, and the tins are filled in the sequence

1st scoopful: 1, 2, 3, 4
2nd scoopful: 2, 3, 4, 1
3rd scoopful: 3, 4, 1, 2, etc.

so that no tin is filled exclusively with material from a particular section of the scoopful.

A simple and useful sample divider is the riffle. This apparatus usually takes the form of a box divided into a number of compartments with bottoms

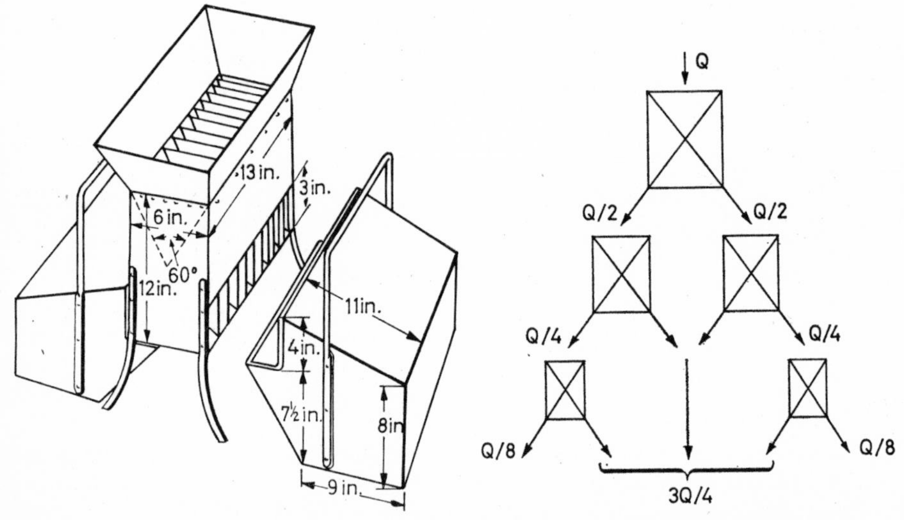

Figure 9.2. A riffle sample divider. (*From* B.S. 1796[2])

Figure 9.3. Battery of riffles used for reduction of a large quantity of material

sloping about 60° to the horizontal, the slopes of alternate compartments being directed towards opposite sides of the box. Thus, when the bulk sample is poured through the riffle it is divided into two equal portions. One type of riffle[2] is shown in *Figure 9.2.* The dimensions of a riffle depend on the size of the particles and the feed rate of material—an even flow of solids must be spread over the whole inlet area. When very large bulk samples have to be reduced to small test quantities, a battery of riffles decreasing in size may be employed, as shown diagrammatically in *Figure 9.3.* This arrangement is also suitable for the continuous or intermittent sampling of materials flowing out of hoppers or other items of process plant.

LABORATORY SIEVING

Standard Sieves

Standard sieves used for particle size testing usually consist of a circular brass frame, 8 or 12 in. in diameter, with a wire cloth rigidly mounted in the bottom. The bottom rim of each sieve fits closely inside the upper rim of another sieve in the series, thus forming a dust seal. A top lid and bottom receiving pan are generally provided for the nest of sieves. Each sieve is given a number, the mesh number, N, which refers to the number of wires per inch, which is the same thing as the number of apertures per inch (see *Figure 9.4*). Mesh numbers range from about 5 to 400, depending on the scale employed. The basic dimensions of a wire screen are the wire diameter,

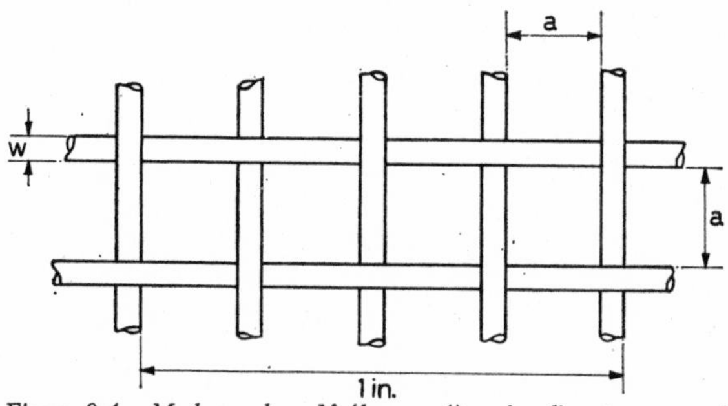

Figure 9.4. Mesh number, N (here = 4), wire diameter, w, and mesh aperture, a

w, the sieve aperture, a, and the fractional open area, A. The relationships between these quantities, a and w being expressed in inches, are

$$N = \frac{1}{a + w} \tag{1}$$

$$A = N^2a^2 = (1 - Nw)^2 \tag{2}$$

A varies from about 0·3 to 0·45, i.e. 30 to 45 per cent of the screen consists of open spaces. Therefore, as the mesh number increases, a decreases and w must also decrease. For example, a 10-mesh screen employs a wire of diameter about 0·04 in. whereas 100 and 300-mesh screens have wire diameters of about 0·004 and 0·001 in., respectively. The wires may be made of brass, phosphor bronze or occasionally of mild steel; they have to be of regular circular cross-section. The screens must be plain woven, except cloths finer than 200-mesh which may be twilled.

Unfortunately many different standard sieve scales are in existence. Thus, whilst all 100-mesh sieves, for example, have 100 apertures and 100 wires to the inch, the width of the apertures will vary according to the scale, because each scale employs a slightly different diameter wire. The two most widely used standard scales are the American Tyler and the British Standard[3]. In the latter scale the sieves have wires graded according to the Standard Wire Gauge (S.W.G.). The Tyler and British Standard scales are compared in *Table 9.1.*

The Tyler scale, which has gained world-wide acceptance, is based on the standard 200-mesh screen with 0·0021-in. diam. wires, giving an aperture of 0·0029 in. The 20- and 100-mesh screens in this series have also been standardized by the U.S. Bureau of Standards. The aperture of any sieve in the main Tyler scale is equal to that of the next smaller sieve number multiplied by $\sqrt{2}$. Thus, proceeding in the order of decreasing mesh number, the area of the aperture is doubled at each successive sieve. For closer sizing purposes, the complete Tyler screen range consists of a series of sieve apertures that increase by the factor $\sqrt[4]{2}$.

Table 9.1. Comparison Between the Tyler and British Standard Fine-mesh Sieve Scales

Tyler			*British Standard*		
Mesh No.	*Aperture*		*Mesh No.*	*Aperture*	
	in.	micron†		in.	micron†
*6	0·131	3,327	5	0·1320	3,340
7	0·110	2,794	6	0·1107	2,810
*8	0·093	2,362	7	0·0949	2,410
9	0·078	1,981	8	0·0810	2,050
*10	0·065	1,651	10	0·0660	1,670
12	0·055	1,397	12	0·0553	1,400
*14	0·046	1,168	14	0·0474	1,200
16	0·039	991	16	0·0395	1,000
*20	0·0328	833	18	0·0336	850
24	0·0276	701	22	0·0275	700
*28	0·0232	589	25	0·0236	600
32	0·0195	495	30	0·0197	500
*35	0·0164	417	36	0·0166	421
42	0·0138	351	44	0·0139	353
*48	0·0116	295	52	0·0116	295
60	0·0097	246	60	0·0099	252
*65	0·0082	208	72	0·0083	211
80	0·0069	175	85	0·0070	177
*100	0·0058	147	100	0·0060	152
115	0·0049	124	120	0·0049	125
*150	0·0041	104	150	0·0041	105
170	0·0035	88	170	0·0035	88
*200	0·0029	74	200	0·0030	76
250	0·0024	61	240	0·0026	65
*270	0·0021	53	300	0·0021	53
325	0·0017	43	350	0·0017	44
*400	0·0015	37	—		

* Sieves of the main Tyler series † 1 micron = 10^{-3} mm

The wire diameters in the Tyler scale have been chosen to guarantee this $\sqrt[4]{2}$ relationship. In the British Standard scale, because of the use of S.W.G. wires, the multiplying factor is only approximately $\sqrt[4]{2}$. Another standard scale, the U.S. (A.S.M.T.), based on the 1,000 μ (0·0394 in.) aperture for the number 18-mesh, also employs the $\sqrt[4]{2}$ factor, but as it permits the use of a range of wire diameters for any given sieve, can be made interchangeable with the Tyler series.

Each Standard Specification lays down a permissible aperture tolerance. The aperture is measured in both the warp and weft directions of the cloth, and the average aperture width must comply with the limits imposed. The tolerances of the British Standard are $\pm$ 3 per cent for sieves up to 16-mesh, $\pm$ 5 per cent for 18- to 44-mesh, $\pm$ 6 per cent for 52- to 120-mesh, and $\pm$ 8 per cent for meshes $>$ 120. Therefore, Tyler and British Standard meshes numbered 10, 12, 14, 16, 60, 100, 150, 170 and 200 may be considered to be approximately equivalent. Throughout the rest of the scales the mesh numbers do not coincide.

In view of the tolerance limits, it is quite possible for one set of Standard sieves to differ appreciably from another set in the series. For some purposes these differences are serious, and a special calibration must be made. This can be done by one of several methods. An image, of definite magnification, of the wire cloth can be projected onto a screen, and direct measurements of the aperture in different directions and in different areas of the sieve can be made. A mean aperture may thus be calculated[3]. STAIRMAND[4] has suggested another method in which the screens are calibrated by sieving graded particles through them; the overall effective aperture of the sieve can then be found (see p. 230).

The largest particles that can be graded on the B.S. fine mesh scale are about $3\frac{1}{3}$ mm (5-mesh); on the Tyler scale the upper limit is about $6\frac{1}{2}$ mm (3-mesh). For the grading of coarser particles, medium and coarse wire mesh sieve scales are available. The Tyler scale ranges from $\frac{1}{4}$ to 3 in. nominal aperture, the B.S. scale (*Table 9.2*) from $\frac{1}{32}$ to $\frac{1}{2}$ in., perforated plates being recommended for coarser screens[3].

Table 9.2. British Standard Medium Mesh Scale[3]

Nominal aperture		*Wire diameter*	
in.	mm	in.	mm
$\frac{1}{32}$	0·79	0·021	0·53
$\frac{1}{16}$	1·60	0·038	0·97
$\frac{1}{8}$	3·18	0·072	1·83
$\frac{3}{16}$	4·76	0·080	2·03
$\frac{1}{4}$	6·35	0·092	2·34
$\frac{3}{8}$	9·53	0·104	2·64
$\frac{1}{2}$	12·7	0·128	3·25

Sieve Tests

Briefly, a sieve test is commenced by placing a weighed quantity of the solid material on to the top sieve in a nest which comprises a number of sieves, arranged in order of decreasing aperture, mounted on a bottom collecting pan. A top lid may be fitted to minimize loss of dust. The nest is shaken for a given period and the sieves are then removed, one by one, and shaken individually either for a specified time or until the rate at which particles pass through the sieve falls below a certain specified rate. The procedure is repeated for each of the sieves in turn. All the collected fractions are then weighed. The loss of material for accurate work should not exceed 0·5 per cent; for routine analyses in control laboratories up to 2 per cent loss may be permitted.

The preparation of the test sample is an important preliminary operation. The weight of the material must not change during the sieving operation. Damp materials must be dried in an oven, then be allowed to cool in the atmosphere before sieving. If the material is hygroscopic the sample must be oven-dried, cooled in a desiccator and sieved with minimum exposure to the atmosphere. For 8-in. diam. B.S. sieves the recommended test sample weights[2] are

25 g for materials of S.G. $<$ 1·2
50 g for materials of S.G. 1·2–3·0
100 g for materials of S.G. $>$ 3·0

These sample weights may be modified in special cases. For coarse solids, e.g. where the bulk of the particles is retained on 52-mesh, the above weights may be increased, provided that the weight retained on any one sieve does not exceed 50 per cent of the test sample weight. Again, the sample weight may be increased to better the accuracy of determining small percentages of relatively coarse particles. Other recommendations[1] for the weight of a test sample are

Limiting particle size mm	*Test sample* g
4·0 to 2·0	1,000
2·0 to 1·0	500
1·0 to 0·5	200
0·5 to 0·25	100
$>$ 0·25	50

For materials which contain a high percentage of fine particles, a preliminary removal of dust may be made before carrying out the main sieve test. This is done with a nest consisting of a medium screen, e.g. B.S. 25- or 30-mesh, the finest sieve to be used in the main test, and the bottom collecting pan; the reason for the medium mesh is to protect the fine mesh from the abrasive action of the coarse particles. The sample is sieved, as described below, for a suitable period, and the fractions are removed and weighed. The material retained on the medium and fine meshes can then be taken as the test sample for the main test using the complete nest of fine sieves.

Sieve testing can be carried out by hand or on a machine designed to hold the nest of sieves and perform the shaking, rotating and tapping operations. Many different types of such machines are available from suppliers of laboratory equipment. Hand sieving may be carried out as follows[2]. Hold the nest of sieves, together with the lid and collector pan, in the left hand, inclining the sieve surfaces downwards towards the left at an angle of 30° to the horizontal. Tap the higher side of the sieve frame six or eight times with the hand or a piece of wood. Whilst maintaining the inclination of the sieve, shake the nest to and fro several times, also rotating it in the plane of the gauze through an angle of about 60°. These alternate tapping and shaking operations are repeated for 5 minutes. The top sieve is removed, inverted over a clean piece of paper, and tapped gently to remove the material retained on it. The bottom surface of the gauze, i.e. the surface that is now uppermost while the sieve is held in its inverted position, may be brushed gently to aid the cleaning operation. All the discharged material is transferred back to the cleaned sieve mounted on a collector pan and submitted to a sieving operation, as described above, for 2 minutes. If the amount of material passing through the sieve in this 2-minute period is less than 0·2 per cent of the original test sample weight, sieving on this sieve may be considered complete; if the amount passing exceeds 0·2 per cent, the procedure is repeated until the 'end point' is achieved. All the weighed material from the collector pan is then transferred on to the material retained on the top sieve of the remaining nest which is submitted to another 5-minute sieving operation. The 'end point' rate test is again applied to the top sieve, and the whole procedure is repeated until all the sieves in the nest have been dealt with.

For machine sieving the following procedure may be adopted. The complete nest is shaken on the machine for 5 minutes. After this period the top sieve is removed and submitted to the cleaning process and 'end point' test *by hand*, as described above. The reduced nest, with the top sieve containing the material which passed through the former top sieve during the end point test, is again shaken on the machine for 5 minutes, and so on.

Wet sieving may be used for very fine powders which pass through gauzes finer than about 150-mesh, or in cases where the particles are fragile and will not withstand the vigorous shaking motions of dry sieving. Small quantities of the solid particles are thoroughly mixed, with an added wetting agent if necessary, in the chosen liquid, e.g. water, alcohol or light petroleum fraction. The slurry is then washed on to a fine screen held near the surface of the liquid in a bowl. The load on the sieve is swirled and jigged until all the fine particles have passed through the gauze. The oversize fraction can be dried and submitted to a dry sieve test, as described above. The fine

Table 9.3. Tabular Representation of Sieve Tests

(a) Fractional percentages			*(b) Cumulative percentages*		
B.S. Mesh		Per cent *by weight*	*B.S. Mesh*	*On (oversize)*	*Through (undersize)*
Through	*On*				
—	44	1·0	44	1·0	99·0
44	60	20·0	60	21·0	79·0
60	85	40·5	85	61·5	38·5
85	120	30·8	120	91·5	8·5*
120	—	8·5*			
		100·0			

solids which pass through the gauze into the liquid in the bowl can either be removed, dried and weighed, or submitted to a further sizing test, e.g. by sedimentation or elutriation.

The results of a sieve test can be reported, in tabular form, as shown in *Table 9.3*, for fractional percentages and for cumulative percentages, distinguishing between oversize and undersize. The quantity marked with an asterisk (*) should be annotated 'including a loss, during test, of *x* per cent'. The method of sieving should also be stated.

MacCalman[5] published one of the most comprehensive accounts of the manufacture and use of test sieves. He discusses in considerable detail the potential errors encountered in this method of size analysis.

Methods of Analysis[6]

Once a sieve test has been performed, and the results tabulated in an approved manner, there remains the task of assessing the size characteristics of the tested material and of extracting the maximum amount of information from the data. Whilst a table may record all the measured quantities, this form of expression is not always the best one; the magnitudes of the various quantities may be readily visualized, but certain trends may be completely obscured in a mass of figures. The real significance of a sizing test can most readily be judged when the data are expressed graphically. From such a pictorial representation trends in the data are easily detected, and the prediction of the expected behaviour of the material on sieves other than those used in the test can often be made with a reasonable degree of accuracy.

Many different forms of graphical expression may be employed, and the use and applicability of some of these methods will be demonstrated with reference to the results of the sieve test given in *Table 9.4*. These data are chosen for their fairly wide spread over the size range of approximately 100

Table 9.4. Sieve Test Data[6] used for the Construction of Figures 9.5 to 9.11

B.S. Mesh	*Sieve aperture* μ*	*Fractional weight* per cent *retained*	*Cumulative weight* per cent *oversize*	*Cumulative weight* per cent *undersize*
7	2,410	1·2	1·2	98·8
10	1,670	2·9	4·1	95·9
14	1,200	18·8	22·9	77·1
18	850	28·8	51·7	48·3
25	600	22·0	73·7	26·3
36	421	11·1	84·8	15·2
52	295	6·0	90·8	9·2
72	211	3·9	94·7	5·3
100	152	1·8	96·5	3·5
150	105	1·3	97·8	2·2
>150	—	2·2	—	—

* 1 micron, $\mu = 10^{-3}$ mm

to 2,500 μ to illustrate the scope of the methods of plotting. The data are listed in the three recommended ways, viz. the percentage by weight of the fractions retained on each sieve, and the cumulative percentages by weight of oversize and undersize material. Alternate sieves in the B.S. series have been used, i.e. meshes with aperture widths decreasing approximately in the ratio $\sqrt{2}:1$. The equivalent B.S. mesh numbers are translated into aperture widths in microns.

Four different types of graph paper may be used, depending on the sort of information that is required: (*a*) the ordinary squared or arithmetic, (*b*) the log-linear or semi-log, where one of the axes is marked off on a log scale and the other on an arithmetic scale, (*c*) the log-log, where both axes are marked off on a logarithmic scale, and (*d*) the arithmetic-probability, where one axis is marked off on a probability scale, the intervals being based on the probability integral.

Arithmetic Graphs

In *Figure 9.5a* the weight percentages of the fractions retained on each successive sieve used in the test are plotted against the widths of the sieve apertures (in microns). The lines joining the points have no significance;

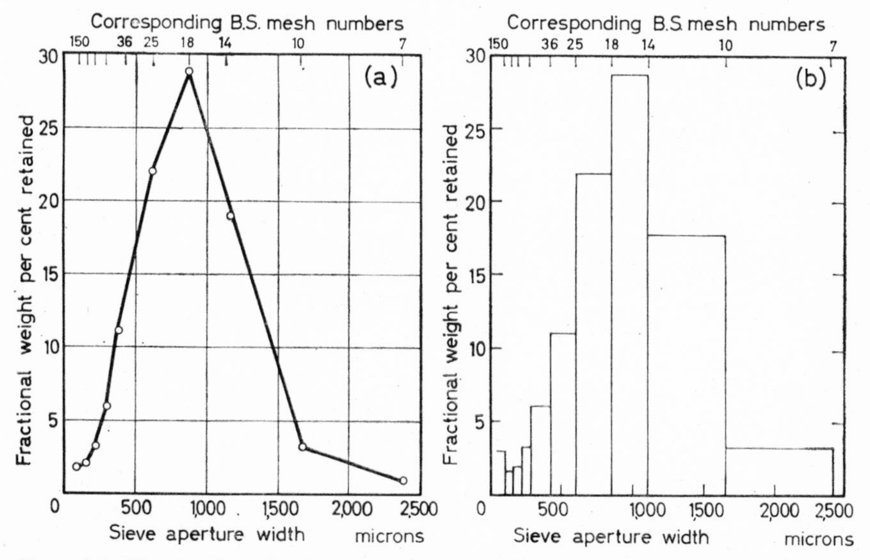

Figure 9.5. Sieve test data plotted on arithmetic graph paper—percentage by weight of the fractions retained between two given sieves in the B.S. series: (a) the frequency polygon; (b) the frequency histogram

they merely complete the graph or frequency polygon. The sharp peak in the distribution curve occurs at 18-mesh (850 μ). This point, however, represents the fraction which passes 14-mesh (1,200 μ) and is retained on 18-mesh, so it could be plotted at the 'mean' size $(1,200 + 850)/2 = 1,025\ \mu$. Alternatively, the results may be represented in the form of a frequency histogram, as shown in *Figure 9.5b*, depicting the size range of each collected fraction. For example, 18·8 per cent of the test sample passed through the 10-mesh and was retained on the 14-mesh (see *Table 9.4*). Thus, the 18·8 per cent horizontal is drawn between the 1,200 and 1,670 μ positions on the histogram.

From both diagrams in *Figure 9.5* the general picture of the overall spread of particle size can be seen quite clearly. However, the simple arithmetic method of plotting suffers from the disadvantage of producing a congested picture in the regions of the fine mesh sieves. If all the available sieves in the B.S. range had been used, i.e. those belonging to the $\sqrt[4]{2}$ series, the conditions in the small-size region of this type of plot would have become quite obscure. The fine mesh data, of course, could be re-plotted on an enlarged scale, but the overall picture would then be lost.

The cumulative weight percentages are plotted against aperture size in *Figure 9.6*. This type of plot often permits the drawing of smooth curves through the plotted points, although the accuracy with which the curves can be drawn can be very poor if not all the sieves in the range have been used.

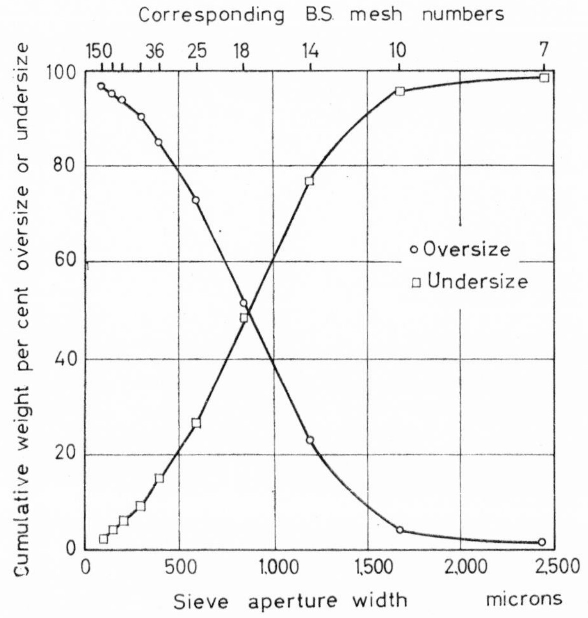

Figure 9.6. Sieve test data plotted on arithmetic graph paper—cumulative oversize and undersize percentages

The two curves in *Figure 9.6* are mirror images of one another. The advantage of this method of plotting is that estimates can be made of the percentages of the material that would be retained on or pass through any sieve, standard or non-standard, with apertures in the range tested. It could be predicted, for example, that about 87 per cent of the original material would pass through a B.S. 12-mesh (1,400 μ) or that about 40 per cent would be retained on a B.S. 16-mesh (1,000 μ). However, as in *Figure 9.5*, the data in the region of the small aperture sieves are congested, and interpolation is difficult.

Semi-log Graphs

When the test data are plotted on semi-log paper (*Figure 9.7a*), with the aperture widths recorded on the logarithmic scale, the points in the coarse sieve region are brought closer together, and those in the fine sieve region located further apart than in the corresponding simple arithmetic plot. In fact, the successive points are more or less equally spaced along the horizontal scale. (If the B.S. sieves followed the $\sqrt{2}$ rule exactly, as the Tyler scale does, the points would be equidistant.) The frequency histogram (*Figure 9.7b*)

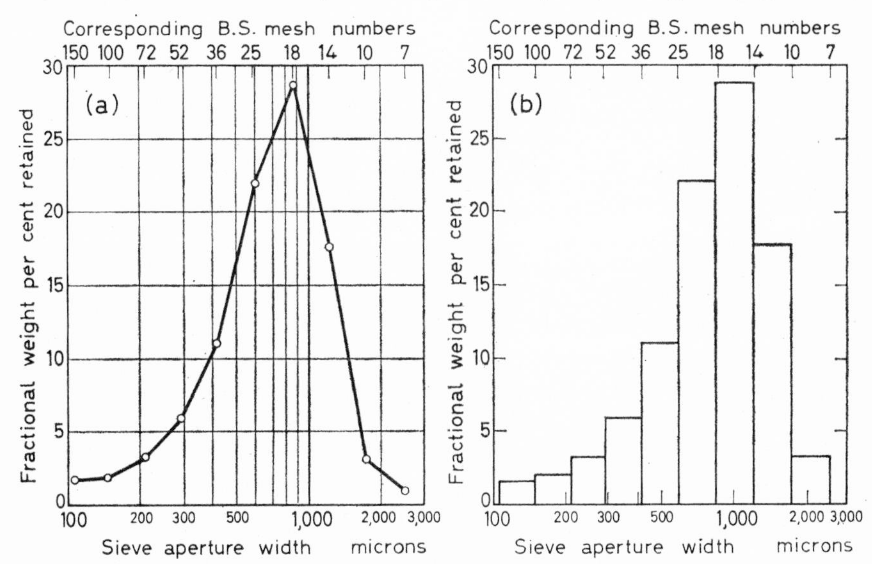

Figure 9.7. Sieve test data plotted on semi-log graph paper—percentage by weight of fractions retained between two sieves in the B.S. series: (a) the frequency polygon; (b) the frequency histogram

is composed of columns of approximately equal widths. The picture of the size spread in the fine sieve region is thus clarified, and data obtained on the complete $\sqrt[4]{2}$ series of sieves can be plotted on graphs of this type without congestion.

The semi-log graphs for the cumulative oversize and undersize are plotted in *Figure 9.8*; here again the curves are mirror images. Interpolation in the fine sieve region is facilitated due to the even spread of the plotted points. It can be estimated, for instance, that about 87 per cent of the original material would be retained on a 44-mesh (353 μ) or that about 7 per cent would pass through a 60-mesh (252 μ).

Log–log Graphs

In *Figure 9.9* the weight percentages of the fractions retained on the given sieves are plotted against the aperture size on a log–log basis. This type of plot usually tends to bring the points in the fine sieve region into a straight

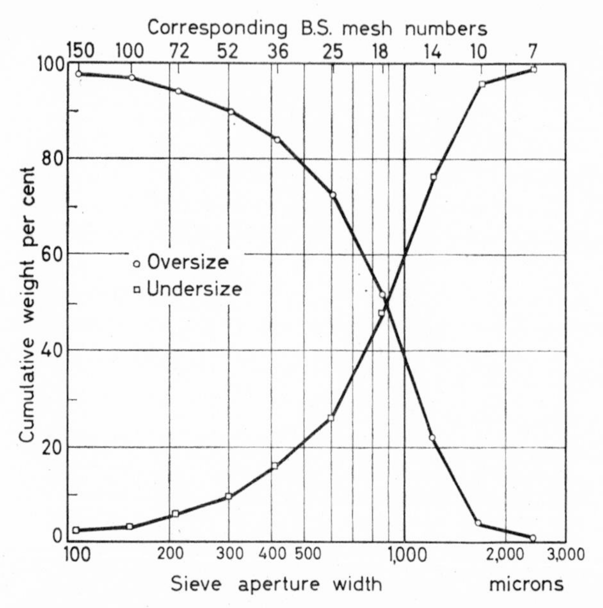

Figure 9.8. Sieve test data plotted on semi-log graph paper—cumulative oversize and undersize percentages

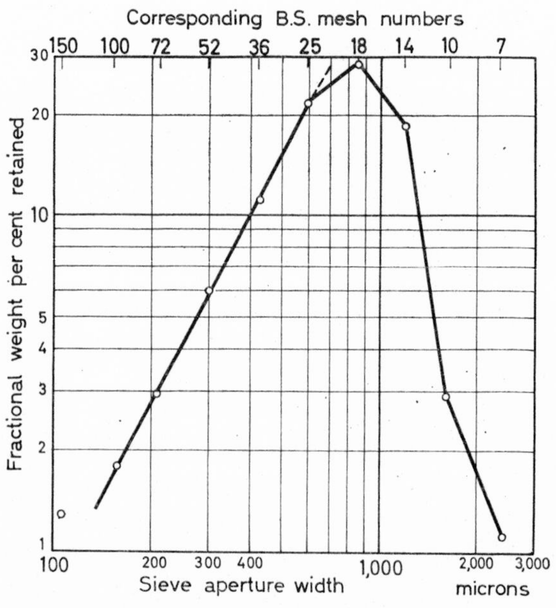

Figure 9.9. Sieve test data plotted on log-log graph paper—fractional weight percentages retained between two sieves in the B.S. series

line. The data in *Figure 9.9* are fitted by a straight line between about 600 to 150 μ (25- to 100-mesh). A linear correlation of the data indicates the application of an equation of the type

$$p = c \, . \, a^n \tag{3}$$

where p is the weight percentage of material retained on a sieve of aperture a after passing through the next larger sieve used in the test. The constant c, mathematically speaking, is the value of p when $a = 1$, but this cannot be interpreted literally because the linear relationship would not be valid in the region of 1 μ particle size. Therefore, c is just a constant depending on the material; the higher the value of c, the finer is the material. The exponent n, the slope of the line, gives a measure of the particle size spread; for close-sized particles n tends to infinity.

The cumulative percentages of oversize and undersize particles are plotted against aperture width on a log–log basis in *Figure 9.10*. In this type

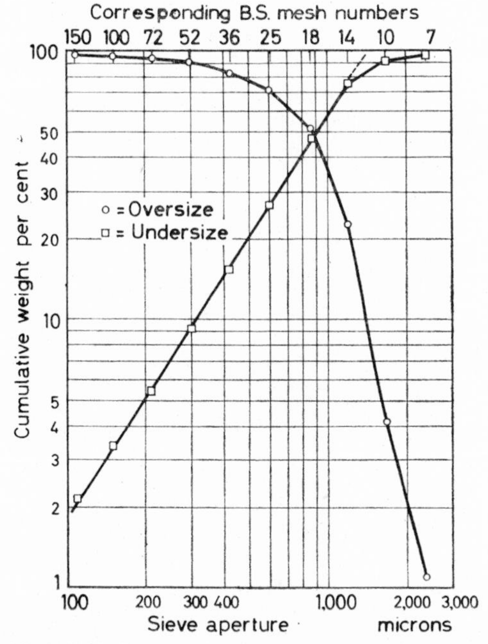

Figure 9.10. Sieve test data plotted on log–log graph paper—cumulative oversize and undersize percentages

of plot the cumulative undersize data tend to lie on a straight line over a wide range of particle size, about 100 to 1,200 μ in this case. The undersize and oversize curves are clearly not mirror images, and oversize data are rarely correlated on this basis. For the undersize line the relationship

$$P = b \, . \, a^m \tag{4}$$

holds over the linear region. P is the total percentage of the original material that would pass through a screen of aperture a. The exponent m, the slope of the line, gives a measure of the spread; the lower the value of m, the greater the size distribution. The constant b indicates the fineness or coarseness of the particles: the higher the value of b, the finer the material. The log–log method of plotting of undersize data is extremely useful because rough checks may be made on the size distribution by the use of only two, or possibly three, test sieves. Material of the type considered in *Table 9.4* could be size-checked with, say, Nos. 25 and 72 B.S. mesh sieves, or Nos. 18, 36 and 72 if a more reliable result was required. Sieves of any standard series can be used, provided that their apertures are accurately known, and the data can readily be translated into any other required sieve range.

Many products of comminution exhibit a linear size distribution relationship when plotted in the above manner, and certain selected finely-crushed materials can be used to calibrate a set of sieves. The method reported by STAIRMAND[4], mentioned above, is based on this behaviour. Briefly, the method is as follows. A cumulative undersize log–log plot is drawn from the

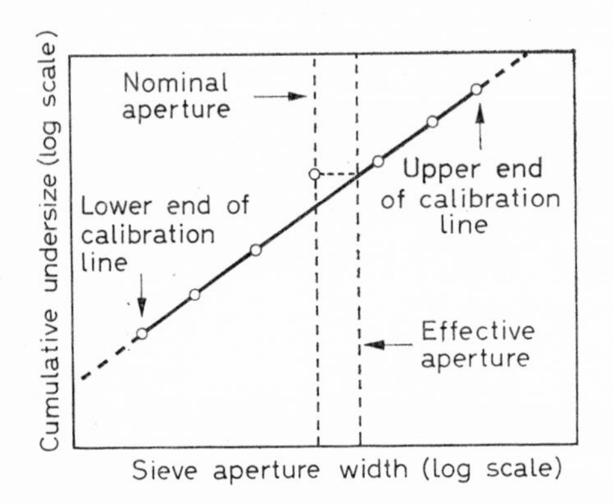

Figure 9.11. Calibration of test sieves

data obtained from the sieve test on the powdered material, using the nominal aperture of the sieves. The best straight line is drawn through the points, which should lie on or very close to this line. For all points lying off the straight line, the nominal aperture of the corresponding sieve is corrected, as shown by the grossly exaggerated example in *Figure 9.11*, and the effective aperture is used in all work in which this sieve is subsequently used.

Probability Graphs

Several different probability plots have been suggested for particle size analysis[1, 7], particularly in connection with the assessment of crushing and grinding processes. The following method, devised by POWERS[8] for use in the sugar industry, has much to be said in its favour for the size specification of crystalline materials. This method, which employs arithmetic-probability graph paper, was suggested as a means of eliminating the confusion caused by the use of different standard sieve scales; the size distribution is recorded in terms of two numbers only—the *mean aperture*, *MA*, and a statistical quantity called the *coefficient of variation*, *CV*, expressed as a percentage. The significance of these terms is as follows. A result expressed as *MA* 600, *CV* 30, or just 600/30, indicates that 50 per cent by weight of the material will pass through a sieve of 600 μ aperture, and that the standard deviation $\times$ 100 divided by the mean aperture = 30 per cent. From a knowledge of these two numbers the percentages of the material passing through any standard sieve can be specified with a reasonable degree of accuracy. Incidentally, Powers suggested that *MA* values should be given in inches, i.e. 0·0197/30 instead of 600/30, but it is felt that the use of microns yields a simpler *MA*/*CV* value.

The use of the *MA*/*CV* method is demonstrated in *Figure 9.12* with the data from *Table 9.4*. The cumulative undersizes are plotted on the probability scale, the sieve aperture widths on the arithmetic scale. If the data between

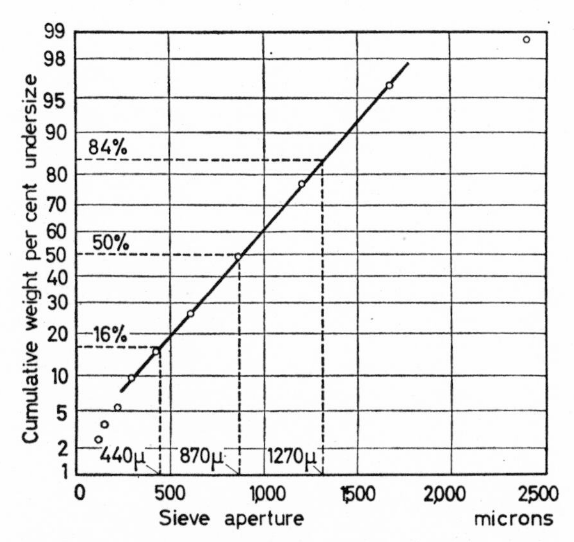

Figure 9.12. Sieve test data plotted on arithmetic/probability graph paper (cumulative undersize percentages), illustrating the use of the MA/CV method of analysis

about 10 and 90 per cent lie on a straight line, the *MA*/*CV* method can be applied. The data in *Figure 9.12* comply with this requirement. Thus, the mean aperture corresponding to 50 per cent is 870 μ. The standard deviation σ can be deduced as follows.

If the area enclosed under a normal probability curve between sieve aperture $a = 0$ to ∞ is taken as unity, the area enclosed between $a = 0$ and $a = \bar{a} + \sigma$, where $\bar{a}$ is the mean aperture *MA*, is 0·8413. This value is obtained from tables of the normal probability function. Therefore, the area enclosed between $a = \bar{a} + \sigma$ and $a = \infty$ is $1 - 0{\cdot}8413 = 0{\cdot}1587$. The value of σ can be obtained from the arithmetic probability diagram by reading the value of a at 84·13 per cent (84 per cent is accurate enough for this purpose) and subtracting the value of $\bar{a}$. Alternatively, the value of a at 15·87 (or 16) per cent can be subtracted from $\bar{a}$, i.e.

$$\sigma = a_{84\%} - \bar{a} = \bar{a} - a_{16\%}$$

These two values of σ may not coincide, so a mean value can be taken as

$$\sigma = \frac{a_{84\%} - a_{16\%}}{2} \tag{5}$$

The coefficient of variation, as a percentage, is given by

$$CV = \frac{100\sigma}{\bar{a}} \tag{6}$$

or

$$CV = \frac{100(a_{84\%} - a_{16\%})}{2MA} \tag{7}$$

From *Figure 9.12*

$$CV = \frac{100(1{,}270 - 440)}{2(870)} = 48 \text{ per cent}$$

and the size distribution can be specified, in terms of MA/CV, as 870/48.

If the crystalline material being produced is known to obey this straight-line rule, routine analysis of the product is greatly simplified. Only two sieves, carefully standardized, need be used when making checks on the product specification. It is important to understand that the size distribution does not have to be Gaussian for the MA/CV method to apply. Many skew distributions also give the necessary linear relationship between about 10 to 90 per cent, but in these cases the MA will not coincide with the modal diameter at the peak of the distribution curve.

Particle Size and Surface Area

It is not possible to measure or define absolutely the size of an irregular particle, and perfectly regular crystalline solids are rarely, if ever, encountered. The terms length, breadth, thickness or diameter applied to irregular particles are meaningless unless accompanied by further definition, because so many different values of these quantities can be measured. The only precise properties that can be defined for a single solid particle are the volume and surface area, but even the measurement of these quantities may present insuperable experimental difficulties. All particle size measurements are made by indirect methods: some property of the solid body which can be related to size is measured.

Despite these difficulties of definition and measurement it is most convenient, for classification purposes, if a single-length parameter can be ascribed to an irregular solid particle. The most frequent expression used in connection with particle size is the 'equivalent diameter', i.e. the diameter of a sphere which behaves in an identical manner to the given particle when submitted to the same experimental procedure. Several of these equivalent diameters may be defined. For example, a particle which just passes through a sieve aperture is classified according to the diameter of a sphere which would also just pass through. The term equivalent sieve aperture diameter, d_a, is usually applied in this case.

The present chapter is concerned primarily with sieving operations, which are almost exclusively used for grading commercial crystals, but a few brief references can be made to some of the other methods that can be employed for size classification in the sub-sieve range. Sedimentation and elutriation methods, for instance, are based on Stokes' law which relates the free falling velocity u of a particle in a fluid medium of density ρ_f and viscosity η to the diameter d_s of a solid sphere of the same density ρ_s as the particle, by means of the equation

$$d_s = \sqrt{\frac{18\eta u}{(\rho_s - \rho_f)g}} \tag{8}$$

This equivalent diameter, d_s, is usually referred to as the Stokes' diameter. Water or any other suitable liquid may be used as the fluid medium in sedimentation processes, liquids or gases can be used as elutriating media. Both operations can be carried out fractionally to get a picture of the range of size distribution in the original sample.

Microscopic techniques can also be used for the size estimation of very fine particles. The actual size recorded is the diameter of a circle of the same area as the projected image of the particle viewed in a direction perpendicular to its plane of maximum stability. The particle image is compared with graduated circles on an eyepiece or graticule. The symbol d_{pa} may be ascribed to this 'projected area' diameter. The approximate useful size ranges for the above methods are

Sieving	d_a:	3,000 to 50 μ
Elutriation	d_s:	50 to 5 μ
Gravity sedimentation	d_s:	20 to 1 μ
Visible light microscopy	d_{pa}:	100 to 0·2 μ

As a rough guide in the comparison of size analyses by these various methods the following relationship may be used

$$d_a : d_s : d_{pa} \sim 1 : 0{\cdot}94 : 1{\cdot}4$$

For a detailed account of the scope and methods of particle size measurement in industrial practice, reference should be made to specialized publications devoted to this subject[9–13]. Details of the microscopic methods of particle size analysis for particles in the sub-sieve range are given in a British Standards specification[14].

The terms 'fine' and 'coarse' are so frequently used in industrial practice, usually without definition, to describe crystals and crystalline powders that it is worth recording the recommendations of the British Pharmacopœia[15] on the grading of powdered materials:

Coarse:	all passes 10-mesh, $\not>$ 40 per cent passes 44-mesh
Moderately coarse:	all passes 22-mesh, $\not>$ 40 per cent passes 60-mesh
Moderately fine:	all passes 44-mesh, $\not>$ 40 per cent passes 85-mesh
Fine:	all passes 85-mesh
Very fine:	all passes 120-mesh

The mesh numbers refer to the British Standard scale described above. It is appreciated, of course, that these definitions will not be regarded as suitable for all purposes, but at least they give some definite meaning to these rather loosely applied terms.

In many cases the degree of fineness of a particulate mass is better expressed in terms of the available surface area of the particles rather than as an equivalent diameter. Particle area is a very important factor to be considered when chemical reactions or other mass transfer operations are to be performed with the solid substance. For very fine particles, surface area measurements may be made by turbidimetric, adsorption, dissolution, permeability and many other techniques[11, 12].

The permeability method has proved invaluable for the determination of the degree of fineness of powdered substances such as Portland cement; a permeability cell for this purpose is described in a British Standards specification[16]. The method has also been suggested for the size analysis of fine crystalline materials, e.g. icing sugar, and a laboratory apparatus suitable for routine analysis has been described by HILL and MULLER[17]. A known quantity of air is forced through a small bed of the fine solids under a constant pressure drop, and the flow time is recorded. The theory is based on the laminar flow of fluids through porous beds, and the specific surface S (cm^2/g) of the material is calculated from the Kozeny equation

$$S^2 = \frac{\Delta P}{k u \eta L \rho^2} \cdot \frac{\varepsilon^3}{(1 - \varepsilon)^2} \tag{9}$$

where ΔP = pressure drop across the bed; ε = voidage of the bed; L = depth of the bed; η = viscosity of the air; u = empty-tube velocity; ρ = density of the solid material; k is a constant (Kozeny's constant) which has a value equal to about 5·0 for granular solids.

For particles in the sieve size range, an estimate of the surface area may be made from the results of a sieve analysis, but this estimate may be liable to gross error. A precise calculation of the volume or surface area of a solid body of regular geometric shape can only be made when its length, breadth and thickness are known. For crystals, or indeed particulate solids in general, these three dimensions can never be precisely measured. Therefore, before giving a brief account of some of the methods of calculation available, a word of warning is necessary. It must be fully appreciated that the precision of calculation is always far greater than that of measurement of the various quantities used in the mathematical expressions. An equation, especially a complex one, always has a look of absolute dependability, but in this particular connection it most certainly leads to a sense of false security. All calculated volume or surface area data must be used with caution.

Most calculation methods are based on one dimension of the particle, usually the equivalent diameter. If this dimension is obtained from a sieve

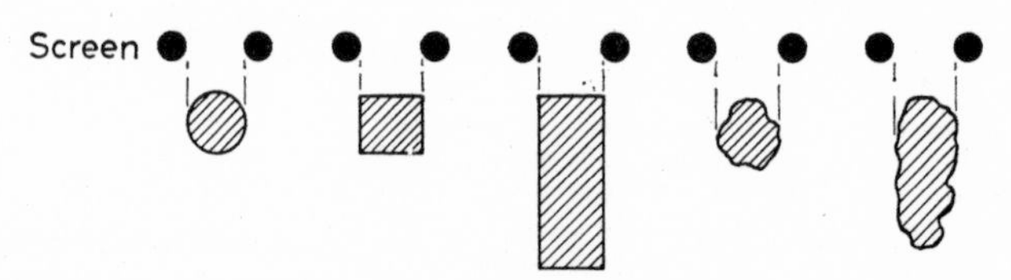

Figure 9.13. Various particle shapes that would all be classified under the same sieve-aperture diameter

analysis it will be the sieve aperture diameter, d_a, but as crystals are never true spheres, this diameter will be the second largest dimension of the particle. *Figure 9.13* demonstrates some particle shapes that would, in a sieve analysis, all yield the same value for d_a. One source of error is thus clearly seen.

For a single particle, the size of which is defined by some length parameter or diameter, d, the following relationships can be applied

$$\text{volume} \qquad v = f_v d^3 \tag{10}$$

$$\text{mass} \qquad m = f_v \rho d^3 \tag{11}$$

$$\text{surface area} \qquad s = f_s d^2 \tag{12}$$

The constants f_v and f_s may be called volume and surface shape factors, respectively. For spherical (diameter $= d$) and cubical (length of side $= d$) particles

$$f_v = \tfrac{\pi}{6} \text{ (sphere) and 1 (cube)}$$
$$f_s = \pi \text{ (sphere) and 6 (cube)}$$

From equations 10 and 12 two basic ratios may be defined

$$\text{surface: volume} \qquad \frac{s}{v} = \frac{f_s d^2}{f_v d^3} = \frac{F}{d} \tag{13}$$

$$\text{surface: mass} \qquad \frac{s}{m} = \frac{f_s d^2}{f_v \rho d^3} = \frac{F}{\rho d} \tag{14}$$

Equation 14 defines the important quantity known as the specific surface, i.e. the surface area per unit mass of solid. Some authors, unfortunately, have also called the surface:volume ratio a specific surface, but this definition is not widely adopted. The constant F $(= f_s/f_v)$ may be called the overall, surface–volume or specific surface shape factor. For spheres and cubes, $F = 6$, for other shapes $F > 6$. Values of $F \sim 10$ are frequently encountered in comminuted solids, and much higher values may be found for flakes and plate-like crystals. If the particles are elongated or needle-shaped, their volume and surface area may be calculated on the assumption that they are cylindrical; length and diameter may be measured microscopically, or the diameter can be taken as the equivalent sieve aperture diameter.

In a total mass, M, of uniform particles, each of mass m and equivalent diameter d, the number of particles, n, is given by

$$n = \frac{M}{m} = \frac{M}{f_v \rho d^3} \tag{15}$$

and the total surface area, Σs, by

$$\Sigma s = ns = \frac{f_s M d^2}{f_v \rho d^3} = \frac{FM}{\rho d} \tag{16}$$

However, before equations 15 and 16 can be applied to masses of non-uniform particles, some average value of the equivalent diameter must be defined. A few of the many suggested methods are described below.

The simplest of all average diameters is the arithmetic mean. If sieving has been carried out between two sieves of aperture a_1 and a_2, the average particle diameter, $\bar{d}$, is given by

$$\bar{d} = (a_1 + a_2)/2 \tag{17}$$

This description is quite adequate for two consecutive sieves in the $\sqrt[4]{2}$ series, but it can be absolutely meaningless for two sieves at extreme ends of the mesh range. Another simple average diameter is the geometric mean, defined by

$$\bar{d} = \sqrt{a_1 a_2} \tag{18}$$

Values of $\bar{d}$ calculated from equation 18 are smaller than those given by equation 17, but for two close sieves the difference is not great.

When the surface area of the particles is an important property the surface mean diameter can be employed, defined by

$$\bar{d} = \frac{\Sigma nd^3}{\Sigma nd^2} = \frac{\Sigma M}{\Sigma(M/d)} \tag{19}$$

where n and M are the number and mass, respectively, of all particles of equivalent diameter d. The root mean square diameter is also frequently used when surface properties are important. This statistical quantity is defined by

$$\bar{d} = \sqrt{\frac{\Sigma nd^2}{\Sigma n}} = \sqrt{\frac{\Sigma(M/d)}{\Sigma(M/d^3)}} \tag{20}$$

Values of $\bar{d}$ calculated from equations 19 and 20 can differ considerably, yet for a mass of particles with a wide size distribution there is no general agreement as to the preferred method.

Two other statistical diameters are often encountered, viz. the modal and median diameters; both are determined from frequency plots (size interval versus number of particles in each interval). The modal diameter is the diameter at the peak of the frequency curve, whereas the median diameter defines a mid-point in the distribution—half the total number of particles are smaller than the median, half are larger. If the distribution curve obeys the Gaussian or Normal Error law the median and modal diameters coincide.

In connection with particle size measurement more than twenty different 'average' diameters have been proposed, and whilst several have certain points in their favour in special cases, none has yet been found to be generally satisfactory. Therefore, all calculations based on an average diameter are prone to appreciable error, and it is recommended that such calculated quantities should be clearly annotated with the method of calculation so that the results of different workers can be compared.

INDUSTRIAL SCREENING

Crystalline products are generally marketed in graded sizes, and consequently dried crystals are frequently submitted to a screening process before final packaging. One of the functions of the classifying crystallizers, described in

Chapter 8, is to produce directly a product of the required size, and thus to eliminate the necessity for screening which proves to be a wasteful, costly bottleneck process. But even the products from these crystallizers may require a gentle screening to remove pieces of scale, or fine dust resulting from attrition during filtration, drying, handling and other post-crystallization operations.

The objects of most crystal-screening processes are to make a 'cut' in the original material, to remove the roughs and fines and to leave as the main product a mass of fairly regular granular crystals. There are cases where regularity of crystal size is not desired. Drugs and other fine chemicals required for tabletting purposes, for instance, must have a reasonable spread of particle size to reduce the quantity of air entrapped in the mass when it is compressed in the die of the tabletting machine; a certain proportion of fines is required to fill the voids of the larger particles. In some industrial processes, complicated screening followed by blending operations may be necessary to produce the final product.

Screening on an industrial scale is quite different from the laboratory procedures described above for sieve testing purposes. In the latter operation screening is continued to an end point, i.e. until no more, or very little, of the material passes through the given screen. In industrial practice there is neither the time nor indeed the necessity to approach this degree of perfection. The operation is usually continuous, feed material flowing at a steady rate on to the shaking or vibrating screen and remaining on the screening surface for a relatively short time. The passage of particles through the sieve apertures is impeded by the motion of the screen and by the presence of other particles. Particle interference coupled with the short residence time of material on the screen lead to imperfect separation. The size of the sieve apertures is also an important factor; sieves finer than about 100-mesh are rarely used industrially on account of their low throughput and liability to clogging or 'binding'.

Types of Screens

Industrial screens may be fabricated in mild steel, stainless steel, brass, phosphor bronze, Monel metal and many other alloys to suit special requirements. Stainless steel is probably the most widely favoured, as it is corrosion-resistant and does not suffer severely from abrasion. Fine screens in a sieving machine are usually supported on a more open screen with stronger wires to prevent distortion of the fine mesh by the weight of material flowing over it. The presence of a supporting mesh, however, restricts the effective screening area.

The nominal aperture of a screen determines the diameter of a spherical particle that will just pass through or be retained on the screen. This condition, however, only applies if the screen is laid horizontally and the particle is presented to the aperture in a vertically downward direction. For industrial screening purposes many factors have to be considered before a mesh size can be specified for a given grading duty. For instance, the screens are usually laid in a sloping position, and the effective aperture of the screen can be much less than the nominal aperture. Again, crystals are not spherical, nor are they presented to the apertures in a vertically downward direction; passage through the screen is effected by a combination of jostling and pushing. Elongated or irregular particles can easily block the sieve apertures, rendering the screen ineffective. A similar state of affairs is encountered when the feed material contains a high proportion of particles of near-mesh size.

Most wire screens have square apertures, but specially woven screens with elongated apertures are occasionally employed. These latter types are useful for the sieving of needle crystals as they are less prone to blinding; they are not suitable, however, for the grading of tabular or platy crystals. The throughput or capacity of an elongated aperture screen is greater than that of an equivalent square aperture screen, but the sharpness of size grading that can be effected is generally inferior.

Screening Equipment

No attempt will be made here to describe in any detail the vast number of screening units that are employed in industrial size grading. For an account of the construction and operation of these machines reference should be made to handbooks dealing with materials handling equipment. The treatment of this subject given by TAGGART[18] is most comprehensive. Broadly speaking, industrial screens may be classified according to the motion, if any, of the screening surface, and it is on this basis that the following notes are made.

Stationary screens, punched and slotted plates, and parallel bars (grizzlies) are used for the coarse grading of particles larger than about $\frac{1}{2}$ in. They may be laid in a horizontal or sloping position, and the material is passed on to or down the screening surface. The passage of the material through the screen may be assisted by raking, and rough lumps may be broken up on the more robust assemblies.

Revolving screens, or trommels, consist of a horizontal cylindrical screen, of wire or perforated plate, which rotates within a casing. Feed material enters at one end of the cylinder, fines pass through the screen into the collector casing, and the coarse material leaves at the other end of the cylinder. The longitudinal axis of rotation usually slopes towards the discharge end to facilitate the movement of material through the trommel. The meshes or holes may increase in size at set intervals along the cylindrical screen, and several product grades can be collected. Alternatively, two or more concentric screens of different aperture may be used. Conical trommels, with horizontal axes of rotation, are also available. Rotating screens of nominal aperture smaller than about $\frac{1}{8}$ in. are rarely used.

Shaking, gyrating and vibrating screens are most frequently employed for the size grading of crystalline products, and literally dozens of different machines are in common use. Whilst their operating mechanisms differ considerably, they are all aimed at producing a jerky but continuous movement of material over the whole available screening surface. The screens are usually laid in a sloping position, about 15 to 30° to the horizontal, but some, particularly those on the circular gyrators, may be laid almost horizontally. Several decks of screens may be mounted above one another in the same unit, thus permitting the separation of a number of different size fractions. The vibrating screens, in which the vibrations (about 10 to 50 per sec) may be produced electrically or mechanically, e.g. by an off-balance flywheel, are less prone to blinding than are the shakers and gyrators, and generally give larger throughputs per unit area of screen surface.

Screening Capacity and Efficiency

The material fed to a screen can be considered to be composed of two fractions, an oversize fraction consisting of particles that are too large to pass through the screen apertures, and an undersize fraction consisting of particles that are too small to be retained on the screen. The screening efficiency, or effectiveness of separation, therefore, should indicate the degree of success obtained in the segregation of these two fractions. In an industrial screening operation a 'clean' separation is never achieved; undersize particles are invariably left in the oversize fraction mainly because the material does not remain on the sieve for a sufficiently long period, and oversize particles may be found in the undersize fraction if the screen mesh is non-uniform, punctured or inadequately sealed around its edges.

The capacity of a screen is the feed rate at which it performs the specified duty. In general, other factors remaining constant, the capacity, i.e. throughput, decreases as the required degree of separation, i.e. efficiency, increases. Some compromise, therefore, has to be made between capacity and efficiency. Various expressions of capacity are used. For a given screen, the specification of lb./min or ton/h may be quite adequate. Capacity may also be quoted as mass per unit time per unit area of screening surface, but care must be taken here because the length and width of the screen may be independent factors. The specification of capacity in terms of, say, lb./h (or lb./h · ft.^{2}) per unit of sieve aperture enables a rough estimation to be made of the capacity of another screen, because this quantity should remain fairly constant for a given screening unit operating on a given feedstock.

The term 'efficiency' applied to screening processes is not easy to define. In fact, there is no generally accepted definition of the term, and various industries adopt the one which most simply and adequately meets their needs. The following analysis indicates a few of the expressions commonly employed. The sieve-test data shown diagrammatically in *Figure 9.14* point out the differences between perfect and actual screening operations. *Figure 9.14a* gives the sieve analysis (cumulative oversize fractions plotted against sieve aperture) of the feed material. Therefore, for an effective screen aperture a^*, fraction O represents the oversize particles, fraction U represents the undersize. For perfect separation, the sieve analyses of these two fractions would be as shown in *Figure 9.14b*; no particles smaller than a^* appear in the oversize fraction, none larger than a^* in the undersize fraction. In practice, however, unwanted particle sizes do appear in the undersize and oversize flow streams (*Figure 9.14c*).

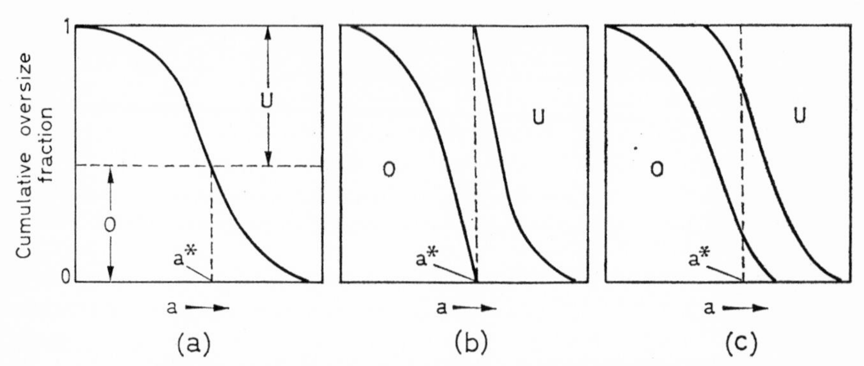

Figure 9.14. Cumulative oversize diagrams: (a) feedstock; (b) perfect screening; (c) actual screening

A screening operation can be considered as the separation of a feedstock F into an oversize top product O and an undersize bottom product U, i.e.

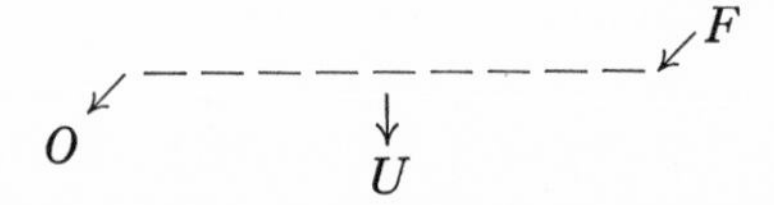

Therefore, if F, O and U represent the masses of these flow streams, an overall mass balance gives

$$F = O + U \tag{21}$$

If $x_F{}^o$, $x_O{}^o$ and $x_U{}^o$ represent the mass fractions of oversize material, i.e. particles larger than a^*, in the feed, overflow and underflow, respectively, and $x_F{}^u$, $x_O{}^u$ and $x_U{}^u$ represent the corresponding undersize mass fractions, then a balance on the oversize material gives

$$Fx_F{}^o = Ox_O{}^o + Ux_U{}^o \tag{22}$$

and a balance on the undersize material gives

$$Fx_F{}^u = Ox_O{}^u + Ux_U{}^u \tag{23}$$

Thus, from equations 21 to 23

$$\frac{O}{F} = \frac{x_F{}^o - x_U{}^o}{x_O{}^o - x_U{}^o} = \frac{x_U{}^u - x_F{}^u}{x_U{}^u - x_O{}^u} \tag{24}$$

and

$$\frac{U}{F} = \frac{x_O{}^o - x_F{}^o}{x_O{}^o - x_U{}^o} = \frac{x_F{}^u - x_O{}^u}{x_U{}^u - x_O{}^u} \tag{25}$$

For perfect screening, therefore,

$$Ox_O{}^o = Fx_F{}^o \tag{26}$$

and

$$Ux_U{}^u = Fx_F{}^u \tag{27}$$

For actual screening, two efficiencies can be defined:

$$E_O = \frac{Ox_O^o}{Fx_F^o} \tag{28}$$

and

$$E_U = \frac{Ux_U^u}{Fx_F^u} \tag{29}$$

Equation 28 gives a measure of the success of recovering oversize particles in the overflow stream, equation 29 of undersize material in the underflow stream. For perfect screening both E_O and E_U will be unity. Equations 28 and 29 require values of the mass flow rates F, O and U, but substitution from equations 24 and 25 can eliminate these quantities

$$E_O = \frac{x_O^o(x_F^o - x_U^o)}{x_F^o(x_O^o - x_U^o)} \tag{30}$$

$$= \frac{(1 - x_O^u)(x_U^u - x_F^u)}{(1 - x_F^u)(x_U^u - x_O^u)} \tag{30a}$$

and

$$E_U = \frac{x_U^u(x_F^u - x_O^u)}{x_F^u(x_U^u - x_O^u)} \tag{31}$$

$$= \frac{(1 - x_U^o)(x_O^o - x_F^o)}{(1 - x_F^o)(x_O^o - x_U^o)} \tag{31a}$$

An overall screen effectiveness, E, can be defined[19] as the product of E_O and E_U. *Table 9.5* illustrates the calculation of these efficiencies.

Table 9.5. Sieve Analyses (effective screen aperture $a^* = 460\ \mu$)

B.S. Mesh	*Aperture μ*	*Cumulative weight fraction oversize*		
		Feedstock	*Overflow*	*Underflow*
18	850	0·02	0·06	—
22	700	0·12	0·29	—
25	600	0·26	0·52	0·02
30	500	0·45	0·78	0·08
36	421	0·68	0·90	0·22
44	353	0·81	0·96	0·46
52	292	0·90	1·00	0·68
72	211	0·98	—	0·84
<72	—	1·00	—	1·00

These data are plotted in *Figure 9.15*, where it can be seen that the values of x_F^o, x_O^o and x_U^o at $a^* = 460\ \mu$ are 0·58, 0·86 and 0·14, respectively. Therefore, from equations 30 and 31

$$E_O = \frac{0{\cdot}86(0{\cdot}58 - 0{\cdot}14)}{0{\cdot}58(0{\cdot}86 - 0{\cdot}14)} = 0{\cdot}91$$

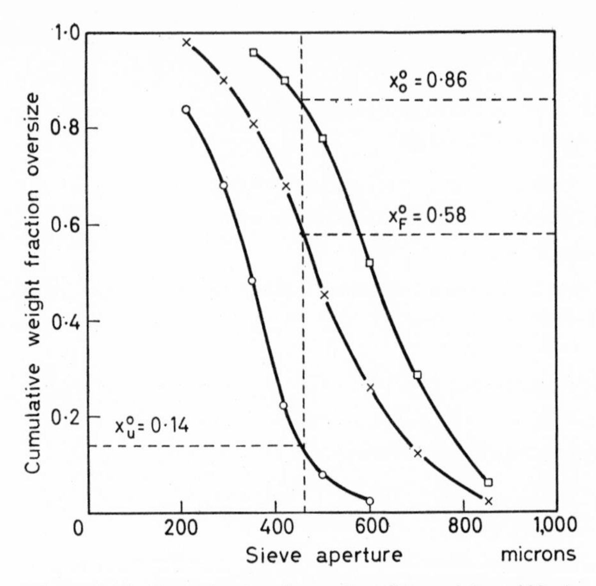

Figure 9.15. Determination of screening efficiency ($a^* = 460\ \mu$)

and

$$E_U = \frac{(1 - 0{\cdot}14)\ (0{\cdot}86 - 0{\cdot}58)}{(1 - 0{\cdot}58)\ (0{\cdot}86 - 0{\cdot}14)} = 0{\cdot}80$$

and the overall effectiveness $E = E_O E_U = 0{\cdot}91 \times 0{\cdot}80$

$= 0{\cdot}728$ or 73 per cent

A simpler form of equation 30 is often used to express the recovery of true undersize material in the overflow fraction, on the assumption that all the underflow stream consists of undersize material, i.e. $x_U{}^u = 1$

$$E_O' = \frac{x_F{}^u - x_O{}^u}{x_F{}^u(1 - x_O{}^u)} \tag{32}$$

or

$$E_O' = \frac{x_O{}^o - x_F{}^o}{x_O{}^o(1 - x_F{}^o)} \tag{32a}$$

The above assumption, of course, is not always valid.

Formulae of the above types have been criticized because they do not take into consideration the difference between an easy separation duty and a difficult one. For example, consider the sieving of two feedstocks *A* and *B* through, say, an 18-mesh. In both cases, all the particles are smaller than 18-mesh, but *A* contains a high proportion of particles smaller than say 60-mesh, *B* a high proportion in the range 18–22-mesh. Clearly, *A* can be sieved with ease whereas the sieving of *B* is a difficult operation. TAGGART[18] gives an account of several efficiency calculations based on the near-mesh particle contents of various flow streams.

Some of the factors that can affect the capacity of a screen and the efficiency of separation are

(*a*) *Feedstock Properties:* particle shape and size; bulk density; moisture content; abrasion resistance

(*b*) *Screen Characteristics:* percentage open area; aperture size and shape; length and width of screen; angle of inclination; material of construction; uniformity of mesh

(*c*) *Operating Conditions:* feed rate; method of feeding; depth of layer on screen; frequency of vibration; amplitude of vibration; direction of vibration.

Several authors[18–20] have discussed the influence of these variables on screening operations, but there is still ample scope for further investigation in this relatively unexplored field.

REFERENCES

1 HASSIALIS, M. D. and BEHRE, H. A., Sampling and testing (Section 19 in reference 18)

2 *Methods for the Use of B.S. Fine-mesh Test Sieves*, B.S. 1796, 1952. London; British Standards Institution

3 *Test Sieves*, B.S. 410, 1943. London; British Standards Institution

4 STAIRMAND, C. J., Some practical aspects of particle size analysis in industry (in reference 12)

5 MACCALMAN, D., The accuracy of sieving tests, *Industr. Chem. Mfr* 13 (1937) 464, 507; 14 (1938) 64, 101, 143, 197, 231, 306, 363, 386, 498; 15 (1939) 161, 184, 247, 290

6 MULLIN, J. W., The analysis of sieve test data, *Industr. Chem. Mfr* 36 (1960) 272

7 AUSTIN, J. B., Methods of representing the distribution of particle size, *Industr. Engng Chem.* (*Anal. ed.*) 11 (1939) 334

8 POWERS, H. E. C., Determination of the grist of sugars, *Intern. Sugar J.* 50 (1948) 149

9 HEYWOOD, H., Measurement of the fineness of powdered materials, *Proc. Instn mech. Engrs* 140 (1938) 257

10 HEYWOOD, H., Numerical definitions of particle size and shape, *Chem. & Ind.* 56 (1937) 149

11 *Symposium on New Methods for Particle Size Determination in the Sub-sieve Range* (8 papers), 1941. American Society for Testing Materials

12 Symposium on Particle Size Analysis (12 papers), *Trans. Instn chem. Engrs., Lond.* 25 (1947) *Special Supplement*

13 HERDAN, G., *Small Particle Statistics*, 1960. London; Butterworths

14 *Determination of Particle Size of Powders* (*Microscopic Methods*), British Standards Institution, London, *to be published*

15 *British Pharmacopœia*, 1958. London; General Medical Council, Pharmaceutical Press

16 *Portland Cement*, B.S. 12, 1947. London; British Standards Institution

17 HILL, S. and MULLER, E. G., Routine control of the fineness of icing sugar, *Intern. Sugar J.* 60 (1958) 194

18 TAGGART, A. F., *Handbook of Mineral Dressing*, 1945. New York; Wiley

19 BROWN, G. G. and associates, *Unit Operations*, 1950. New York; Wiley

20 FOWLER, R. T. and LIM, S. C., Influence of various factors on the effectiveness of separation on a vibrating screen, *Chem. Engng Sci.* 10 (1959) 163

APPENDIX

Table A.1. International Atomic Weights (1954)

Name	Symbol	At. No.	International At. Wt. 1954	Valency
Actinium	Ac	89	227	—
Aluminium	Al	13	26·98	3
Antimony	Sb	51	121·76	3, 5
Argon	A	18	39·944	0
Arsenic	As	33	74·91	3, 5
Astatine	At	85	[210]	1, 3, 5, 7
Barium	Ba	56	137·36	2
Beryllium	Be	4	9·013	2
Bismuth	Bi	83	209·00	3, 5
Boron	B	5	10·82	3
Bromine	Br	35	79·916	1, 3, 5, 7
Cadmium	Cd	48	112·41	2
Calcium	Ca	20	40·08	2
Carbon	C	6	12·011	2, 4
Cerium	Ce	58	140·13	3, 4
Cesium	Cs	55	132·91	1
Chlorine	Cl	17	35·457	1, 3, 5, 7
Chromium	Cr	24	52·01	2, 3, 6
Cobalt	Co	27	58·94	2, 3
Copper	Cu	29	63·54	1, 2
Dysprosium	Dy	66	162·51	3
Erbium	Er	68	167·27	3
Europium	Eu	63	152·0	2, 3
Fluorine	F	9	19·00	1
Francium	Fr	87	[223]	1
Gadolinium	Gd	64	157·26	3
Gallium	Ga	31	69·72	2, 3
Germanium	Ge	32	72·60	4
Gold	Au	79	197·0	1, 3
Hafnium	Hf	72	178·50	4
Helium	He	2	4·003	0
Holmium	Ho	67	164·94	3
Hydrogen	H	1	1·0080	1
Indium	In	49	114·82	3
Iodine	I	53	126·91	1, 3, 5, 7
Iridium	Ir	77	192·2	3, 4
Iron	Fe	26	55·85	2, 3
Krypton	Kr	36	83·8	0
Lanthanum	La	57	138·92	3
Lead	Pb	82	207·21	2, 4
Lithium	Li	3	6·940	1
Lutecium	Lu	71	174·99	3
Magnesium	Mg	12	24·32	2
Manganese	Mn	25	54·94	2, 3, 4, 6, 7
Mercury	Hg	80	200·61	1, 2
Molybdenum	Mo	42	95·95	3, 4, 6

Name	Symbol	At. No.	International At. Wt. 1954	Valency
Neodymium	Nd	60	144·27	3
Neon	Ne	10	20·183	0
Nickel	Ni	28	58·71	2, 3
Niobium	Nb	41	92·91	3, 5
Nitrogen	N	7	14·008	3, 5
Osmium	Os	76	190·2	2, 3, 4, 8
Oxygen	O	8	16·000	2
Palladium	Pd	46	106·4	2, 4
Phosphorus	P	15	30·975	3, 5
Platinum	Pt	78	195·09	2, 4
Polonium	Po	84	210	—
Potassium	K	19	39·100	1
Praseodymium	Pr	59	140·92	3
Promethium	Pm	61	[145]	3
Protactinium	Pa	91	231	—
Radium	Ra	88	226·05	2
Radon	Rn	86	222	0
Rhenium	Re	75	186·22	—
Rhodium	Rh	45	102·91	3
Rubidium	Rb	37	85·48	1
Ruthenium	Ru	44	101·1	3, 4, 6, 8
Samarium	Sm(Sa)	62	150·35	2, 3
Scandium	Sc	21	44·96	3
Selenium	Se	34	78·96	2, 4, 6
Silicon	Si	14	28·09	4
Silver	Ag	47	107·880	1
Sodium	Na	11	22·991	1
Strontium	Sr	38	87·63	2
Sulphur	S	16	32·066 ± 0·003	2
Tantalum	Ta	73	180·95	5
Technetium	Tc	43	[99]	6, 7
Tellurium	Te	52	127·61	2, 4, 6
Terbium	Tb	65	158·93	3
Thallium	Tl	81	204·39	1, 3
Thorium	Th	90	232·05	4
Thulium	Tm	69	168·94	3
Tin	Sn	50	118·70	2, 4
Titanium	Ti	22	47·90	3, 4
Tungsten	W	74	183·86	6
Uranium	U	92	238·07	4, 6
Vanadium	V	23	50·95	3, 5
Xenon	Xe	54	131·3	0
Ytterbium	Yb	70	173·04	2, 3
Yttrium	Y	39	88·92	3
Zinc	Zn	30	65·38	2
Zirconium	Zr	40	91·22	4

Substance	Solubility product	Temperature °C	Substance	Solubility product	Temperature °C
Aluminum hydroxide	4×10^{-13}	15	Lead sulphide	$3{\cdot}4 \times 10^{-28}$	18
Aluminium hydroxide	$1{\cdot}1 \times 10^{-15}$	18	Lithium carbonate	$1{\cdot}7 \times 10^{-3}$	25
Barium carbonate	7×10^{-9}	16	Magnesium ammonium phosphate	$2{\cdot}5 \times 10^{-13}$	25
Barium chromate	$1{\cdot}6 \times 10^{-10}$	18	Magnesium carbonate	$2{\cdot}6 \times 10^{-5}$	12
Barium fluoride	$1{\cdot}7 \times 10^{-6}$	18	Magnesium fluoride	$7{\cdot}1 \times 10^{-9}$	18
Barium iodate ($2H_2O$)	$6{\cdot}5 \times 10^{-10}$	25	Magnesium hydroxide	$1{\cdot}2 \times 10^{-11}$	18
Barium oxalate ($2H_2O$)	$1{\cdot}2 \times 10^{-7}$	18	Magnesium oxalate	$8{\cdot}57 \times 10^{-5}$	18
Barium oxalate ($\frac{1}{2}H_2O$)	$2{\cdot}18 \times 10^{-7}$	18	Manganese hydroxide	4×10^{-14}	18
Barium sulphate	$0{\cdot}87 \times 10^{-10}$	18	Manganese sulphide	$1{\cdot}4 \times 10^{-15}$	18
Cadmium sulphide	$3{\cdot}6 \times 10^{-29}$	18	Mercuric sulphide	4×10^{-53} to 2×10^{-49}	18
Calcium carbonate (calcite)	$0{\cdot}99 \times 10^{-8}$	15	Mercurous bromide	$1{\cdot}3 \times 10^{-21}$	25
Calcium fluoride	$3{\cdot}4 \times 10^{-11}$	18	Mercurous chloride	2×10^{-18}	25
Calcium oxalate (H_2O)	$1{\cdot}78 \times 10^{-9}$	18	Mercurous iodide	$1{\cdot}2 \times 10^{-28}$	25
Calcium sulphate	$1{\cdot}95 \times 10^{-4}$	10	Nickel sulphide	$1{\cdot}4 \times 10^{-24}$	18
Calcium tartrate ($2H_2O$)	$0{\cdot}77 \times 10^{-6}$	18	Potassium acid tartrate $[K^+][HC_4H_4O_6^-]$	$3{\cdot}8 \times 10^{-4}$	18
Cobalt sulphide	3×10^{-26}	18	Silver bromate	$3{\cdot}97 \times 10^{-5}$	20
Cupric iodate	$1{\cdot}4 \times 10^{-7}$	25	Silver bromide	$4{\cdot}1 \times 10^{-13}$	18
Cupric oxalate	$2{\cdot}87 \times 10^{-8}$	25	Silver carbonate	$6{\cdot}15 \times 10^{-12}$	25
Cupric sulphide	$8{\cdot}5 \times 10^{-45}$	18	Silver chloride	$1{\cdot}56 \times 10^{-10}$	25
Cuprous bromide	$4{\cdot}15 \times 10^{-8}$	18–20	Silver chromate	9×10^{-12}	25
Cuprous chloride	$1{\cdot}02 \times 10^{-6}$	18–20	Silver cyanide $[Ag^+][Ag(CN)^-_2]$	$2{\cdot}2 \times 10^{-12}$	20
Cuprous iodide	$5{\cdot}06 \times 10^{-12}$	18–20	Silver hydroxide	$1{\cdot}52 \times 10^{-8}$	20
Cuprous sulphide	2×10^{-47}	16–18	Silver iodide	$1{\cdot}5 \times 10^{-16}$	25
Cuprous thiocyanate	$1{\cdot}6 \times 10^{-11}$	18	Silver sulphide	$1{\cdot}6 \times 10^{-49}$	18
Ferric hydroxide	$1{\cdot}1 \times 10^{-36}$	18	Silver thiocyanate	$0{\cdot}49 \times 10^{-12}$	18
Ferrous hydroxide	$1{\cdot}64 \times 10^{-14}$	18	Strontium carbonate	$1{\cdot}6 \times 10^{-9}$	25
Ferrous oxalate	$2{\cdot}1 \times 10^{-7}$	25	Strontium fluoride	$2{\cdot}8 \times 10^{-9}$	18
Ferrous sulphide	$3{\cdot}7 \times 10^{-19}$	18	Strontium oxalate	$5{\cdot}61 \times 10^{-8}$	18
Lead carbonate	$3{\cdot}3 \times 10^{-14}$	18	Strontium sulphate	$3{\cdot}81 \times 10^{-7}$	17·4
Lead chromate	$1{\cdot}77 \times 10^{-14}$	18	Zinc oxalate	$1{\cdot}35 \times 10^{-9}$	18
Lead fluoride	$3{\cdot}2 \times 10^{-8}$	18	Zinc sulphide	$1{\cdot}2 \times 10^{-23}$	18
Lead iodate	$1{\cdot}2 \times 10^{-13}$	18			
Lead iodide	$7{\cdot}47 \times 10^{-9}$	15			
Lead oxalate	$2{\cdot}74 \times 10^{-11}$	18			
Lead sulphate	$1{\cdot}06 \times 10^{-8}$	18			

Table A.3. Solubilities of Inorganic Salts in Water (g *of anhydrous Compounds per* 100 g *of Water*)

Compound	*Formula*	*Solubility* (°C)								*Stable hydrates* 0–25° C
		0	10	20	30	40	60	80	100	
Aluminium chloride	$AlCl_3$		46		47					6
sulphate	$Al_2(SO_4)_3$	31·3	33·5	36·2	40·4	46·1	59·2	73·0	89·1	18
nitrate	$Al(NO_3)_3$	60	68	74	82	89	106	132	160	9
Ammonium alum	$(NH_4)_2Al_2(SO_4)_4$	2·1	5·0	7·7	11·0	14·9	26·7		109·7 (95°)	24
bicarbonate	NH_4HCO_3	12	16	21	27	35	decomp.			—
bromide	NH_4Br	60·6	68·0	75·5	83·5	91·0	108	126	146	—
chloride	NH_4Cl	29·7	33·4	37·2	41·4	45·8	55·2	65·6	77·3	—
dihydrogen phosphate	$NH_4H_2PO_4$	22·0	28·0	36·5	45·8	56·6				—
iodide	NH_4I	154	163	172	181	191	209	230	250	—
nitrate	NH_4NO_3	118	150	192	242	297	421	580	870	—
oxalate	$(NH_4)_2C_2O_4$	2·1	3·1	4·4	6·0	8·0	14			1
sulphate	$(NH_4)_2SO_4$	71·0	73·0	75·4	78·0	81·0	88·0	95·3	103·3	—
thiocyanate	NH_4CNS	121		162						—
vanadate	NH_4VO_3			4·8	8·4	13·2				—
Barium acetate	$Ba(C_2H_3O_2)_2$	58	63	72	75	79	74	74	74	3
bromide	$BaBr_2$	98	100	104	107	112	124	140	160	2
chlorate	$Ba(ClO_3)_2$	20·3	27·0	33·8	41·7	49·6	66·8	84·8	105	1
chloride	$BaCl_2$	31·6	33·2	35·7	38·2	40·7	46·4	52·4	58·3	2
hydroxide	$Ba(OH)_2$	1·6	2·5	3·9	5·6	8·2	21	101		8
iodide	BaI_2	170	186	203	220	232	247	261	272	6
nitrate	$Ba(NO_3)_2$	5·0	7·0	9·2	11·6	14·2	20·3	27·0	34·2	—
Beryllium chloride	$BeCl_2$	68		73	77	79				4
nitrate	$Be(NO_3)_2$	98		107	110		177			4
sulphate	$BeSO_4$	35	37	39	41	44	54	67	85	4
Boric acid	H_3BO_3	2·7	3·6	5·0	6·6	8·7	14·8	23·8	40·3	—
Cadmium bromide	$CdBr_2$	56·3	75·5	96·5	128	152			160	4
chloride	$CdCl_2$	90	135	134	132	135	136	140	147	2½
iodide	CdI_2	80	83	86	90				128	—
nitrate	$Cd(NO_3)_2$	120		140		220	400		660	4
sulphate	$CdSO_4$	76·5	76·0	76·6		78·5	83·7		60·8	3

Table A.3.—continued

Compound	*Formula*	*Solubility* (°C)								*Stable hydrates* 0–25° C
		0	10	20	30	40	60	80	100	
Caesium chloride	$CsCl$	161	175	187	197	208	230	250	271	—
chlorate	$CsClO_3$	2·46	3·8	6·2	9·5	13·8	26·2	45·0	79·0	—
nitrate	$CsNO_3$	9·3	14·9	23·0	33·9	47·2	83·8	134	197	—
perchlorate	$CsClO_4$	0·1	1·0	1·6	2·6	4·0	7·3	14·4	30·0	—
sulphate	Cs_2SO_4	167	173	179	184	190	200	210	220	—
Calcium acetate	$Ca(C_2H_3O_2)_2$	37·4	36·0	34·7	33·8	33·2	32·7	33·5	29·7	2
bicarbonate	$Ca(HCO_3)_2$	16·2	16·4	16·6	16·8	17·1	17·5	18·0	18·4	—
chloride	$CaCl_2$	59·5	65·0	74·5	102		137	147	159	6
iodide	CaI_2	192	196	204	220	240			430	8
nitrate	$Ca(NO_3)_2$	102	115	129	153	196		359	363	4
sulphate	$CaSO_4$	0·18	0·19	0·20	0·21	0·21	0·20	0·18	0·16	2
Cobalt ammonium sulphate	$Co(NH_4)_2SO_4$	6·0	9·2	12·6	17·5	21·8	32·7	49·0		6
bromide	$CoBr_2$	92		110		156	226		257	6
chloride	$CoCl_2$	42	46	50	56		92	97	104	6
iodide	CoI_2	138	160	185	234	300		400		1
nitrate	$Co(NO_3)_2$	85	89	97	110	126	167	211		6
sulphate	$CoSO_4$	25·5	30·0	36·2	41·8	48	60	70	83	7
Copper(-ic) chloride	$CuCl_2$	69	71	74	76	81			98	2
nitrate	$Cu(NO_3)_2$	81·8	95·3	125		160	179	208	250	6
sulphate	$CuSO_4$	14·3	17·4	20·7	25·0	28·5	40·0	55·0	75·4	5
Ferric ammonium sulphate	$Fe_2(SO_4)_3(NH_4)_2SO_4$			43						24
chloride	$FeCl_3$	74·4	81·9	91·8				526	540	6
Ferrous ammonium sulphate	$FeSO_4(NH_4)_2SO_4$	12·5		26·4		32·9	45·7			6
bromide	$FeBr_2$	102		115	122	128	144	160	177	6
chloride	$FeCl_2$	61	64	68	73	77	89	100	106	6, 4
potassium sulphate	$FeSO_4K_2SO_4$	20	25	32	39	45	59			6
sulphate	$FeSO_4$	15·6	20·5	26·5	32·9	40·2				7
Lead acetate	$Pb(C_2H_3O_2)_2$	19·7	29·2	44·1	69·5	116				3
bromide	$PbBr_2$	0·45	0·62	0·85	1·2	1·5	2·4	3·3	4·8	—
chloride	$PbCl_2$	0·67	0·81	1·0	1·2	1·5	2·0	2·6	3·3	—
nitrate	$Pb(NO_3)_2$	39	48	57	66	75	95	115	139	—

Table A.3.—continued

Compound	Formula	Solubility (°C)								Stable hydrates 0–25° C
		0	10	20	30	40	60	80	100	
Lithium bromide	$LiBr$	143	160	177	191	205	224	245	266	2
carbonate	Li_2CO_3	1·54	1·43	1·33	1·25	1·17	1·01	0·85	0·72	—
chloride	$LiCl$	64	70	80	90		102	112	125	3
hydroxide	$LiOH$	12·6	12·7	12·8	13·0	13·2	13·9	15·4	17·5	1
iodide	LiI	151	158	165	172	180			480	3
nitrate	$LiNO_3$	48	60	76					227	3
sulphate	Li_2SO_4	35	35	34		33			29	1
Magnesium bromide	$MgBr_2$	92·0	95·0	96·5	99·2	101·6	107·5	113·7	120·6	6
chloride	$MgCl_2$	52·8	53·5	54·5	56·0	57·5	61·0	66·0	73·0	6
iodide	MgI_2	120		140		174				8
nitrate	$Mg(NO_3)_2$	66·5				84·7			137 (90°)	6
sulphate	$MgSO_4$		30·9	35·5	40·8	45·5	55·1	64·2	74	7
Manganous chloride	$MnCl_2$	63·4	68·1	73·9	80·7	88·6	109	113	115	4
nitrate	$Mn(NO_3)_2$	50·5	54·1	58·8	67·4					6
sulphate	$MnSO_4$	53·2	60·0	64·5	66·4	68·8	55·0	48·0	34·0	7
Mercuric bromide	$HgBr_2$	0·3	0·4	0·6	0·7	1·0	1·7	2·8	4·9	—
chloride	$HgCl_2$	4·66	5·43	6·59	8·14	10·2	17·4	30·9	58·3	—
Mercurous perchlorate	$Hg_2(ClO_4)_2$	282		368	420	457	500	540	600	4
Nickel ammonium sulphate	$Ni(NH_4)_2(SO_4)_2$	1·6	4·0	6·5	9·0	12·0	17·5			6
bromide	$NiBr_2$	112	122	131	138	144	152	154	155	6
chloride	$NiCl_2$	54	60	64	69	73	82	87	88	6
iodide	NiI_2	124	135	147	157	174	184	187		6
nitrate	$Ni(NO_3)_2$	80	88	96	109	122	163		77	6
sulphate	$NiSO_4$	26	32	37	42		55	63		7
Potassium nitrate	$KC_2H_3O_2$	217	234	256	284	323	350	380		1½
aluminium sulphate	$K_2Al_2(SO_4)_4$	3·0	4·0	5·9	8·4	11·7	24·8	71·0		24
bicarbonate	$KHCO_3$	22·5	27·7	33·2	39·1	45·4	60·0			—
bisulphate	$KHSO_4$	36·3		51·4		67·3			121·6	—
bromate	$KBrO_3$	3·1	4·0	6·8	10·0	13·1	22·5	33·9	50·0	—
bromide	KBr	53·5	58·0	64·6	70·0	74·2	84·5	96·0	102·0	—
carbonate	K_2CO_3	106	108	110	114	117	127	140	156	1½
chlorate	$KClO_3$	3·3	5·0	7·0	10·5	14·0	24·5	38·5	57	—
chloride	KCl	27·6	31·0	34·0	37·0	40·0	45·5	51·1	56·7	—
chromate	K_2CrO_4	58·2	60·0	61·7	63·4	65·2	68·6	72·1	75·6	—
dichromate	$K_2Cr_2O_7$	5	7	12	20	26	43	61	80	—

Compound	Formula	Solubility (°C)								Stable hydrates 0–25° C
		0	10	20	30	40	60	80	100	
Potassium ferricyanide	$K_3Fe(CN)_6$	31	36	43	50	60	66	72	81	—
ferrocyanide	$K_4Fe(CN)_6$	14	21	27	34	40	54	69	86	3
hydroxide	KOH	97	103	112	126				178	2
hydrogen tartrate	$KHC_4H_4O_6$	0·32	0·40	0·53	0·90	1·3	2·5	4·6	6·9	—
hydrogen oxalate	KHC_2O_4	2·2	3·4	5·2	7·4	11·0		34·7	51·0	—
iodate	KIO_3	4·7	6·2	8·1	10·3	12·2	18	25	32	—
iodide	KI	128	135	144	150	160	175	190	210	—
nitrate	KNO_3	13·3	20·9	31·6	45·8	63·9	110	169	247	—
nitrite	KNO_2	280	290	300	310	330			413	—
oxalate	$K_2C_2O_4$	25·9	30·2	34·7	39·2	43·8		63·4	75	1
perchlorate	$KClO_4$	0·75	1·1	1·8	2·6	4·4	9·0	14·8	21·8	—
permanganate	$KMnO_4$	2·8	4·4	6·3	9·0	12·6	22·2			—
platinichloride	K_2PtCl_6	0·74	0·90	1·12	1·41	1·76	2·64	3·79	5·18	—
sulphate	K_2SO_4	7·4	9·2	10·9	13·0	14·8	18·2	21·4	24·2	—
thiocyanate	$KCNS$	176	189	242						—
Rubidium bromide	$RbBr$	89		110			150	175	190	—
chlorate	$RbClO_3$	2·1	3·4	5·4	8·0	11·5	22·3	38·2	65	—
chloride	$RbCl$	70·6	77·4	83·6	89·5				128	—
nitrate	$RbNO_3$	13·3	22·6	36·5	55·5	79·0	136	211	305	—
perchlorate	$RbClO_4$		0·64	0·98	1·5	2·4	4·9	9·3	18	—
sulphate	Rb_2SO_4	34·2	39·7	45·0	50·3	55·2			79·5	—
Silver acetate	$AgC_2H_3O_2$	0·72	0·88	1·04	1·21	1·41	1·89	2·52		—
nitrate	$AgNO_3$	122	170	222	300	376	525	669	952	—
sulphate	Ag_2SO_4	0·57	0·70	0·80	0·89	0·98	1·15	1·30	1·41	—
Sodium acetate	$NaC_2H_3O_2$	36·3	40·8	46·5	54·5	65·5	139	153	170	3
bicarbonate	$NaHCO_3$	6·9	8·2	9·6	11·1	12·7	16·4	decomp.	—	—
borate (tetra)	$Na_2B_4O_7$	1·5	1·8	2·7	3·9	6·0	20·3	31·5	52·5	10
bromate	$NaBrO_3$	27	30	35	42	50	63	76	91	—
bromide	$NaBr$	79·5	83·8	90·5	97·2	105		118	121	2
carbonate	Na_2CO_3	7·1	12·5	21·4	38·8	48·5	46·4	45·8	45·5	10
chlorate	$NaClO_3$	80	89	101	113	126	155	189	233	—
chloride	$NaCl$	35·7	35·8	36·0	36·3	36·6	37·3	38·4	39·8	—
chromate	Na_2CrO_4	31·7	50·2	88.7	88·7	96·0	115	125	126	10
dichromate	$Na_2Cr_2O_7$	163	170	178	196	220	275	380	430	2
dihydrogen phosphate	NaH_2PO_4	58	70	85	107	138	179	207	247	2

Table A.3.—continued

Compound	*Formula*	*Solubility* (°C)								*Stable hydrate* 0–25° C
		0	10	20	30	40	60	80	100	
Sodium ferrocyanide	$Na_4Fe(CN)_6$			18		30		59	63	—
hydrogen arsenate	Na_2HAsO_4	7·3	15·5	26·5	37	47	65	85		12
hydrogen phosphate	Na_2HPO_4	1·7	3·6	7·7	20·8	51·8	82·9	92·4	102	12
hydroxide	$NaOH$	42·0	51·5	109	119	129	174		340	4, 3½
iodate	$NaIO_3$	2·5	5·6	9·1	13·2		23	27	34	1½
iodide	NaI	159	169	179	196	210	250		302	2
nitrate	$NaNO_3$	73	80	88	96	104	124	148	180	—
nitrite	$NaNO_2$	72	78	85	92	98		133	163	—
oxalate	$Na_2C_2O_4$			3·7					6·33	—
phosphate	Na_3PO_4	1·5	4	11	20	31	55	81	108	12
pyrophosphate	$Na_4P_2O_7$	3·2	3·9	6·2	10·0	13·5	21·8	30·0	40·3	10
sulphate	Na_2SO_4	4·8	9·0	19·4	40·8	48·8	45·3	43·7	42·5	10
sulphide	Na_2S		15·4	18·8	22·5	28·5	39	49		9
sulphite	Na_2SO_3	14·4	20·0	26·5	36	28	28	28		7
thiosulphate	$Na_2S_2O_3$	52	61	70	84	103	207	250	266	5
Stannous chloride	$SnCl_2$	84								—
iodide	SnI_2			1·0	1·2	1·4	2·1	3·0	4·0	—
sulphate	$SnSO_4$			19						—
Strontium acetate	$Sr(C_2H_3O_2)_2$	36·9	41·6	42	39·5			36·1	36·4	4, ½
bromide	$SrBr_2$	85	93	102	112	124	150	182	223	6
chloride	$SrCl_2$	43·5	47·7	52·9	58·7	65·3	81·8	90·5	101	6
hydroxide	$Sr(OH)_2$	0·9	1·2	1·7	2·6	3·8			91·2	8
iodide	SrI_2	164		179		198			370	6
nitrate	$Sr(NO_3)_2$	40	54	70	89	90	94	98	101	4
Thallium chlorate	$TlClO_3$	2		4				37	57	—
chloride	$TlCl$	0·21	0·25	0·33	0·42	0·52	0·8	1·2	1·8	—
hydroxide	$TlOH$	25·4			39·9	49·5	73·8	106	148	—
nitrate	$TlNO_3$	3·91	6·22	9·55	14·3	20·9	46·2	111	414	—
sulphate	Tl_2SO_4	2·7	3·7	4·9	6·2	7·5	10·9	14·6	18·4	—
Uranyl nitrate	$UO_2(NO_3)_2$	97·5	110	125	143	169	252			6
Zinc bromide	$ZnBr_2$	390	420	440			620	640	670	2
chlorate	$ZnClO_3$	145	153	200	209	223				6, 4
nitrate	$Zn(NO_3)_2$	95		118		207				6
sulphate	$ZnSO_4$	42	47	54	61	70		87	81	7

Table A.4. *Solubilities of Organic Solids in Water* (g *of Anhydrous Compound per* 100 g *of Water*)

Compound	*Formula*	*Solubility* (°C)								*Anhydrous melting point* °C
		0	10	20	30	40	60	80	100	
Acetamide	$CH_3 \cdot CONH_2$	138	175	230	310	440	850			81
Acetanilide	$C_6H_5NH \cdot COCH_3$		0·48	0·52	0·63	0·87	2·1	4·7		114
Adipic acid	$(CH_2)_4(COOH)_2$	0·8	1·0	1·9	3·0	5·0	18	70	160	153
Alanine (*d*)	$CH_3CH \cdot NH_2 \cdot COOH$	12·7	14·2	15·8	17·6	19·6	24·3	30·0	37·3	~ 300 (*d*)
Alanine (*dl*)	$CH_3CH \cdot NH_2 \cdot COOH$	12·1	13·8	15·7	17·9	20·3	26·3	33·9	44·0	~ 300 (*d*)
o-Aminophenol	$C_6H_4 \cdot OH \cdot NO_2$	1·7	1·9	2·0	2·2	2·4	2·7	3·0	7·0	173
m-Aminophenol	$C_6H_4 \cdot OH \cdot NO_2$		2·0	2·7	3·8	5·6	21	280	950	123
p-Aminophenol	$C_6H_4 \cdot OH \cdot NO_2$	1·1	1·3	1·6	1·9	2·3	3·6	7·9	37	184 (*d*)
Anthranilic acid (*o*-)	$C_6H_4 \cdot NH_2 \cdot COOH$		0·3	0·35	0·6	0·9			95	145
Benzamide	$C_6H_5 \cdot CONH_2$		0·6	1·0	1·3	1·6	5	200	800	130
Benzoic acid	$C_6H_5 \cdot COOH$	0·17	0·20	0·29	0·40	0·56	1·16	2·72	5·88	122
Cinnamic acid	$C_6H_5 \cdot CH{:}CH \cdot COOH$			0·05					0·59	133
Citric acid*	$C_3H_4 \cdot OH \cdot (COOH)_3$	96	118	146	183	215	277	372	526	153
Dicyandiamide	$NH_2 \cdot C({:}NH) \cdot NH \cdot CN$	1·3	1·9	3·2	5·0	7·8	19	38		208
Fructose	$CH_2OH(CHOH)_3COCH_2OH$	75		80		85	90			95–105
Fumaric acid (*trans*-)	$HOOC \cdot CH{:}CH \cdot COOH$	0·23	0·35	0·50	0·72	1·1	2·3	5·2	9·8	287
Glucose (dextrose)*	$C_6H_{12}O_6$	46	70	92	120	160	280	440		146
Glutamic acid (*d*)	$COOH(CH_2)_2CH \cdot NH_2 \cdot COOH$	0·34	0·50	0·72	1·0	1·5	3·2	6·5	14·0	198 (*d*)
Glycine	$CH_2 \cdot NH_2 \cdot COOH$	14·2	18·0	22·5	27	33	45	57	70	235 (*d*)
p-Hydroquinone	$C_6H_4(OH)_2$	4·0	5·4	7·2	9·6	13	35	88	198	170
o-Hydroxybenzoic acid	$C_6H_4 \cdot OH \cdot COOH$	0·13	0·15	0·20	0·28	0·42	0·91	2·26	8·12	159
m-Hydroxybenzoic acid	$C_6H_4 \cdot OH \cdot COOH$	0·35	0·55	0·86	1·3	2·0	4·5	12·4	58·7	200
p-Hydroxybenzoic acid	$C_6H_4 \cdot OH \cdot COOH$	0·25	0·35	0·53	0·80	1·25	4·29	13·7	49·9	215
Lactose*	$C_{12}H_{22}O_{11}$	12·2	15·0	19·5	25·2	33·3	57·5	102	153	202
Maleic acid	$HOOC \cdot CH{:}CH \cdot COOH$	39·3	50	70	90	115	178	283		130

Table A.4.—continued

Compound	*Formula*	*Solubility* (°C)								*Anhydrous melting point* °C
		0	10	20	30	40	60	80	100	
Malic acid (*dl*)	$CH\cdot OH\cdot CH_2(COOH)_2$	89	105	126	150	180	270	460		128
Malonic acid	$CH_2(COOH)_2$	108	128	153	180	212	292	455	810	135 (*d*)
Maltose*	$C_{12}H_{22}O_{11}$	57	65	78	93	110	175	300		
Mannitol (*d*)	$(CH_2OH)_2(CHOH)_4$	10·4	13·7	18·6	25·2	34·6	64·4	115	197	166
Melamine	$C_3N_3(NH_2)_3$	0·12	0·18	0·27	0·42	0·71	1·5	2·8	5·0	~250
Oxalic acid†	$(COOH)_2$	3·5	6·0	9·5	14·5	21·6	44·3	84·4		189
Pentaerythritol	$C(CH_2OH)_4$	4	5	6	8	13	22	40	100	262
Phenacetin (*p*-)	$C_2H_5O\cdot C_6H_4\cdot NHCOCH_3$			0·07					1·43	135
Phthalic acid (*o*-)	$C_6H_4(COOH)_2$	0·23	0·36	0·56	0·8	1·2	2·8	6·3	18·0	208
Picric acid (2,4,6)	$C_6H_2\cdot OH\cdot (NO_2)_3$	1·0	1·1	1·2	1·5	1·9	3·1	4·6	7·2	122
o-Pyrocatechol	$C_6H_4(OH)_2$			45·1		172	412	1,120	8,360	104
Raffinose‡	$C_{18}H_{32}O_{16}$	3·4	6·6	13·6	27·1	49·9	86·9	153·8		118
Resorcinol (*m*-)	$C_6H_4(OH)_2$	66·2	85	123	170	225	390	634	1,060	111
Salicylic acid (*o*-)	$C_6H_4\cdot OH\cdot COOH$	0·13	0·15	0·20	0·28	0·42	0·91	2·26	8·12	159
Succinic acid	$(CH_2\cdot COOH)_2$	2·8	4·4	6·9	10·5	16·2	35·8	70·8	127	183
Succinimide	$(CH_2CO)_2NH$	10	16	26	48	83	140	213		125
Sucrose	$C_{12}H_{22}O_{11}$	179	190	204	219	238	287	362	487	170–186 (*d*)
Sulphanilic acid (*p*-)	$C_6H_4\cdot NH_2\cdot SO_3H$	0·45	0·80	1·12		2·03	3·01	4·51	6·67	> 280 (*d*)
Tartanic acid (*d* or *l*)	$(CHOH\cdot COOH)_2$	115	126	139	156	176	220	273	343	170
Tartaric acid (racemic)*	$(CHOH\cdot COOH)_2$	8·2	12·3	18·0	25·2	37·0	64·5	98·1	138	205
Taurine	$NH_2CH_2CH_2SO_3H$	3·9	6·0	8·8	12·4	16·8	27·4	38·4	45·7	~330 (*d*)
Thiourea	$NH_2\cdot CS\cdot NH_2$	4·9	8·0	13·6	20·1	30·8	71	138	238	181
Urea	$NH_2\cdot CO\cdot NH_2$	67	80	105	135	164	250	400	730	133
Uric acid	$C_5H_4O_3N_4$	0·002	0·004	0·006	0·009	0·012	0·023	0·039	0·062	(*d*)

* Crystallizes from water with $1H_2O$ † Crystallizes from water with $2H_2O$ ‡ Crystallizes from water with $5H_2O$

Table A.5. Heats of Solution of Inorganic Salts in Water at Approximately Room Temperature and Infinite Dilution

A positive value indicates an exothermic, a negative value an endothermic heat of solution

Substance	*Formula*	*Heat of solution* kcal/mole
Aluminium chloride	$AlCl_3$	+ 77·9
chloride	$AlCl_3 \cdot 6H_2O$	+ 13·1
sulphate	$Al_2(SO_4)_3$	+ 120
sulphate	$Al_2(SO_4)_3 \cdot 18H_2O$	+ 7·0
Ammonium bicarbonate	NH_4HCO_3	− 6·7
bromide	NH_4Br	− 4·5
chloride	NH_4Cl	− 3·8
iodide	NH_4I	− 3·6
nitrate	NH_4NO_3	− 6·5
oxalate	$(NH_4)_2C_2O_4$	− 8·0
oxalate	$(NH_4)_2C_2O_4 \cdot H_2O$	− 11·5
sulphate	$(NH_4)_2SO_4$	− 2·5
bisulphate	$(NH_4)_2HSO_4$	+ 0·56
Barium acetate	$Ba(C_2H_3O_2)_2$	+ 6·0
bromide	$BaBr_2$	+ 5·2
bromide	$BaBr_2 \cdot 2H_2O$	− 3·9
chlorate	$Ba(ClO_3)_2$	− 6·7
chlorate	$Ba(ClO_3)_2 \cdot H_2O$	− 11·0
chloride	$BaCl_2$	+ 2·4
chloride	$BaCl_2 \cdot 2H_2O$	− 4·5
hydroxide	$Ba(OH)_2$	+ 11·4
hydroxide	$Ba(OH)_2 \cdot 8H_2O$	− 14·5
iodide	BaI_2	+ 10·3
iodide	$BaI_2 \cdot 6H_2O$	− 6·6
nitrate	$Ba(NO_3)_2$	− 9·6
sulphate	$BaSO_4$	− 4·6
Beryllium bromide	$BeBr_2$	+ 62·6
chloride	$BeCl_2$	+ 51·1
iodide	BeI_2	+ 72·6
sulphate	$BeSO_4$	+ 18·1
sulphate	$BeSO_4 \cdot 4H_2O$	+ 1·1
Boric acid	H_3BO_3	− 5·4
Cadmium bromide	$CdBr_2$	+ 0·43
bromide	$CdBr_2 \cdot 4H_2O$	− 7·3
chloride	$CdCl_2$	+ 3·11
chloride	$CdCl_2 \cdot 2\frac{1}{2}H_2O$	− 3·0
nitrate	$Cd(NO_3)_2 \cdot 4H_2O$	− 5·1
sulphate	$CdSO_4$	+ 10·7
Caesium chloride	$CsCl$	− 4·6
nitrate	$CsNO_3$	− 9·6
sulphate	Cs_2SO_4	− 4·9
Calcium acetate	$Ca(C_2H_3O_2)_2$	+ 7·0
acetate	$Ca(C_2H_3O_2)_2 \cdot H_2O$	+ 5·9
chloride	$CaCl_2$	+ 4·9
chloride	$CaCl_2 \cdot 6H_2O$	− 4·1
iodide	CaI_2	+ 28·0
iodide	$CaI_2 \cdot 8H_2O$	+ 1·7
nitrate	$Ca(NO_3)_2$	+ 4·1
nitrate	$Ca(NO_3)_2 \cdot 4H_2O$	− 8·0
sulphate	$CaSO_4$	+ 5·2
sulphate	$CaSO_4 \cdot 2H_2O$	− 0·18

Table A.5.—continued

Substance	*Formula*	*Heat of solution* kcal/mole
Cobalt bromide	$CoBr_2$	+ 18·4
bromide	$CoBr_2 \cdot 6H_2O$	− 1·3
chloride	$CoCl_2$	+ 18·5
chloride	$CoCl_2 \cdot 6H_2O$	− 2·9
nitrate	$Co(NO_3)_2$	+ 11·9
nitrate	$Co(NO_3)_2 \cdot 6H_2O$	− 4·9
sulphate	$CoSO_4$	+ 15·0
sulphate	$CoSO_4 \cdot 7H_2O$	− 3·6
Copper acetate	$Cu(C_2H_3O_2)_2$	+ 2·5
nitrate	$Cu(NO_3)_2$	+ 10·4
nitrate	$Cu(NO_3)_2 \cdot 6H_2O$	− 10·7
sulphate	$CuSO_4$	+ 15·9
sulphate	$CuSO_4 \cdot 5H_2O$	− 2·86
Ferric chloride	$FeCl_3$	+ 31·7
chloride	$FeCl_3 \cdot 6H_2O$	+ 5·6
Ferrous chloride	$FeCl_2$	+ 17·9
chloride	$FeCl_2 \cdot 4H_2O$	+ 2·7
sulphate	$FeSO_4$	+ 14·9
sulphate	$FeSO_4 \cdot 7H_2O$	− 4·4
Lead acetate	$Pb(C_2H_3O_2)_2$	+ 1·4
acetate	$Pb(C_2H_3O_2)_2 \cdot 3H_2O$	− 5·9
chloride	$PbCl_2$	− 6·5
nitrate	$Pb(NO_3)_2$	− 7·6
Lithium bromide	$LiBr$	+ 11·5
bromide	$LiBr \cdot 2H_2O$	+ 2·1
carbonate	Li_2CO_3	+ 3·1
chloride	$LiCl$	− 8·7
chloride	$LiCl \cdot 3H_2O$	− 2·0
hydroxide	$LiOH$	+ 4·5
hydroxide	$LiOH \cdot H_2O$	+ 9·6
iodide	LiI	+ 14·9
iodide	$LiI \cdot 3H_2O$	− 0·17
nitrate	$LiNO_3$	+ 0·47
nitrate	$LiNO_3 \cdot 3H_2O$	− 7·9
sulphate	Li_2SO_4	+ 6·7
sulphate	$Li_2SO_4 \cdot H_2O$	+ 3·8
Magnesium bromide	$MgBr_2$	+ 43·5
bromide	$MgBr_2 \cdot 6H_2O$	+ 19·8
chloride	$MgCl_2$	+ 36·0
chloride	$MgCl_2 \cdot 6H_2O$	+ 3·1
iodide	MgI_2	+ 50·2
nitrate	$Mg(NO_3)_2 \cdot 6H_2O$	− 3·7
sulphate	$MgSO_4$	+ 21·1
sulphate	$MgSO_4 \cdot 7H_2O$	− 3·18
Manganous chloride	$MnCl_2$	+ 16·0
chloride	$MnCl_2 \cdot 4H_2O$	+ 1·5
nitrate	$Mn(NO_3)_2$	+ 12·7
nitrate	$Mn(NO_3)_2 \cdot 6H_2O$	− 6·1
sulphate	$MnSO_4$	+ 13·8
sulphate	$MnSO_4 \cdot 7H_2O$	− 1·7
Mercuric bromide	$HgBr_2$	− 2·4
chloride	$HgCl_2$	− 3·3

Table A.5.—continued

Substance	*Formula*	*Heat of solution* kcal/mole
Nickel bromide	$NiBr_2$	+ 19·0
chloride	$NiCl_2$	+ 19·3
chloride	$NiCl_2 \cdot 6H_2O$	− 1·15
iodide	NiI_2	+ 19·4
nitrate	$Ni(NO_3)_2$	+ 11·7
nitrate	$Ni(NO_3)_2 \cdot 6H_2O$	− 7·5
sulphate	$NiSO_4$	+ 15·1
sulphate	$NiSO_4 \cdot 7H_2O$	− 4·2
Potassium acetate	$KC_2H_3O_2$	+ 3·5
aluminium sulphate	$K_2Al_2(SO_4)_4$	+ 48·5
aluminium sulphate	$K_2Al_2(SO_4)_4 \cdot 12H_2O$	− 10·1
bicarbonate	$KHCO_3$	− 5·1
bisulphate	$KHSO_4$	− 3·1
bromate	$KBrO_3$	− 10·1
bromide	KBr	− 5·1
carbonate	K_2CO_3	+ 6·9
carbonate	$K_2CO_3 \cdot 1\frac{1}{2}H_2O$	− 0·45
chlorate	$KClO_3$	− 10·3
chloride	KCl	− 4·4
chromate	K_2CrO_4	− 4·8
cyanide	KCN	− 2·9
dichromate	$K_2Cr_2O_7$	− 17·8
hydroxide	KOH	+ 13·0
iodate	KIO_3	− 6·9
iodide	KI	− 5·2
nitrate	KNO_3	− 8·6
oxalate	$K_2C_2O_4$	− 4·6
oxalate	$K_2C_2O_4 \cdot H_2O$	− 7·5
perchlorate	$KClO_4$	− 13·0
permanganate	$KMnO_4$	− 10·5
sulphate	K_2SO_4	− 6·3
sulphite	K_2SO_3	+ 1·8
sulphite	$K_2SO_3 \cdot H_2O$	+ 1·4
thiocyanate	$KCNS$	− 6·1
thiosulphate	$K_2S_2O_3$	− 4·5
Rubidium bromide	$RbBr$	− 6·0
chloride	$RbCl$	− 4·2
nitrate	$RbNO_3$	− 8·8
sulphate	Rb_2SO_4	− 6·7
Silver acetate	$AgC_2H_3O_2$	− 5·4
nitrate	$AgNO_3$	− 5·4
Sodium acetate	$NaC_2H_3O_2$	+ 4·1
acetate	$NaC_2H_3O_2 \cdot 3H_2O$	− 4·7
arsenate	Na_3AsO_4	+ 31·0
arsenate	$Na_3AsO_4 \cdot 12H_2O$	− 12·6
bicarbonate	$NaHCO_3$	− 4·2
borate (tetra)	$Na_2B_4O_7$	+ 10·3
borate	$Na_2B_4O_7 \cdot 10H_2O$	− 25·8
bromide	$NaBr$	− 0·58
bromide	$NaBr \cdot 2H_2O$	− 4·6
carbonate	Na_2CO_3	+ 5·6
carbonate	$Na_2CO_3 \cdot 10H_2O$	− 16·2
chlorate	$NaClO_3$	− 5·4
chloride	$NaCl$	− 1·2

Table A.5.—continued

Substance	*Formula*	*Heat of solution* kcal/mole
Sodium chromate	Na_2CrO_4	+ 2·4
chromate	$Na_2CrO_4 \cdot 10H_2O$	− 15·8
cyanide	$NaCN$	− 0·50
cyanide	$NaCN \cdot 2H_2O$	− 4·4
hydrogen phosphate	NaH_2PO_4	+ 5·6
hydroxide	$NaOH$	+ 10·2
iodide	NaI	+ 1·5
iodide	$NaI \cdot 2H_2O$	− 3·9
nitrate	$NaNO_3$	− 5·0
nitrite	$NaNO_2$	− 3·6
oxalate	$Na_2C_2O_4$	− 5·5
perchlorate	$NaClO_4$	− 3·6
phosphate	Na_3PO_4	+ 13·0
phosphate	$Na_3PO_4 \cdot 12H_2O$	− 15·0
pyrophosphate	$Na_4P_2O_7$	+ 12·0
pyrophosphate	$Na_4P_2O_7 \cdot 10H_2O$	− 12·0
sulphate	Na_2SO_4	+ 0·28
sulphate	$Na_2SO_4 \cdot 10H_2O$	− 18·7
sulphide	Na_2S	+ 15·2
sulphide	$Na_2S \cdot 9H_2O$	− 16·7
tartrate	$Na_2C_4H_4O_6$	− 1·1
tartrate	$Na_2C_4H_4O_6 \cdot 2H_2O$	− 5·9
thiocyanate	$NaCNS$	− 1·8
thiosulphate	$Na_2S_2O_3$	+ 1·8
thiosulphate	$Na_2S_2O_3 \cdot 5H_2O$	− 11·4
Stannous bromide	$SnBr_2$	− 1·7
chloride	$SnCl_2$	+ 0·36
iodide	SnI_2	− 5·8
Strontium acetate	$Sr(C_2H_3O_2)_2$	+ 6·2
acetate	$Sr(C_2H_3O_2)_2 \cdot \frac{1}{2}H_2O$	+ 5·9
bromide	$SrBr_2$	+ 16·0
bromide	$SrBr_2 \cdot 6H_2O$	− 6·4
chloride	$SrCl_2$	− 11·5
chloride	$SrCl_2 \cdot 6H_2O$	− 7·1
hydroxide	$Sr(OH)_2$	+ 10·3
hydroxide	$Sr(OH)_2 \cdot 8H_2O$	− 14·3
iodide	SrI_2	+ 20·4
iodide	$SrI_2 \cdot 6H_2O$	− 4·5
nitrate	$Sr(NO_3)_2$	− 4·8
nitrate	$Sr(NO_3)_2 \cdot 4H_2O$	− 12·4
Thallium chloride	$TlCl$	− 10·0
hydroxide	$TlOH$	− 3·2
nitrate	$TlNO_3$	− 10·0
sulphate	Tl_2SO_4	− 8·3
Uranyl nitrate	$UO_2(NO_3)_2$	+ 19·0
nitrate	$UO_2(NO_3)_2 \cdot 6H_2O$	− 5·5
sulphate	$UO_2SO_4 \cdot 3H_2O$	+ 5·0
Zinc acetate	$Zn(C_2H_3O_2)_2$	+ 9·8
acetate	$Zn(C_2H_3O_2)_2 \cdot 2H_2O$	+ 4·0
bromide	$ZnBr_2$	+ 15·0
chloride	$ZnCl_2$	+ 15·7
iodide	ZnI_2	+ 11·6
nitrate	$Zn(NO_3)_2 \cdot 6H_2O$	− 5·9
sulphate	$ZnSO_4$	+ 18·5
sulphate	$ZnSO_4 \cdot 7H_2O$	− 4·3

Table A.6. Heats of Solution of Organic Solids in Water at Approximately Room Temperature and Infinite Dilution

A positive value indicates an exothermic, a negative value an endothermic heat of solution

Substance	*Formula*	*Heat of solution* kcal/mole
Acetamide	CH_3CONH_2	− 2·0
Aconitic acid	$C_3H_3(COOH)_3$	− 4·2
Benzoic acid	C_6H_5COOH	− 6·5
Chloral hydrate	$CCl_3CH(OH)_2$	− 0·90
Chloroacetic acid	$CH_2ClCOOH$	− 3·4
Citric acid	$C_3H_4OH(COOH)_3$	− 5·4
Fumaric acid	$HOOC \cdot CH{:}CH \cdot COOH$	− 5·9
Hydroquinone	$C_6H_4(OH)_2$	− 4·4
o-Hydroxybenzoic acid	$C_6H_4OHCOOH$	− 6·3
m-Hydroxybenzoic acid	$C_6H_4OHCOOH$	− 6·2
p-Hydroxybenzoic acid	$C_6H_4OHCOOH$	− 5·8
Itaconic acid	$HOOC \cdot C({:}CH_2)CH_2COOH$	− 5·9
Lactose	$C_{12}H_{22}O_{11}$	+ 2·5
Lactose (hydrate)	$C_{12}H_{22}O_{11} \cdot H_2O$	− 3·7
Maleic acid	$HOOC \cdot CH{:}CH \cdot COOH$	− 4·4
Malic acid (*dl*)	$CH \cdot OH \cdot CH_2(COOH)_2$	− 3·3
Malonic acid	$CH_2(COOH)_2$	− 4·5
Mannitol (*d*)	$(CH_2OH)_2(CHOH)_4$	− 5·3
p-Nitroaniline	$C_6H_4NH_2NO_2$	− 3·7
o-Nitrophenol	$C_6H_5OHNO_2$	− 6·3
m-Nitrophenol	$C_6H_5OHNO_2$	− 5·2
p-Nitrophenol	$C_6H_5OHNO_2$	− 4·5
Oxalic acid	$(COOH)_2$	− 2·3
Oxalic acid (hydrate)	$(COOH)_2 \cdot 2H_2O$	− 8·5
Oxamic acid	$NH_2CO \cdot COOH$	− 6·9
Phloroglucinol (1,3,5)	$C_6H_3(OH)_3$	− 1·7
Phloroglucinol (hydrate)	$C_6H_3(OH)_3 \cdot H_2O$	− 6·7
Phthalic acid	$C_6H_4(COOH)_2$	− 4·8
Picric acid (2,4,6)	$C_6H_2OH(NO_2)_3$	− 7·1
Pyrocatechol	$C_6H_4(OH)_2$	− 3·5
Pyrogallol (1,2,3)	$C_6H_3(OH)_3$	− 3·7
Raffinose	$C_{18}H_{32}O_{16}$	+ 8·4
Raffinose (hydrate)	$C_{18}H_{32}O_{16} \cdot 5H_2O$	− 9·7
Resorcinol	$C_6H_4(OH)_2$	− 3·9
Salicylic acid	$C_6H_4OHCOOH$	− 6·3
Succinic acid	$(CH_2COOH)_2$	− 6·4
Sucrose	$C_{12}H_{22}O_{11}$	− 1·3
Tartaric acid (*d*-)	$(CHOH \cdot COOH)_2$	− 3·5
Tartaric acid (*dl*-)	$(CHOH \cdot COOH)_2$	− 5·4
Thiourea	NH_2CSNH_2	− 5·3
Urea	NH_2CONH_2	− 3·6
Urea nitrate	$NH_2CONH_2 \cdot HNO_3$	− 10·8

Table A.7. Heats of Fusion of Organic Substances

Substance	*Formula*	*Melting point* °C	*Heat of fusion* cal/g
o-Aminobenzoic acid	$C_6H_4NH_2COOH$	145	35·5
m-Aminobenzoic acid	$C_6H_4NH_2COOH$	180	38·0
p-Aminobenzoic acid	$C_6H_4NH_2COOH$	189	36·5
Anthracene	$C_{14}H_{10}$	217	38·7
Anthraquinone	$(C_6H_4)_2(CO)_2$	282	37·5
Benzene	C_6H_6	5·5	30·1
Benzoic acid	C_6H_5COOH	122	33·9
Benzophenone	$(C_6H_5)_2CO$	48	23·5
t-Butyl alcohol	C_4H_9OH	25	21·9
Cetyl alcohol	$CH_3(CH_2)_{14}CH_2OH$	49	33.8
Cinnamic acid	$C_6H_5CH{:}CH \cdot COOH$	133	36·5
p-Cresol	$CH_3 \cdot C_6H_4OH$	35	26·3
Cyanamide	NH_2CN	43	49·5
Cyclohexane	C_6H_{12}	6·7	7·57
p-Di-bromobenzene	$C_6H_4Br_2$	86	20·6
p-Di-chlorobenzene	$C_6H_4Cl_2$	53	29·7
p-Di-iodobenzene	$C_6H_4I_2$	129	16·2
o-Di-nitrobenzene	$C_6H_4(NO_2)_2$	117	32·3
m-Di-nitrobenzene	$C_6H_4(NO_2)_2$	90	24·7
p-Di-nitrobenzene	$C_6H_4(NO_2)_2$	174	40·0
Diphenyl	$C_{12}H_{10}$	69	28·8
Hexamethylbenzene	$C_{12}H_{18}$	166	30·5
Hydroquinone	$C_6H_4(OH)_2$	172	58·8
Naphthalene	$C_{10}H_8$	80	35·6
α-Naphthol	$C_{10}H_7OH$	95	38·9
β-Naphthol	$C_{10}H_7OH$	121	31·3
o-Nitroaniline	$C_6H_4NH_2NO_2$	70	27·9
m-Nitroaniline	$C_6H_4NH_2NO_2$	113	41·0
p-Nitroaniline	$C_6H_4NH_2NO_2$	147	36·5
Nitrobenzene	$C_6H_5NO_2$	6	22·5
Phenanthrene	$C_{14}H_{10}$	97	24·5
Phenol	C_6H_5OH	41	29·0
Pyrocatechol	$C_6H_4(OH)_2$	104	49·4
Resorcinol	$C_6H_4(OH)_2$	110	46·2
Stearic acid	$C_{17}H_{35}COOH$	68	47·6
Stilbene	$C_{14}H_{12}$	124	40·0
o-Toluic acid	$CH_3C_6H_4COOH$	104	35·4
m-Toluic acid	$CH_3C_6H_4COOH$	109	27·6
p-Toluic acid	$CH_3C_6H_4COOH$	180	40·0
p-Toluidine	$CH_3C_6H_4NH_2$	43	39·9
Thymol	$C_{10}H_{13}OH$	50	27·5
o-Xylene	$C_6H_4(CH_3)_2$	− 25	30·6
m-Xylene	$C_6H_4(CH_3)_2$	− 48	26·1
p-Xylene	$C_6H_4(CH_3)_2$	13	38·5

AUTHOR INDEX

AUTHOR INDEX

SUBJECT INDEX